Current Algebras and Groups

PLENUM MONOGRAPHS IN NONLINEAR PHYSICS

Series Editor: Hedley C. Morris

Department of Mathematics
San Jose State University
San Jose, California

CURRENT ALGEBRAS AND GROUPS
Jouko Mickelsson

Current Algebras and Groups

Jouko Mickelsson

University of Jyväskylä
Jyväskylä, Finland

Springer Science+Business Media, LLC

Library of Congress Cataloging in Publication Data

Mickelsson, Jouko.
Current algebras and groups / Jouko Mickelsson.
p. cm. — (Plenum monographs in nonlinear physics)
Includes bibliographical references.

1. Algebra of currents. 2. Group theory. I. Title. II. Series.
QC174.52.A43M53 1989 89-37698
530.1'2'015125 — dc20 CIP

ISBN 978-1-4757-0297-2 ISBN 978-1-4757-0295-8 (eBook)
DOI 10.1007/978-1-4757-0295-8

Originally published by Plenum Press, New York in 1989.

PREFACE

Let M be a smooth manifold and G a Lie group. In this book we shall study infinite-dimensional Lie algebras associated both to the group $Map(M, G)$ of smooth mappings from M to G and to the group of diffeomorphisms of M. In the former case the Lie algebra of the group is the algebra $M\mathbf{g}$ of smooth mappings from M to the Lie algebra $\mathbf{g}$ of G. In the latter case the Lie algebra is the algebra $Vect\, M$ of smooth vector fields on M. However, it turns out that in many applications to field theory and statistical physics one must deal with certain extensions of the above mentioned Lie algebras.

In the simplest case M is the unit circle S^1, G is a simple finite-dimensional Lie group and the central extension of $Map(S^1, \mathbf{g})$ is an *affine Kac-Moody algebra.* The highest weight theory of finite-dimensional Lie algebras can be extended to the case of an affine Lie algebra. The important point is that $Map(S^1, \mathbf{g})$ can be split to positive and negative Fourier modes and the finite-dimensional piece $\mathbf{g}$ corresponding to the zero mode. In a highest weight representation there is a vector v, the vector of highest weight, such that it is annihilated by the positive Fourier components and the positive roots of the finite-dimensional algebra $\mathbf{g}$, v is an eigenvector for the elements in the Cartan subalgebra of $\mathbf{g}$ and any vector in the representation space can be reached from v by an action of the negative Fourier components and the negative roots of $\mathbf{g}$. There is a similar picture for the highest weight representations of the Virasoro algebra, which is a central extension of $Vect\, S^1$.

Both the affine Kac-Moody algebras and the Virasoro algebras play an important role in physics. They are typical examples of *current algebras* for a quantized system in one space and one time dimension. The group G is the group of symmetries of the system at a given space-time point. In the Hamiltonian picture the maps $f : S^1 \to \mathbf{g}$ are the *charge densities* associated to one-parameter families of symmetries in $Map(S^1, G)$. In a two-dimensional system in statistical physics the same current algebras appear in a different way. Assuming that the system is *conformally invariant*, meaning that the conformal transformations of the plane parametrized by the complex coordinate $z = x + iy$ are symmetries of the partition function, one can decompose the stress energy tensor (and other observables) to a sum of a holomorphic piece $T(z)(dz)^2$ and an antiholomorphic piece $T(\overline{z})(d\overline{z})^2$ which commute with each other. It turns out that both pieces satisfy the commutation relations of a Virasoro algebra.

The theory of Virasoro algebra and the affine algebras is intertwined through the *Sugawara formula*, which defines the generators of the former as quadratic expressions in the basis of the latter. Originally the Sugawara formula was introduced in quantum field theory for the purpose of describing a physical system completely in terms of currents; for example, the time evolution is determined by a Hamiltonian (which is an element of the Virasoro algebra) which is a sum of squares of the charge densities. However, the formula is important also in conformally invariant statistical physics.

In a standard approach to quantum field theory the states of the system are vectors in a Fock space. The structure of the Fock space depends on the type of statistics obeyed by the particles. In the bosonic case the components of a quantized field satisfy canonical commutation relations (CCR) and the Fock space is the symmetric tensor algebra formed from a "one-particle Hilbert space" H. In the fermionic case we have canonical anticommutation relations and the Fock space is an infinite-dimensional version of the exterior algebra based on H. The Hilbert space H has some predefined splitting $H = H_+ \oplus H_-$ to positive and negative energy states. The physical meaning of a vacuum in the Fock space is that it is a state such that all the negative energy levels are filled and the positive energy levels are empty. One can interpret the vacuum in terms of *semi-infinite cohomology*: The vacuum is a "differential form" $f_0 \wedge f_1 \wedge f_2 \wedge \ldots$ with the standard action of wedge product and contraction by *finite* forms. Taking a wedge product by a basis element f_i corresponds to the action of a creation operator a_i^* and the contraction by f_i is an annihilation operator a_i.

The semi-infinite forms can be interpreted as determinants of some infinite matrices in the same way as an ordinary exterior form in the dual space, when evaluated for a sequence of vectors in the underlying vector space, is a determinant formed from the components of the vectors. In fact, the vectors in the Fock space are sections of a complex line bundle DET_1^*, the dual *determinant bundle*, over an infinite-dimensional Grassmannian manifold Gr_1. The points in the Grassmannian are infinite-dimensional planes in H which are obtained from H_+ by a linear transformation g such that the off-diagonal blocks of g, with respect to the splitting $H = H_+ \oplus H_-$, are Hilbert-Schmidt operators; the group of all these linear transformations is denoted by GL_1.

When the dimension of the space is one the group $Map(M, G)$ can be embedded in GL_1. The latter acts projectively in the Fock space via a representation of the *Bogoliubov automorphisms*: An automorphism g of the CAR algebra defines a linear operator $T(g)$ in the Fock space such that $g(x) = T(g)xT(g)^{-1}$ for all creation and annihilation operators

x if and only if g is in GL_1. The operator $T(g)$ is uniquely defined up to a phase. The projective representation T corresponds to a true representation of a central extension $\widehat{GL}_1$. In the case of $G = SU(N)$ one obtains the so-called basic highest weight representation of the affine group associated to the loop group $Map(S^1, G)$, by the restriction to the subgroup $Map(S^1, G) \subset GL_1$.

The first problem when one tries to generalize the above picture to higher space dimensions is that the group $Map(M, G)$ is no longer contained in GL_1 in any natural way. For this reason one cannot really construct a representation of $Map(M, G)$ (or some of its extension) in the Fock space. However, the concept of the determinant bundle over the Grassmannian can be generalized in an appropriate way. First, one must deal with a bigger group GL_p, $2p = \dim M + 1$, which consists of the bounded invertible operators such that the off-diagonal blocks are in the *Schatten ideal* L_{2p}; an operator A is in L_{2p} if $(A^*A)^p$ has a trace. The Grassmannian Gr_p consists of planes which can be reached from H_+ by an action of an element in GL_p. The definition of the holomorphic determinant bundle generalizes to this setting.

The second problem is that there is no central extension of GL_p which lifts the action of GL_p on the base space of the determinant bundle to the total space DET_p. There is an extension $\widehat{GL}_p$ by an Abelian group which acts in DET_p. The Abelian normal subgroup of $\widehat{GL}_p$ is the group of smooth functions on Gr_p taking values in the multiplicative group of nonzero complex numbers. However, the extension does not act holomorphically on DET_p. In the one-dimensional case the Fock space can be identified as the space of holomorphic sections of the dual determinant bundle. When $p > 1$ the space of holomorphic sections is strictly smaller than the Fock space; anyway, for constructing representations of the current algebra, the space of holomorphic sections is not the right object to study when $p > 1$ because in general a holomorphic section is mapped to a nonholomorphic section by an element of GL_p [or the subgroup $Map(M, G)$]. One has to consider the action of $\widehat{GL}_p$ in the space of all smooth sections of DET_p^*.

In Yang-Mills theory the group $Map(M, G)$ is the group of gauge transformations. The Lie algebra $Map(M, \mathrm{g})$ is the algebra of time components of the gauge currents. When the gauge field is coupled to chiral fermions the algebra is modified by the *chiral anomaly.* The chiral anomaly measures the change of the determinant of the Dirac operator under gauge transformations. Without the anomaly the determinant of the Dirac operator would be a function on the space $\mathcal{A}/\mathcal{G}$, vector potentials modulo gauge transformations, but because of the anomaly the determinant must be interpreted as a section in a complex line bundle over $\mathcal{A}/\mathcal{G}$. In the Hamiltonian formalism the time-independent gauge transformations should act in the space Γ of sections of the *Dirac determinant bundle.* In the case of $1+1$ space-time dimensions there is indeed a projective representation of the gauge group in Γ. On the Lie algebra level the projective factor determines a *Schwinger term* in the commutation relations. The Schwinger term is precisely what is needed to define an affine Kac-Moody algebra.

In higher dimensions there are operator valued Schwinger terms: Instead of the central extension we have an extension by the Abelian ideal consisting of functions in the space of vector potentials. This extension is equivalent to the extension obtained from $\widehat{GL_p}$ by an embedding of the group of gauge transformations into GL_p.

The aim of this book is to introduce the reader to the mathematical techniques of modern theory of current algebras and groups, and explain a few physical applications as an illustration of the use of current algebra methods. I have tried to make the book as self-contained as possible. The mathematical background needed consists of some familiarity with the basic concepts of differentiable manifolds; for a brush up of these things I have included the Section 3.0. In addition, the reader is assumed to be familiar with the calculus of linear operators in Hilbert spaces. This book describes mainly the mathematics of current algebras and strictly speaking no background knowledge in physics is needed. However, in order to put the physical applications in a proper perspective (this concerns mainly Chapters 9, 10, 12, and Section 4.4) some familiarity with quantum field theory is helpful.

I shall now describe briefly the contents. Chapter 1 is an introduction to the representation theory of finite-dimensional semisimple Lie algebras. No a priori knowledge of Lie algebras is assumed. Many of the results are given without proofs and the reader willing to fill the holes is referred to standard textbooks on Lie algebras, e.g., Humphreys [1980].

The affine Kac-Moody algebras are an infinite-dimensional analog of the simple finite-dimensional algebras. Chapter 2 contains the classification of affine Lie algebras and the basic results about the structure of their root systems. After that we shall study the highest weight representations and their character formula.

Chapter 3 is an introduction to the theory of principal bundles. This is needed for understanding the structure of the groups associated to current algebras as well as the geometry of the anomalies of Dirac operators.

In Chapter 4 we shall discuss the geometry of current groups. Section 4.1 contains first a description of the group cohomology needed in the construction of the groups associated to the extensions of $Map(M,G)$. In Section 4.2 we shall study the case $M = S^1$, which corresponds to the affine Lie algebras. In Section 4.3 we construct the current group in three space dimensions and its canonical connection. In Section 4.4 we explain a connection between spin and statistics in the Wess-Zumino-Witten model as a consequence of the structure of the group extension.

Chapter 5 contains first the basic definitions and constructions of a Clifford algebra, spin bundle, and the Dirac operator. Then we study the construction of the determinant of the Dirac operator and the geometry and topology of the Dirac determinant bundle over the space $\mathcal{A}/\mathcal{G}$. The connection between curvature of the determinant bundle and Schwinger terms is explained.

In Chapter 6 we construct the bundles DET_p and the group extensions $\widehat{GL_p}$ which we have mentioned above. We also explain the construction of the (generalized) spin bundle and the Dirac operator on the Grassmannian Gr_1. In Section 6.5 we construct an invariant Hermitian form in the space of sections of the dual determinant bundle DET_2^*. (The case $p = 2$ corresponds to the physically interesting case of space-time dimension four.) However, the Hermitian form is not positive (semi)definite. In fact, according to a recent result [Pickrell, 1988], the group $\widehat{GL_p}$ does not have any interesting *unitary* representations. This result makes it very plausible that the current algebra (including the Schwinger terms) in $3+1$ dimensions does not have any faithful unitary representations. It is quite probable that the basic principles of quantum field theory in higher dimensions than two must be formulated in an essentially different way than in the $1+1$ dimensional case.

In Chapter 7 we explain the Sugawara construction of the Virasoro algebra from a representation of an affine Kac-Moody algebra. The unitary highest weight representations of Virasoro algebra consist of a discrete series parametrized by a discrete set of values of the *central charge* c in the range $0 < c < 1$ and of the continuous series $1 \leq c$. The discrete series representations correspond to certain lattice models in two-dimensional statistical mechanics, like the Ising model, the tricritical Ising model and the 3-state Potts model. In continuum quantum field theory (and in string theory) one normally meets representations with $c \geq 1$. There are a number of interesting generalizations of Virasoro algebra of which we shall briefly discuss two: The first is the algebra of meromorphic vector fields on a Riemann surface (this is relevant for the formulation of conformal field theories on Riemann surfaces) and its central extension. The second is the algebra of vector fields on a higher dimensional manifold and its Abelian extensions. Despite the potential importance of this algebra in field theory in higher dimensions almost nothing is known about its representations.

Chapter 8 is concerned with the relations between bosonic and fermionic field theories in $1+1$ dimensions. The vertex operator construction of fermionic fields from bosons is explained in the contexts of the representation theory of the infinite-dimensional linear Lie algebra $\mathbf{gl}_1$. In a sense, this is a "universal construction" because the current algebra in one space dimension can always be thought of as a subalgebra of $\mathbf{gl}_1$.

In Chapter 9 we shall study some applications of the representation theory of affine algebras and groups to constructions of quantum theories of moving one-dimensional compact objects, "strings."

In Chapter 10 we explain an explicit construction of vacuum vectors in a quantized Wess-Zumino-Witten model in $1+1$ dimensions. We also study a symplectic formulation of the WZW model in $3+1$ dimensions.

In Chapter 11, as another application of the Fock space approach to highest weight representations of affine Lie algebras, we shall study the soliton solutions of the nonlinear Kadomtsev-Petviashvili equation and associated nonlinear differential equations. We also describe another construction of the vertex operator which associates fermion field operators to certain solitonic structures on a circle. In the light of the results of Section 4.4 (where we show that solitons on S^3 behave like fermions) this seems to indicate that there should be a generalization of the vertex operator construction to $3+1$ dimensions.

In the final Chapter 12 we examine the structure of the fermionic *Fock bundle* over a family of Dirac operators. We shall use the theory of infinite-dimensional Grassmannians. The vector potentials are parametrized by infinite-dimensional planes, the space of positive energy solutions of the one-particle Dirac equation. The Fock bundle is then a vector bundle over a Grassmannian Gr_p, where p is again related to the space-time dimension, $d+1 = 2p$. Sections of the Fock bundle are the Schrödinger wave functions of a system of fermions coupled to a gauge field. The following examples are studied: massless fermions in two dimensions coupled to Yang-Mills field, massless fermions coupled to metrics, fermions coupled to Yang-Mills field in four dimensions. In the last case there are many more open problems than solutions; the difficulties are associated with the nonexistence of unitary representations of the appropriate current algebra.

There is no attempt to give a complete picture of the historical development of current algebra methods and it is clear that the present book reflects the author's personal view on the topic. The references have been selected accordingly: The choice has been made on the basis of how the author has learned the subject matter and no attempt at objective evaluation of the relative importance of the contributions is made; there are certainly important omissions which are due to the ignorance of the author. However, I hope that there is a sufficiently representative collection for the reader to learn more about different aspects of current algebras and groups. For Kac-Moody algebras and some of their applications there is a fairly extensive bibliography [Benkart, 1986].

Despite the fact that current algebras play a role in quantum electrodynamics, the method became widely popular in the 1960s in the study of strong interactions of elementary particles. The reason was that there was no widely accepted model of strong interactions and people tried to extract information from the commutation relations of currents which were assumed to be to a certain extent model independent. Today there is a popular model of strong interactions, the *quantum chromodynamics,* abbreviated as QCD. However, it has been extremely difficult to compute reliable results from QCD which could be compared with experiments. Therefore people have turned again to current algebra methods, assuming that at least in a low energy limit the current commutation relations give information about a more complete theory. For the older approach we refer to Adler and Dashen [1968]. To get a fuller picture of the modern developments we refer to Treiman et al. [1985].

I wish to thank S. Rajeev for many fruitful discussions which have influenced many parts of this book; it is a pleasure to thank also H.D. Doebner, R. Jackiw, and H. Römer for hospitality and discussions during my stay in 1986/88 at the Arnold-Sommerfeld Institute (Clausthal-Zellerfeld), MIT (Cambridge, MA), and the University of Freiburg, made possible by financial support from the Alexander von Humboldt Foundation and the Fulbright/ASLA Foundation, as well as M. Bowick, C. Devchand, M. Forger, and I. Singer for illuminating discussions. I am also grateful to A. Lehtonen for introducing me to the TEXnicalities of mathematical typesetting, V. Lappalainen for valuable help in those matters, and Tuula Blåfield for help in correcting errors in the manuscript.

CONTENTS

CHAPTER 1 SEMISIMPLE LIE ALGEBRAS

1.1. Lie algebras and homomorphisms

Let $\mathbf{F}$ be the field of real or complex numbers. A *Lie algebra* is a vector space $\mathbf{g}$ over $\mathbf{F}$ with a *Lie product* (or *commutator*) $[\cdot,\cdot] : \mathbf{g} \times \mathbf{g} \to \mathbf{g}$ such that

(1) $x \mapsto [x, y]$ is linear for any $y \in \mathbf{g}$,
(2) $[x, y] = -[y, x]$,
(3) $[x, [y, z]] + [y, [z, x]] + [z, [x, y]] = 0$.

The last condition is called the *Jacobi identity* . From (1) and (2) it follows that also $y \mapsto [x, y]$ is linear for any $x \in \mathbf{g}$. In this chapter we shall consider only finite-dimensional Lie algebras. In any vector space $\mathbf{g}$ one can always define a *trivial Lie product* $[x, y] \equiv 0$. A Lie algebra with this commutator is *Abelian.* Some nontrivial examples follow:

Example 1.1.1. Let $\mathbf{o}(n)$ denote the space of all real antisymmetric $n \times n$ matrices. The commutator of a pair of matrices is defined by

$$[x, y] = xy - yx$$

(ordinary matrix multiplication in xy). Since $(xy)^t = y^t x^t$, where x^t denotes the transpose of the matrix x, the commutator of two antisymmetric matrices is again antisymmetric. The commutator clearly satisfies (1) and (2); (3) is checked by a simple computation. The dimension of the real vector space $\mathbf{o}(n)$ is $\frac{1}{2}n(n-1)$.

The matrix Lie algebras, like $\mathbf{o}(n)$ above, are closely related to groups of matrices. Let $O(n)$ denote the group of all orthogonal $n \times n$ matrices A, $A^t A = 1$. Then the Lie algebra $\mathbf{o}(n)$ consists precisely of those matrices x for which $A(s) = \exp sx \in O(n)$ for all $s \in \mathbf{R}$. Namely, taking the derivative of $A(s)^t A(s)$ at $s = 0$ one gets $x^t + x$. So $A(s) \in O(n)$ implies $x \in \mathbf{o}(n)$. On the other hand if $x \in \mathbf{o}(n)$ then $(\exp sx)^t = \exp sx^t = \exp(-sx) = (\exp sx)^{-1}$, so $A(s) \in O(n)$.

Example 1.1.2. The real vector space $\mathbf{u}(n)$ consisting of *anti-Hermitian* $n \times n$ matrices $x, x^* = -x$, where $x^* = \overline{x}^t$ and the bar means complex conjugation, is a Lie algebra with respect to the matrix commutator. Its dimension is n^2. Denoting by $U(n)$ the group of *unitary* matrices A, $A^* A = 1$, one can prove as in the case of orthogonal matrices that $\exp sx \in U(n) \, \forall s \in \mathbf{R}$ iff $x \in \mathbf{u}(n)$.

Example 1.1.3. The traceless anti-Hermitian $n \times n$ matrices form a Lie algebra to be denoted by $\mathbf{su}(n)$ and it corresponds to the group $SU(n) = \{A \in U(n) \mid \deg A = 1\}$. The dimension of $\mathbf{su}(n)$ is $n^2 - 1$.

Example 1.1.4. Let J be the antisymmetric $2n \times 2n$ matrix

$$\begin{pmatrix} 0 & 0 & \dots & 0 & -1 & 0 & \dots & 0 \\ 0 & 0 & \dots & 0 & 0 & -1 & \dots & 0 \\ \vdots & \vdots & \ddots & \vdots & \vdots & \vdots & \ddots & \vdots \\ 0 & 0 & \dots & 0 & 0 & 0 & \dots & -1 \\ 1 & 0 & \dots & 0 & 0 & 0 & \dots & 0 \\ 0 & 1 & \dots & 0 & 0 & 0 & \dots & 0 \\ \vdots & \vdots & \ddots & \vdots & \vdots & \vdots & \ddots & \vdots \\ 0 & 0 & \dots & 1 & 0 & 0 & \dots & 0 \end{pmatrix}.$$

Since $\deg J = (-1)^{n+1} \neq 0$ the form $\langle x, y \rangle = x^t J y$ is nondegenerate (the vectors x, y are written as column matrices). Define $\mathbf{sp}(2n)$ to consist of all real $2n \times 2n$ matrices x such that $x^t J + J x = 0$. This is a Lie algebra and one can associate to $\mathbf{sp}(2n)$ the group $Sp(2n)$ consisting of real matrices A such that $A^t J A = J$, or equivalently such that A preserves the form $\langle u, v \rangle = u^t J v$, $\langle Au, Av \rangle = \langle u, v \rangle$ for all $u, v \in \mathbf{R}^{2n}$. $Sp(2n)$ is the *symplectic group* defined by J.

Exercise 1.1.5. Find a basis for $\mathbf{sp}(2n)$ and show that $\dim \mathbf{sp}(2n) = 2n^2 + n$.

One can analogously define the complex orthogonal Lie algebra $\mathbf{o}(n, \mathbf{C})$ and the complex symplectic Lie algebra $\mathbf{sp}(2n, \mathbf{C})$.

Let $\{X_1, X_2, \dots, X_n\}$ be a vector space basis of a Lie algebra $\mathbf{g}$. We define the *structure constants* c_{ij}^k by

$$[X_i, X_j] = c_{ij}^k X_k$$

(sum over the repeated index k; we shall use the same summation convention also later). From the defining properties (1) and (2) follows that the commutator $[X, Y]$ for arbitrary $X, Y \in \mathbf{g}$ is determined by the structure constants. The Jacobi identity can be written as

$$c_{ij}^l c_{lk}^m + c_{jk}^l c_{li}^m + c_{ki}^l c_{lj}^m = 0$$

$\forall i, j, k, m$. By the antisymmetry of the Lie product we have $c_{ij}^k = -c_{ji}^k$.

Let $\mathbf{g}$ and $\mathbf{g}'$ be Lie algebras. A linear map $\phi : \mathbf{g} \to \mathbf{g}'$ is a *homomorphism* if

$$\phi([x, y]) = [\phi(x), \phi(y)]$$

$\forall x, y \in \mathbf{g}$. An invertible homomorphism is an *isomorphism* . The inverse of an isomorphism is also an isomorphism. An isomorphism of $\mathbf{g}$ into itself is an *automorphism* of the Lie algebra $\mathbf{g}$.

A linear subspace $\mathbf{k} \subset \mathbf{g}$ is a *subalgebra* of $\mathbf{g}$ if $[x, y] \in \mathbf{k}\, \forall x, y \in \mathbf{k}$. A subalgebra is a Lie algebra in its own right.

Exercise 1.1.6. Let $\phi : \mathbf{g} \to \mathbf{g}'$ be a homomorphism. Show that the *kernel* $\ker\phi = \{x \in \mathbf{g} \mid \phi(x) = 0\} \subset \mathbf{g}$ and the *image* $\mathrm{im}\phi = \{\phi(x) \mid x \in \mathbf{g}\} \subset \mathbf{g}'$ are subalgebras.

A subspace $\mathbf{k} \subset \mathbf{g}$ is an *ideal* if $[x, y] \in \mathbf{k}\, \forall x \in \mathbf{g}$ and $y \in \mathbf{k}$. In particular, an ideal is always a subalgebra. If $\mathbf{k} \subset \mathbf{g}$ is an ideal then the quotient space $\mathbf{g}/\mathbf{k}$ is naturally a Lie algebra: The commutator of the cosets $x + \mathbf{k}$ and $y + \mathbf{k}$ is by definition the coset $[x, y] + \mathbf{k}$. If $x' + \mathbf{k} = x + \mathbf{k}$ and $y' + \mathbf{k} = y + \mathbf{k}$ (i.e., $x' - x \in \mathbf{k}$ and $y' - y \in \mathbf{k}$) then $[x', y'] = [x + (x' - x), y + (y' - y)] \equiv [x, y] \bmod \mathbf{k}$ by the ideal property of $\mathbf{k}$; thus $[x', y']$ represents the same element in $\mathbf{g}/\mathbf{k}$ as $[x, y]$ and so the commutator is well-defined in $\mathbf{g}/\mathbf{k}$.

PROPOSITION 1.1.7. *Let $\phi : \mathbf{g} \to \mathbf{g}'$ be a homomorphism which is onto* (i.e., $\mathbf{g}' = \mathrm{im}\phi$). Then the Lie algebras $\mathbf{g}'$ and $\mathbf{g}/\ker\phi$ are isomorphic.

PROOF: Define $\psi : \mathbf{g}/\ker\phi \to \mathbf{g}'$ by $\psi(x + \ker\phi) = \phi(x)$. Obviously ψ is one-to-one and it is a homomorphism by $\psi([x+\ker\phi, y+\ker\phi]) = \psi([x, y]+\ker\phi) = \phi([x, y]) = [\psi(x+\ker\phi), \psi(y+\ker\phi)]$.

1.2. Semisimple Lie algebras

Let $\mathbf{g}$ be a Lie algebra. Set $\mathbf{g}^{(0)} = \mathbf{g}$ and define inductively $\mathbf{g}^{(n+1)} = [\mathbf{g}^{(n)}, \mathbf{g}^{(n)}]$. We say that $\mathbf{g}$ is *solvable* if $\mathbf{g}^{(n)} = 0$ for some n.

Example 1.2.1. The Lie algebra $\mathbf{g}$ consisting of all upper triangular $n \times n$ matrices is solvable. Now $\mathbf{g}^{(0)}$ consists of matrices x with $x_{ij} = 0$ for $j < i$, $\mathbf{g}^{(1)}$ consists of matrices with $x_{ij} = 0$ for $j < i = 1$, $x \in \mathbf{g}^{(2)}$ iff $x_{ij} = 0$ for $j < i + 2$, $x \in \mathbf{g}^{(3)}$ iff $x_{ij} = 0$ for $j < i + 2^2$; in general, $x \in \mathbf{g}^{(k)}$ iff $x_{ij} = 0$ for $j < i + 2^{k-1}$. Thus the process terminates at 0 when $2^{k-1} > n - 1$.

A Lie algebra which has no solvable ideals except the ideal consisting of the zero vector only is said to be *semisimple.* A non-Abelian Lie algebra $\mathbf{g}$ which has no nontrivial ideals (i.e., other than 0 and $\mathbf{g}$) is *simple.* By a *representation* of a Lie algebra we mean any homomorphism of the Lie algebra into the Lie algebra of linear operators in some vector space. The *adjoint representation* of $\mathbf{g}$ is the homomorphism $ad : \mathbf{g} \to \mathrm{End}\,\mathbf{g}$ given by

$$(ad\, x)(y) = [x, y],$$

where End V denotes the space of linear maps of a vector space V into itself. The map *ad* is clearly linear. We check the homomorphism property:

$$(ad\,[x,y])(z) = [[x,y],z] = [x,[y,z]] - [y,[x,z]] = [ad\,x, ad\,y](x)$$

by the Jacobi identity. When $\dim \mathbf{g} < \infty$ we can define

$$(x,y) = \mathrm{tr}(ad\,x \cdot ad\,y)$$

for $x, y \in \mathbf{g}$. Here "tr" stands for the ordinary matrix trace $\mathrm{tr}A = \Sigma A_{ii}$. The trace of a linear operator does not depend on the choice of basis in the vector space since $\mathrm{tr}SAS^{-1} = \mathrm{tr}S^{-1}SA = \mathrm{tr}A$. Clearly the *Killing form* (x,y) is linear in both of the arguments and it is symmetric by $\mathrm{tr}AB = \mathrm{tr}BA$. The Killing form is *associative* in the following sense:

$$([x,y],z) = (y,[z,x]) \ \forall x,y,z \in \mathbf{g}.$$

Namely, $([x,y],z) = \mathrm{tr}(ad[x,y] \cdot ad\,z) = \mathrm{tr}([ad\,x, ad\,y]adz) = \mathrm{tr}(ad\,y \times ad\,z\,ad\,x - ad\,y\,ad\,x\,ad\,z) = \mathrm{tr}(ad\,y \cdot ad[z,x]) = (y,[z,x])$, where we have used the homomorphism property of *ad*. The following theorem is fundamental in the theory of semisimple Lie algebras.

THEOREM 1.2.2.

(1) *A finite-dimensional Lie algebra* $\mathbf{g}$ *is semisimple if and only if its Killing form is nondegenerate [i.e.,* $(x,y) = 0\, \forall y$ *implies* $x = 0$*].*
(2) $\mathbf{g}$ *is semisimple if and only if it is a direct sum of pairwise commuting simple ideals.*

The proof of this theorem can be found in Humphreys [1980], Section 5, for example. For the classification of the semisimple Lie algebras we thus need to know only the simple ones. We shall consider here only the case when the ground field $\mathbf{F}$ is the field of complex numbers. The classification is similar for any algebraically complete field of characteristic zero. We denote by A_l $(l = 1,2,3,\ldots)$ the Lie algebra of complex traceless $(l+1) \times (l+1)$ matrices, by B_l $(l = 1,2,3,\ldots)$ the Lie algebra $\mathbf{o}(2l+1,\mathbf{C})$, by D_l $(l = 3,4,5,\ldots)$ the Lie algebra $\mathbf{o}(2l,\mathbf{C})$ and finally by C_l $(l = 1,2,3,\ldots)$ the complex symplectic Lie algebra $\mathbf{sp}(2l,\mathbf{C})$. We state without proofs: The list of simple Lie algebras over $\mathbf{C}$ (up to a Lie algebra isomorphism) is exhausted by the algebras A_l, B_l, C_l, D_l and five *exceptional* Lie algebras G_2, F_4, E_6, E_7 and E_8, to be discussed in more detail later. For a complete treatment of the classification problem see Humphreys [1980].

Let $\mathbf{g}$ be a semisimple Lie algebra. A subalgebra $\mathbf{h} \subset \mathbf{g}$ is a *Cartan subalgebra* if

(1) $ad\, x : \mathbf{g} \to \mathbf{g}$ is diagonalizable for all $x \in \mathbf{h}$
(2) $\mathbf{h}$ is Abelian
(3) $\mathbf{h}$ is not contained in any strictly larger Abelian subalgebra of $\mathbf{g}$.

Let $\mathbf{h} \subset \mathbf{g}$ be a Cartan subalgebra. Then we can simultaneously diagonalize all operators $ad\, x$ for $x \in \mathbf{h}$. Thus

$$\mathbf{g} = \oplus \mathbf{g}_\alpha$$

where $\mathbf{g}_\alpha \subset \mathbf{g}$ are the eigenspaces labeled by some linear forms $\alpha : \mathbf{h} \to \mathbf{C}$,

$$[h, x] = \alpha(h) x \forall x \in \mathbf{g}_\alpha .$$

Clearly $\mathbf{g}_0 = \mathbf{h}$. The nonzero elements α are called *roots* of the pair $(\mathbf{g}, \mathbf{h})$ and the $\mathbf{g}_\alpha$'s are the corresponding *root spaces*. Next we shall show the existence of Cartan subalgebras for A_l - D_l and we shall study the properties of the root spaces.

<u>The algebra A_l</u>. Denote by e_{ij} the matrix with the only nonzero element at the position (i, j), that being equal to one. Let $\mathbf{h} \subset A_l$ be the subalgebra consisting of diagonal matrices. It is easy to see that the conditions (1) - (3) above are satified and so $\mathbf{h}$ is a Cartan subalgebra of dimension l. Using the commutation relations

$$[e_{ij}, e_{kl}] = \delta_{jk} e_{il} - \delta_{il} e_{kj},$$

where δ_{ij} is the Kronecker δ-symbol,

$$\delta_{ij} = \begin{cases} 1, & i = j \\ 0, & i \neq j \end{cases},$$

we conclude that each e_{ij} with $i \neq j$ spans a one-dimensional root subspace. We denote the corresponding root by α_{ij},

$$[h, e_{ij}] = \alpha_{ij}(h) e_{ij} .$$

A basis of $\mathbf{h}$ is given by the matrices

$$h_i = e_{ii} - \frac{1}{l+1} \sum_{j=1}^{l+1} e_{jj} \, , \; 1 \leq i \leq l .$$

From the commutation relations it follows at once that $\alpha_{ij}(h_k) = \delta_{ik} - \delta_{jk}$. Denote $\Delta = \{\alpha_{12}, \alpha_{23}, \ldots, \alpha_{l\, l+1}\}$. Now each α_{ij} with $i < j$ can be written as a sum of elements in Δ, $\alpha_{ij} = \alpha_{i,i+1} + \alpha_{i+1,i+2} + \ldots \alpha_{j-1,j}$. Since $\alpha_{ij} = -\alpha_{ji}$, each α_{ij} with $j > i$ can be written as *minus* a sum of elements in Δ. Elements of Δ are called the *simple roots* and the system of roots Φ has now been decomposed to *positive roots* $\Phi^+ = \{\alpha_{ij} \mid i < j\}$ and *negative roots* $\Phi^- = \{\alpha_{ij} \mid i > j\} = -\Phi^+$. In particular, $\Delta \subset \Phi^+$.

The algebra B_l. The Lie algebra B_l consists of complex antisymmetric $(2l+1) \times (2l+1)$ matrices. Denote $b_{ij} = \sqrt{-1}(e_{ij} - e_{ji})$ and define

$$h_i = b_{2i-1,2i}\,,\ i = 1, 2, \ldots, l.$$

We shall show that the Abelian algebra $\mathbf{h}$ spanned by the elements h_i is a Cartan subalgebra of B_l. Using the commutation relations

$$[b_{ij}, b_{kl}] = \sqrt{-1}(\delta_{jk} b_{il} + \delta_{il} b_{jk} - \delta_{jl} b_{ik} - \delta_{ik} b_{jl})$$

we observe that each of the following elements spans a one-dimensional eigenspace for the adjoint action of $\mathbf{h}$:

$$\begin{aligned}\rho_{\nu\mu} =& \sqrt{-1} b_{2|\nu|-1,2|\mu|-1} + \frac{\nu}{|\nu|} b_{2|\nu|,2|\mu|-1} \\ &+ \frac{\mu}{|\mu|} b_{2|\nu|-1,2|\mu|} - \sqrt{-1}\frac{\nu\mu}{|\nu\mu|} b_{2|\nu|,2|\mu|}, \\ \rho_\nu =& b_{2|\nu|-1,2l+1} - \sqrt{-1}\frac{\nu}{|\nu|} b_{2|\nu|,2l+1},\end{aligned}$$

where $\nu, \mu = \pm 1, \pm 2, \ldots, \pm l$ and $|\nu| < |\mu|$. By a simple computation,

$$\begin{aligned}[h_i, \rho_{\nu\mu}] &= \delta_{i,|\nu|} \frac{\nu}{|\nu|} \rho_{\nu\mu} + \delta_{i,|\mu|} \frac{\mu}{|\mu|} \rho_{\nu\mu}, \\ [h_i, \rho_\nu] &= \delta_{i,|\nu|} \frac{\nu}{|\nu|} \rho_\nu.\end{aligned}$$

The vectors $\rho_\nu, \rho_{\nu\mu}$ span the complement of $\mathbf{h}$ in B_l and thus from the commutation relations above follows that there are no vectors outside of $\mathbf{h}$ which commute with all vectors in $\mathbf{h}$, so $\mathbf{h}$ is indeed a Cartan subalgebra of B_l. Let $\Delta = \{\alpha_{1,-2}, \alpha_{2,-3}, \ldots \alpha_{l-1,l}, \alpha_l\}$, where $\alpha_{\nu\mu}$ (respectively, α_ν) is the root corresponding to $\rho_{\nu\mu}$ (respectively, ρ_ν). Now

$$\begin{aligned}\alpha_k &= \alpha_{k,-k-1} + \alpha_{k+1,-k-2} + \ldots + \alpha_{l-1,-l}\,,\ 1 \le k \le l, \\ \alpha_{i,-j} &= \alpha_{i,-i-1} + \alpha_{i+1,-i-2} + \ldots + \alpha_{j-1,-j}\,,\ 1 \le i < j \le l, \\ \alpha_{ij} &= \alpha_{i,-i-1} + \ldots \alpha_{j-1,-j} + 2\alpha_{j,-j-1} + \ldots 2\alpha_{l-1,-l} + 2\alpha_l,\end{aligned}$$

$1 \le i < j \le l$. Thus we can define the set of positive roots Φ^+ to consist of the roots listed above, with each element of Φ^+ being a sum of elements in Δ. Again we have the decomposition of roots $\Phi = \Phi^+ \cup \Phi^-$ with $\Phi^- = -\Phi^+$.

The algebra C_l. Let $\{v_i \mid i = \pm 1, \pm 2, \ldots, \pm l\}$ be a basis of $\mathbf{C}^{2l}$ and let J denote the symplectic form which in this basis is represented by the antisymmetric matrix on page 2. Define the $2l \times 2l$ matrices

$$\begin{aligned} c_{ij} &= e_{-i,j} + e_{-j,i}, \\ c_{-i,-j} &= e_{i,-j} + e_{j,-i}, \\ c_{-i,j} &= c_{j,-i} = e_{-i,-j} - e_{ji}\,, \quad 1 \le i,j \le l. \end{aligned}$$

These elements span the symplectic Lie algebra $C_l = \mathbf{sp}(2l, \mathbf{C})$. The commutation relations are

$$[c_{\nu\mu}, c_{\lambda\omega}] = J_{\mu\lambda} c_{\nu\omega} + J_{\nu\omega} c_{\mu\lambda} + J_{\nu\lambda} c_{\mu\omega} + J_{\mu\omega} c_{\nu\lambda} \quad \nu, \mu = \pm 1, \pm 2, \ldots \pm l.$$

Let $\mathbf{h}$ be the commutative subalgebra spanned by the vectors $h_i = c_{i,-i}\,, 1 \le i \le l$. Now

$$\begin{aligned} [h_n, c_{ij}] &= (-\delta_{in} - \delta_{jn}) c_{ij}, \\ [h_n, c_{-i,-j}] &= (\delta_{in} + \delta_{jn}) c_{-i,-j}, \\ [h_n, c_{-i,j}] &= (\delta_{in} - \delta_{jn}) c_{-i,j}. \end{aligned}$$

Thus again the root spaces are one-dimensional and they are spanned by the vectors $c_{ij}, c_{-i,-j}$ and $c_{-i,j}$ with $i \neq j$ in the last case. We can define a set of simple roots to be $\Delta = \{\alpha_{-1,2}, \alpha_{-2,3}, \ldots, \alpha_{-l+1,l}, \alpha_{-l,-l}\}$, where $\alpha_{\nu\mu}$ is the root corresponding to the vector $c_{\nu\mu}$. The set of positive roots [sums of elements in Δ] is $\Phi^+ = \{\alpha_{-i,-j} \mid 1 \le i,j \le l\} \cup \{\alpha_{-i,j} \mid 1 \le i < j \le l\}$ and the set of negative roots is $\Phi^- = -\Phi^+ = \{\alpha_{ij} \mid 1 \le i,j \le l\} \cup \{a_{-i,j} \mid 1 \le j < i \le l\}$.

The algebra D_l. As in the case of B_l the Lie algebra is spanned by the vectors $b_{ij} = -b_{ji}$ but now the range of indices is $1 \le i,j \le 2l$. A basis for the Cartan subalgebra $\mathbf{h}$ is given by the vectors

$$h_i = b_{2i-1,2i}\,, \; i = 1, 2, \ldots, l.$$

The root subspaces are $\mathbf{C}\rho_{\nu\mu}$, where $\nu, \mu = \pm 1, \pm 2, \ldots, \pm l\,, \; |\nu| < |\mu|$, as in the case of B_l; however, the elements ρ_ν are not present now. A set of simple roots can be defined as

$$\Delta = \{\alpha_{1,-2}, \alpha_{2,-3}, \ldots, \alpha_{l-1,-l}, \alpha_{l-1,l}\}.$$

The set of positive roots is $\Phi^+ = \{\alpha_{ij} \mid 1 \le i < j \le l\} \cup \{\alpha_{i,-j} \mid 1 \le i < j \le l\}$ and $\Phi^- = \Phi \setminus \Phi^+ = -\Phi^+$.

We have seen in all of the cases $A_l - D_l$ the following important properties of the root systems:

(1) All root subspaces $\mathbf{g}_\alpha$ corresponding to a nonzero root α are one-dimensional.

(2) There is a subset Δ, the simple roots, in the set Φ of nonzero roots such that $\Phi = \Phi^+ \cup \Phi^-$, where each element of Φ is a sum of vectors in Δ and $\Phi^- = -\Phi^+$. The number of elements in Δ is equal to $\dim \mathbf{h}$.

The properties (1) and (2) hold for an arbitrary semisimple Lie algebra [Humphreys, 1980]. In fact, it is sufficient to prove this only for the exceptional simple Lie algebras $G_2, \ldots, E_8$ since any semisimple Lie algebra is a direct sum of simple Lie algebras.

THEOREM 1.2.3. *The Killing form remains nondegenerate when restricted to a Cartan subalgebra of a semisimple Lie algebra.*

The proof for all assertions in this section can be found in Humphreys [1980]. In any vector space V equipped with a symmetric nondegenerate bilinear form $(\cdot,\cdot) : V \times V \to \mathbf{F}$ one can define a natural isomorphism $\phi : V \to V^*$ (V is the dual vector space) such that

$$\phi(v)w = (v, w).$$

In particular, the Killing form $(\cdot,\cdot)|_{\mathbf{h}\times\mathbf{h}}$ gives an isomorphism $\mathbf{h} \to \mathbf{h}^*$. For $\lambda \in \mathbf{h}^*$ we shall denote the corresponding vector in $\mathbf{h}$ by h_λ. We can now define a bilinear form in $\mathbf{h}^*$ by

$$(\lambda, \mu) = (h_\lambda, h_\mu).$$

Example 1.2.4. Consider the algebra A_l (notation as before). For each root $\alpha_{ij} \in \mathbf{h}^*$ we construct the vector $h_{\alpha_{ij}}$. We can write

$$h_{\alpha_{ij}} = \sum_{k=1}^{l} a_k h_k \, , \; a_k \in \mathbf{C}.$$

Now

$$\begin{aligned}(h_i, h_j) &= \mathrm{tr}(ad\, h_i \cdot ad\, h_j) \\ &= \sum_{m\neq n=1}^{l+1} \alpha_{mn}(h_i)\alpha_{mn}(h_j) \\ &= \sum_{m,n=1}^{l+1} (\delta_{im} - \delta_{in})(\delta_{jm} - \delta_{jn}) = 2(l+1)\delta_{ij} - 2.\end{aligned}$$

On the other hand,

$$\begin{aligned}(h_{\alpha_{ij}}, h_k) &= \alpha_{ij}(h_k) = \delta_{ik} - \delta_{jk} \\ &= (\Sigma a_n h_n, h_k) = \sum_{n=1}^{l} 2a_n[(l+1)\delta_{nk} - 1] \\ &= 2(l+1)a_k - 2\sum_{n=1}^{l} a_n.\end{aligned}$$

We have a linear system of equations for the unknowns a_k. The solution is easily found to be $a_i = 1/2(l+1)$, $a_j = -1/2(l+1)$ and $a_k = 0$ for $k \neq i, j$. Thus

$$h_{\alpha_{ij}} = \frac{1}{2(l+1)}(h_i - h_j).$$

From this we can compute the inner products

$$\begin{aligned}(\alpha_{ij}, \alpha_{mn}) &= \frac{1}{4(l+1)^2}(h_i - h_j, h_m - h_n) \\ &= \frac{1}{2(l+1)}(\delta_{im} + \delta_{jn} - \delta_{jm} - \delta_{in}).\end{aligned}$$

Usually it is sufficient to know the root space structure of a semisimple Lie algebra in terms of the inner products of the roots and an explicit knowledge of a matrix realization of the algebra in question is not needed.

The *rank* of a semisimple Lie algebra is the dimension of its Cartan subalgebra.

THEOREM 1.2.5. *Let* $\mathbf{g}$ *be a semisimple Lie algebra of rank* l, $\mathbf{h} \subset \mathbf{g}$ *a Cartan subalgebra, and* Δ *a system of simple roots for* $(\mathbf{g}, \mathbf{h})$. *Then* Δ *forms a basis of* $\mathbf{h}^*$. *Let* E *denote the real vector space spanned by* Δ. *Then the dual* $(\cdot,\cdot) : \mathbf{h}^* \times \mathbf{h}^* \to \mathbf{C}$ *of the Killing form is a positive definite inner product in* E.

We shall now describe the exceptional Lie algebras in terms of their root systems.

$\underline{G_2}$ Let $\{v_1, v_2, v_3\}$ be the standard basis of $\mathbf{R}^3$ and let E be the plane orthogonal to $v_1 + v_2 + v_3$. A basis of E is given by $\{v_1 - v_2, -2v_1 + v_2 + v_3\} = \Delta$. This is a system of simple roots for G_2. The positive roots are $\Phi^+ = \{v_1 - v_2, -v_1 + v_3, -v_2 + v_3, -2v_1 + v_2 + v_3, v_1 - 2v_2 + v_3, -v_1 - v_2 + 2v_3\}$.

$\underline{F_4}$ Let $E = \mathbf{R}^4$ and $\Delta = \{v_2 - v_3, v_3 - v_4, v_4, \frac{1}{2}(v_1 - v_2 - v_3 - v_4)\}$. The root system of F_4 consists of all *integral* linear combinations α of elements in Δ such that $\|\alpha\|^2 = 1$ or $\|\alpha\|^2 = 2$. One can show that $\Phi = \{\pm v_i\}_{i=1}^4 \cup \{\pm(v_i \pm v_j) \mid i \neq j\} \cup \{\pm\frac{1}{2}(v_1 \pm v_2 \pm v_3 \pm v_4) \mid \text{all signs}\}$. Thus the number of elements in Φ is 48.

Exercise 1.2.6. What is the system of positive roots for F_4?

$\underline{E_8}$ Let $E = \mathbf{R}^8$ and $\Delta = \{\frac{1}{2}(v_1 + v_8) - \frac{1}{2}(v_2 + \ldots + v_7), v_1 + v_2, v_2 - v_1, v_3 - v_2, v_4 - v_3, v_5 - v_4, v_6 - v_5, v_7 - v_6\}$. The root system $\Phi(E_8)$ consists of all integral linear combinations α of elements in Δ such that $\|\alpha\|^2 = 2$. One can show that

$$\Phi = \{\pm(v_i \pm v_j) \mid i \neq j\} \cup \{\frac{1}{2}\sum_{i=1}^{8}(-1)^{\epsilon(i)} v_i \mid \epsilon(i) = 0, 1; \ \sum \epsilon(i) \in 2\mathbf{Z}\}.$$

There are 240 elements in Φ.

$\underline{E_7}$ Δ and Φ are defined here in a similar way as in the case of E_8 except that the last vector $v_7 - v_6$ in Δ is left out. There are 126 roots.

$\underline{E_6}$ Same as above, but now the two last vectors $v_6 - v_5$ and $v_7 - v_6$ are dropped. The number of roots is 72.

1.3. Dynkin diagrams

Let $\mathbf{g}$ be a simple Lie algebra, $\mathbf{h} \subset \mathbf{g}$ a Cartan subalgebra, Φ the system of roots and $\Delta \subset \Phi$ a set of simple roots, $\Delta = \{\alpha_1, \alpha_2, \ldots, \alpha_l\}$. For any $\alpha, \beta \in \Phi$ define

$$\langle \beta, \alpha \rangle = 2\frac{(\beta, \alpha)}{(\alpha, \alpha)}.$$

One can show that the numbers $\langle \beta, \alpha \rangle$ are always integers. For example in the case of A_l, $\langle \alpha, \alpha \rangle = 2$ and $\langle \beta, \alpha \rangle \in \{0, \pm 1\}$ for $\beta \neq \alpha$. In addition, $\langle \alpha_i, \alpha_j \rangle < 0$ for $i \neq j$. The $l \times l$ matrix with entries $\langle \alpha_i, \alpha_j \rangle$ is *the Cartan matrix* of the root system. In the case $\mathbf{g} = A_l$, setting $\alpha_i = \alpha_{i,i+1}$ (notation as in Section 1.2), we conclude from the computation of the inner products in Section 1.2 that the Cartan matrix is

$$\begin{pmatrix} 2 & -1 & 0 & \ldots & 0 \\ -1 & 2 & -1 & \ldots & 0 \\ 0 & -1 & 2 & \ldots & 0 \\ \vdots & \vdots & \vdots & \ddots & \vdots \\ 0 & 0 & 0 & \ldots & 2 \end{pmatrix}.$$

In general, the Cartan matrix is not symmetric. For example,

$$B_2 : \begin{pmatrix} 2 & -2 \\ -1 & 2 \end{pmatrix} \qquad G_2 : \begin{pmatrix} 2 & -1 \\ -3 & 2 \end{pmatrix}.$$

Exercise 1.3.1. Compute the Cartan matrix of D_3 and compare the result with the Cartan matrix of A_3.

The *Dynkin diagram* corresponding to a given Cartan matrix $(\langle\alpha_i, \alpha_j\rangle)$ is constructed as follows. The Dynkin diagram has l nodes labeled by $i \in \{1, 2, \ldots, l\}$. The node i is connected to the node j by p lines where $p = \langle\alpha_i, \alpha_j\rangle\langle\alpha_j, \alpha_i\rangle$. In addition if $\|\alpha_i\| < \|\alpha_j\|$ we set an arrow pointing to the node i; if $\|\alpha_i\| > \|\alpha_j\|$ the arrow points to j. From the Cartan matrix of A_l we get its Dynkin diagram:

A_l 

All the roots have equal length. We have defined the root systems for all the other simple Lie algebras, too. It is a straight-forward (but a bit tedious) computation to find out their Dynkin diagrams. We only state the results.

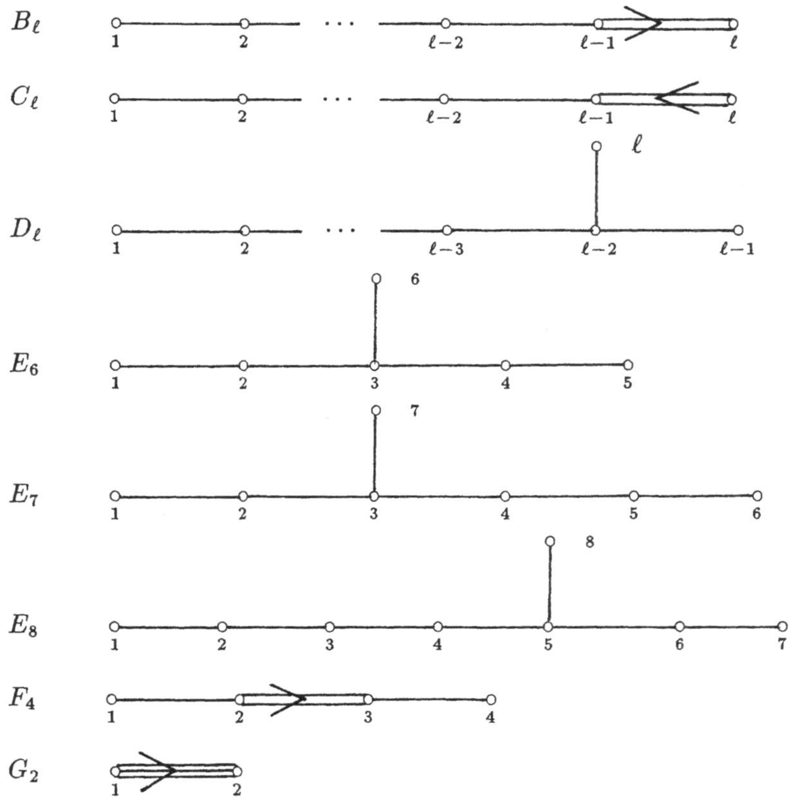

Figure 1 Dynkin diagrams for algebras of the type $B - G$.

The Dynkin diagram determines completely the Cartan matrix. Let p be the number of lines connecting i to the node j. If $p = 0$ then $\langle\alpha_i, \alpha_j\rangle = \langle\alpha_j, \alpha_i\rangle = 0$. If $p = 1$ then $\langle\alpha_i, \alpha_j\rangle = \langle\alpha_j, \alpha_i\rangle = -1$ since $\langle\alpha_i, \alpha_j\rangle$ is always a nonpositive integer for $i \neq j$. If $p = 2, 3$ then we know that one of the numbers $\langle\alpha_i, \alpha_j\rangle, \langle\alpha_j, \alpha_i\rangle$ must be -1 and the other $-p$. But the arrow tells us which one is which: If the arrow is pointing to i then $\langle\alpha_i, \alpha_j\rangle = -1$ and $\langle\alpha_j, \alpha_i\rangle = -p$.

Example 1.3.2. The Cartan matrix corresponding to the diagram F_4 is

$$\begin{pmatrix} 2 & -1 & 0 & 0 \\ -1 & 2 & -2 & 0 \\ 0 & -1 & 2 & -1 \\ 0 & 0 & -1 & 2 \end{pmatrix}.$$

Of course, the matrix can be obtained also directly from our explicit construction of the root system in Section 1.2.

1.4. The enveloping algebra. Linear representations

A Lie algebra is nonassociative, $[x, [y, z]] \neq [[x, y], z]$, except when all the double commutators vanish. Since associative algebras are in many respect easier to handle than nonassociative algebras, the following theorem is very important.

THEOREM 1.4.1. *For any Lie algebra* $\mathbf{g}$ *over* $\mathbf{F}$ *there is an associative algebra* $\mathcal{U}(\mathbf{g})$ *over* $\mathbf{F}$ *and an injective linear map* $i : \mathbf{g} \to \mathcal{U}(\mathbf{g})$ *such that*

$$i([x, y]) = i(x)i(y) - i(y)(x) \quad \forall x \in \mathbf{g}.$$

Furthermore, if A *is any associative algebra (over* $\mathbf{F}$*) and* $\phi : \mathbf{g} \to A$ *is a linear map which also satifies* $\phi([x, y]) = \phi(x)\phi(y) - \phi(y)\phi(x)$ *then there is a uniquely determined homomorphism* $\psi : \mathcal{U}(\mathbf{g}) \to A$ *such that* $\phi = \psi \circ i$. *[A homomorphism of associative algebras is a linear map* ψ *such that* $\psi(uv) = \psi(u)\psi(v)$ *for all* u, v.*]* $\mathcal{U}(\mathbf{g})$ *is called the (universal) enveloping algebra of* $\mathbf{g}$.

We shall not give the complete proof but we shall explain an explicit construction of $\mathcal{U}(\mathbf{g})$. First let us choose a basis $\{x_1, x_2, \ldots\}$ for $\mathbf{g}$. If $\mathbf{g}$ is commutative then $\mathcal{U}(\mathbf{g})$ is just the polynomial algebra (over $\mathbf{F}$) in the "variables" $x_1, x_2, \ldots$. In the general case a basis for the linear space $\mathcal{U}(\mathbf{g})$ consists of the *ordered* monomials

$$x_{i(1)} x_{i(2)} \cdots x_{i(n)},$$

where $i(1) \leq i(2) \leq \cdots \leq i(n)$. A product uv of two ordered monomials u, v is not in general an ordered monomial, but one can write uv as a linear combination of ordered monomials using the commutation relations of $\mathbf{g}$. If $i > j$ we write

$$x_i x_j = x_j x_i + [x_i, x_j] = x_j x_i + \sum_k c_{ij}^k x_k,$$

where the c_{ij}^k's are the structure constants. The rules of the game are best illustrated by an example.

Example 1.4.2. Consider the Lie algebra $\mathbf{g} = A_1$ with an ordered basis $\{x, h, y\}$,

$$x = \begin{pmatrix} 0 & 1 \\ 0 & 0 \end{pmatrix} \qquad h = \begin{pmatrix} 1 & 0 \\ 0 & -1 \end{pmatrix} \qquad y = \begin{pmatrix} 0 & 0 \\ 1 & 0 \end{pmatrix}.$$

The commutation relations are

$$[x, y] = h \qquad [h, x] = 2x \qquad [h, y] = -2y.$$

Let $u = yhx^2$ and $v = y$. Then

$$\begin{aligned} uv &= (yhx)([x,y] + yx) = (yhx)(h + yx) \\ &= yh([x,h] + hx) + yh([x,y] + yx)x \\ &= -2yhx + yh^2x + yh^2x + yh([h,y] + yh)x^2 \\ &= -2yhx + 2yh^2x - 2y^2x^2 + y^2hx^2, \end{aligned}$$

the expression on the left being now a linear combination of ordered monomials.

The map $i : \mathbf{g} \to \mathcal{U}(\mathbf{g})$ is just the identity map.

Let $\mathbf{g}$ be a Lie algebra and V a vector space (both over $\mathbf{F}$). A *linear representation* of $\mathbf{g}$ in V is a linear map ϕ of $\mathbf{g}$ into the space $\operatorname{End} V$ of linear operators in V such that

$$\phi(x)\phi(y) - \phi(y)\phi(x) = \phi([x, y])$$

for all $x, y \in \mathbf{g}$. A linear representation ϕ of $\mathbf{g}$ can always be lifted to a linear representation (to be denoted also by ϕ) of the enveloping algebra $\mathcal{U}(\mathbf{g})$. This follows directly from Theorem 1.4.1 by setting $A = \operatorname{End} V$. (By a representation of an associative algebra we mean a homomorphism of the algebra into an algebra of linear operators.) The reverse is also true: Given a representation $\phi : \mathcal{U}(\mathbf{g}) \to \operatorname{End} V$ of $\mathcal{U}(\mathbf{g})$ we can define a representation $\phi \circ i : \mathbf{g} \to \operatorname{End} V$.

A *left (right) ideal* in an associative algebra A is a linear subspace $I \subset A$ such that $ax \in I$ $(xa \in I)$ for all $a \in A, x \in I$. If $I \subset A$ is a left ideal we can form a canonical representation of A in the quotient space $V = A/I$ as follows. Let $a \in A$ and $[x] = x + I \in V$ be an arbitrary coset. Put

$$\phi(a)[x] = [ax].$$

If $[x] = [x']$, then $x - x' \in I$ and so $a(x - x') \in I$ and $[ax] = [ax']$. Thus $\phi(a) : V \to V$ is a well-defined linear operator. It is simple to see that $\phi : A \to \operatorname{End} V$ is a linear map. Furthermore,

$$\phi(ab)[x] = [abx] = \phi(a)[bx] = \phi(a)\phi(b)[x],$$

and so ϕ is a representation.

A (linear) representation $\phi : \mathbf{g} \to \operatorname{End} V$ of a Lie algebra $\mathbf{g}$ is *irreducible* if there are no invariant subspaces in V other than 0 and V; a subspace $W \subset V$ is *invariant* if $\phi(x)v \in W$ for all $x \in \mathbf{g}$ and $v \in W$. If ϕ is not irreducible then it is *reducible* . The representation is *fully reducible* if V can be written as a direct sum $V = V_1 \oplus V_2 \oplus \cdots \oplus V_n$ of invariant subspaces such that the restriction of ϕ to each V_i is an irreducible representation of $\mathbf{g}$.

From known representations one can build new ones by taking direct sums. If ϕ_i is a representation of $\mathbf{g}$ in V_i $(i \in \Lambda)$, then a representation ϕ of $\mathbf{g}$ in $\oplus \sum_{i \in \Lambda} V_i$ is defined by

$$\phi(x)(v_i)_{i \in \Lambda} = (\phi(x)v_i)_{i \in \Lambda},$$

where $v_i \in V_i$. If ϕ is a representation of $\mathbf{g}$ in a vector space V such that there is an invariant subspace W, then one can construct a representation ψ in the quotient space V/W by setting $\psi(x)(v + W) = \phi(x)v + W$, for any $v \in V$.

THEOREM 1.4.3. *Any finite-dimensional representation of a semisimple Lie algebra is fully reducible.*

Thus for the purpose of classification of finite-dimensional representations of a semisimple Lie algebra it is sufficient to know the irreducible representations.

As an illustration of the enveloping algebra techniques in representation theory, consider the case $\mathbf{g} = A_1$. We use the basis of the example 1.4.2. Let I_λ be the left ideal of $\mathcal{U}(\mathbf{g})$ generated by the elements x and $h - \lambda$, where $\lambda \in \mathbf{C}$ is fixed. I_λ consists of all vectors of the form

$$ux + v(h - \lambda), \qquad \text{with}\, u, v \in \mathcal{U}(\mathbf{g}).$$

A basis element $y^p h^q x^r$ of $\mathcal{U}(\mathbf{g})$ can be written as $\lambda^q y^p \bmod I_\lambda$ if $r = 0$ and $y^p h^q x^r \equiv 0$ if $r > 0$. Thus a basis for $V_\lambda = \mathcal{U}(\mathbf{g})/I_\lambda$ is given by the vectors $v_n = y^n + I_\lambda$, $n = 0, 1, 2, \ldots$. For the canonical representation ϕ of $\mathbf{g}$ in V_λ we have by definition

$$\phi(y)v_n = v_{n+1}.$$

Each v_n is an eigenvector of $\phi(h)$, since

$$\begin{aligned} hy^n &= ([h,y] + yh)y^{n-1} = y(h-2)y^{n-1} \\ &= y^2(h-4)y^{n-2} = \ldots y^n(h-2n) \\ &\equiv (\lambda - 2n)y^n \bmod I_\lambda \end{aligned}$$

and so $\phi(h)v_n = (\lambda - 2n)v_n$. Next we compute the action of $\phi(x)$ in V_λ.

$$\begin{aligned} xy^n &= \sum_{k=0}^{n-1} y^k [x,y] y^{n-k-1} + y^n x \\ &\equiv \sum_{k=0}^{n-1} y^k h y^{n-k-1} \bmod I_\lambda \\ &= n(\lambda - n + 1)y^{n-1}. \end{aligned}$$

Thus $\phi(x)v_n = n(\lambda - n + 1)v_{n-1}$. If $\lambda \notin \mathbf{N}$ then

$$\phi(x)^n v_n = \alpha_n v_0,$$

where $\alpha_n \neq 0$. Using this result one can show that the representation ϕ is irreducible for all $\lambda \notin \mathbf{N}$; in that case the representation is infinite-dimensional. For $\lambda \in \mathbf{N}$ one can construct a finite-dimensional representation as follows. Since now

$$\phi(x)v_{\lambda+1} = 0,$$

the subspace $M_\lambda \subset V_\lambda$ spanned by the vectors $v_{\lambda+1}, v_{\lambda+2}, \ldots$ is invariant. The quotient space

$$L_\lambda = V_\lambda / M_\lambda$$

is spanned by the vectors $v_0 + M_\lambda, v_1 + M_\lambda, \ldots, v_\lambda + M_\lambda$. Since $\phi(x)v_n \neq 0$ for $0 \leq n \leq \lambda$, there are no invariant subspaces in L_λ and the representation is irreducible. Note that L_λ can also be written in the form $\mathcal{U}(\mathbf{g})/J_\lambda$, where J_λ is the left ideal in $\mathcal{U}(\mathbf{g})$ generated by the elements x, $h - \lambda$ and $y^{\lambda+1}$.

In general, a representation $\phi : \mathbf{g} \to \mathrm{End}V$ of $\mathbf{g} = A_1$ is a *highest weight representation* if there is a vector $0 \neq v \in V$ (the *highest weight vector*) such that

(1) $\phi(x)v = 0$
(2) $\phi(h)v = \lambda v$ for some $\lambda \in \mathbf{C}$
(3) $V = \{\phi(u)v \mid u \in \mathcal{U}(\mathbf{g})\}$.

The number λ is the *highest weight* of the representation.

Two representations $\phi : \mathbf{g} \to \mathrm{End}\, V$, $\phi' : \mathbf{g} \to \mathrm{End}\, V'$ of a Lie algebra $\mathbf{g}$ are said to be *equivalent* if there is a linear isomorphism $\alpha : V \to V'$ such that

$$\alpha\phi(x)\alpha^{-1} = \phi'(x)\ \forall x \in \mathbf{g}.$$

We can now prove that an irreducible highest weight representation (ψ, V) of A_1 is uniquely determined, up to an equivalence, by the highest weight λ.

1) Let $\lambda \notin \mathbf{N}$. By (1) - (3) the space V is spanned by the vectors $\{\psi(y)^n v \mid n = 0, 1, 2, \dots\}$. Using the commutation relations as above we get

$$\psi(x)^n \psi(y)^n v = \alpha_n v\,,\ \alpha_n \neq 0.$$

Thus $\psi(y)^n v \neq 0$ for $n = 0, 1, 2, \dots$. The system $\{\psi(y)^n \mid n \in \mathbf{N}\}$ is linearly independent since different vectors correspond to different eigenvalues of the operator $\psi(h)$,

$$\psi(h)\psi(y)^n v = (\lambda - 2n)\psi(y)^n v.$$

We can define a linear isomorphism $\alpha : \mathcal{U}(\mathbf{g})/I_\lambda \to V$ by $\alpha(v_n) = \psi(y)^n v$. We can check that $\alpha\phi(z)\alpha^{-1} = \psi(z)\ \forall z \in \mathbf{g}$. For example,

$$\begin{aligned}\alpha\phi(y)\alpha^{-1}[\psi(y)^n v] &= \alpha\phi(y)v_n = \alpha v_{n+1}\\ &= \psi(y)^{n+1} v = \psi(y)[\psi(y)^n v].\end{aligned}$$

2) The case $\lambda \in \mathbf{N}$. Using the commutation relations we have

$$\begin{aligned}&\psi(x)^n\psi(y)^n v = \alpha_n v \neq 0 \qquad\qquad n = 0, 1, 2, \dots, \lambda\\ &\psi(x)\psi(y)^{\lambda+1} v = 0.\end{aligned}$$

It follows that $\{\psi(y)^n v \mid n = 0, 1, 2, \dots, \lambda\}$ is a basis of V. The rest of the proof goes like in the case 1.

Remark 1.4.4. An irreducible finite-dimensional representation (ψ, V) of A_1 is always a highest weight representation. Let $0 \neq w \in V$ be any eigenvector of $\psi(h)$ in V (which exists because of $\dim V < \infty$). If

$\psi(x)w = 0$ then w is a highest weight vector; otherwise set $v = \psi(x)^n w$, where n is the largest integer such that $\psi(x)^n w \neq 0$. Then w is a highest weight vector. [Because of the irreducibility of the representation, the invariant subspace $\{\psi(u)w \mid u \in \mathcal{U}(\mathbf{g})\}$ must be the whole space V, and so also the condition (3) is satisfied.]

1.5. Highest weight representations of semisimple Lie algebras

We shall now generalize the results obtained in the previous section for A_1, to the case of an arbitrary semisimple Lie algebra $\mathbf{g}$ over $\mathbf{C}$. Note first that a semisimple Lie algebra is always spanned by subalgebras of the type A_1. Namely, let $\mathbf{h} \subset \mathbf{g}$ be a Cartan subalgebra and Φ the system of nonzero roots. If $\alpha \in \Phi$ then also $-\alpha \in \Phi$ (just look at the various root systems listed earlier). Choose $0 \neq x_\alpha \in \mathbf{g}_\alpha$ and $0 \neq y_\alpha \in \mathbf{g}_{-\alpha}$, remembering that $\dim \mathbf{g}_\alpha = 1\, \forall \alpha \in \Phi$. Set $k_\alpha = [x_\alpha, y_\alpha]$. If $h \in \mathbf{h}$ then

$$\begin{aligned}[h, k_\alpha] = [h, [x_\alpha, y_\alpha]] &= [x_\alpha, [y_\alpha, h]] - [y_\alpha, [h, x_\alpha]] \\ &= -[x_\alpha, \alpha(h) y_\alpha] - [y_\alpha, \alpha(h) x_\alpha] = 0.\end{aligned}$$

Since the Cartan subalgebra is a maximal commutative subalgebra of $\mathbf{g}$, we have $k_\alpha \in \mathbf{h}$. Since

$$[k_\alpha, x_\alpha] = \lambda x_\alpha \qquad [k_\alpha, y_\alpha] = -\lambda y_\alpha$$

with $\lambda = \alpha(k_\alpha)$, the subspace spanned by $\{y_\alpha, k_\alpha, x_\alpha\}$ is a subalgebra of $\mathbf{g}$. We want to show that $\lambda \neq 0$.

LEMMA 1.5.1.

(1) *If $\alpha, \beta \in \Phi \cup \{0\}$ and $\alpha + \beta \neq 0$ then $\mathbf{g}_\alpha \perp \mathbf{g}_\beta$ with respect to the Killing form.*

(2) $[x_\alpha, y_\alpha] = (x_\alpha, y_\alpha) h_\alpha\, \forall \alpha \in \Phi$.

PROOF: (1) Let $h \in \mathbf{h}$ such that $(\alpha + \beta)(h) \neq 0$. Choose $0 \neq x_\alpha \in \mathbf{g}_\alpha$ and $0 \neq y_\beta \in \mathbf{g}_\alpha$. Now

$$\begin{aligned}\alpha(h)(x_\alpha, y_\beta) &= ([h, x_\alpha], y_\beta) \\ &= (x_\alpha, [y_\beta, h]) = -\beta(h)(x_\alpha, y_\beta)\end{aligned}$$

and so $(\alpha + \beta)(h)(x_\alpha, y_\beta) = 0$ and $(x_\alpha, y_\beta) = 0$.

(2) Let $h \in \mathbf{h}$. Then

$$\begin{aligned}(h, [x_\alpha, y_\alpha] - (x_\alpha, y_\alpha) h_\alpha) &= (h, [x_\alpha, y_\alpha]) - (x_\alpha, y_\alpha)(h, h_\alpha) \\ &= ([y_\alpha, h], x_\alpha) - (x_\alpha, y_\alpha)\alpha(h) \\ &= \alpha(h)(y_\alpha, x_\alpha) - \alpha(h)(x_\alpha, y_\alpha) = 0.\end{aligned}$$

Thus $\mathbf{h} \perp [x_\alpha, y_\alpha] - (x_\alpha, y_\alpha)h_\alpha$. Since the restriction of the Killing form to $\mathbf{h}$ is nondegenerate, the assertion follows.

Renormalizing the basis by $x = \sqrt{a}x_\alpha$, $y = \sqrt{a}y_\alpha$, $h = ak_\alpha$, where $a = 2/\lambda$ (and $\lambda = (x_\alpha, y_\alpha)\alpha(h_\alpha) = (x_\alpha, y_\alpha)(\alpha, \alpha) \neq 0$) we get the familiar commutation relations $[x, y] = h$, $[h, x] = 2x$, $[h, y] = -2y$ of A_1. We have now proven:

THEOREM 1.5.2. *If $\alpha \in \Phi$ and $0 \neq x_\alpha \in \mathbf{g}_\alpha$, $0 \neq y_\alpha \in \mathbf{g}_{-\alpha}$ then $\{y_\alpha, h_\alpha, x_\alpha\}$ spans a subalgebra isomorphic to A_1.*

A representation $\phi :\rightarrow \mathrm{End} V$ is a *highest weight representation* if there is $0 \neq v \in V$ such that

(1) $\phi(x_\alpha)v = 0 \, \forall \alpha \in \Phi^+$
(2) $\phi(h)v = \lambda(h)v \, \forall h \in \mathbf{h}$
(3) $V = \{\phi(u)v \mid u \in \mathcal{U}(\mathbf{g})\}$

where $\lambda : \mathbf{h} \rightarrow \mathbb{C}$ is some linear form, the *highest weight* of the representation. From now on, when there is no danger of confusion, we shall write shortly zv instead of $\phi(z)v$, when $v \in V$ and $z \in \mathbf{g}$ or $z \in \mathcal{U}(\mathbf{g})$. Consider a finite-dimensional highest weight representation of $\mathbf{g}$ in V, with highest weight vector v. Then for each $\alpha \in \Phi^+$ there has to be $n_\alpha \in \mathbb{N}$ such that $y_\alpha^{n_\alpha+1}v = 0$; otherwise $\{y_\alpha^i v \mid i \in \mathbb{N}\}$ would span an infinite-dimensional subspace because these vectors are linearly independent by the eigenvector property

$$h_\alpha y_\alpha^i v = [\lambda(h_\alpha) - i\alpha(h_\alpha)]y_\alpha^i v.$$

On the other hand, by a similar calculation as was done for A_1,

$$x_\alpha y_\alpha^k v = k(x_\alpha, y_\alpha)\left[\lambda(h_\alpha) - \tfrac{1}{2}(k-1)\alpha(h_\alpha)\right] y_\alpha^{k-1}v.$$

It follows that in the finite-dimensional case (denoting by n_α the smallest number n for which $x_\alpha^{n+1}v = 0$) we must have $\lambda(h_\alpha) = \frac{1}{2}n_\alpha \cdot \alpha(h_\alpha)$. Using the notation of Section 1.3 this relation can be written as

$$n_\alpha = \langle \lambda, \alpha \rangle.$$

In fact, one can prove a stronger result:

THEOREM 1.5.3. *An irreducible highest weight representation of a semisimple Lie algebra is finite-dimensional if and only if the highest weight satisfies $\langle \lambda, \alpha \rangle \in \mathbb{N}$ for all $\alpha \in \Delta$, where $\Delta \subset \Phi^+$ is a system of simple roots. For any linear form $\lambda : \mathbf{h} \rightarrow \mathbb{C}$ which satisfies the above condition* *there is a unique, up to an equivalence, irreducible finite-dimensional representation with highest weight* λ.

We shall give an explicit construction of the irreducible finite-dimensional representations of $\mathbf{g}$. The set of *integral weights* is

$$\Lambda = \{\lambda \in \mathbf{h}^* \mid \langle \lambda, \alpha \rangle \in \mathbf{Z} \, \forall \alpha \in \Delta\}$$

and the set of *dominant integral weights* is

$$\Lambda^+ = \{\lambda \in \mathbf{h}^* \mid \langle \lambda, \alpha \rangle \in \mathbf{N} \, \forall \alpha \in \Delta\}.$$

As an intermediate step we construct for each $\lambda \in \mathbf{h}^*$ the *Verma module*

$$V_\lambda = \mathcal{U}(\mathbf{g})/I_\lambda,$$

where $I_\lambda \subset \mathcal{U}(\mathbf{g})$ is the left ideal generated by all $x_\alpha \in \mathbf{g}_\alpha$ for $\alpha \in \Phi^+$ and by $h - \lambda(h)$ for $h \in \mathbf{h}$. A form $\mu \in \mathbf{h}^*$ is a *weight* of a representation of $\mathbf{g}$ in V if there is $0 \neq v \in V$ such that

$$hv = \mu(h)v \, \forall h \in \mathbf{h}.$$

We denote by $V(\mu) = \{v \in V \mid hv = \mu(h)v \, \forall h \in \mathbf{h}\}$ the corresponding *weight space*.

THEOREM 1.5.4.

(1) *In a Verma module* $\dim V_\lambda(\lambda) = 1$.
(2) *The set of weights of* V_λ *consists of all forms* $\lambda - \sum_{i=1}^{l} k_i \alpha_i$, *where* $k_i \in \mathbf{N}$ *and* $\{\alpha_1, \alpha_2, \ldots, \alpha_l\} = \Delta$.
(3) V_λ *is a direct sum of its weight subspaces* $V_\lambda(\mu)$; *the same holds for each invariant subspace* $W \subset V_\lambda$. *All the weight spaces are finite-dimensional.*

PROOF: Let $v = 1 + I_\lambda \in \mathcal{U}(\mathbf{g})/I_\lambda$. By the definition of I_λ, v is a highest weight vector of weight λ. Let $\Phi^+ = \{\beta_1, \beta_2, \ldots, \beta_n\}$. We can define an ordered basis $\{y_{\beta_1}, \ldots, y_{\beta_n}, h_1, \ldots, h_l, x_{\beta_1}, \ldots, x_{\beta_n}\}$ of $\mathbf{g}$ by choosing the x's and y's as before and choosing a basis $\{h_1, \ldots, h_l\}$ of $\mathbf{h}$. An ordered monomial in $\mathcal{U}(\mathbf{g})$ is written as

$$u = y_{\beta_1}^{r_1} \ldots y_{\beta_n}^{r_n} h_1^{s_1} \ldots h_l^{s_l} x_{\beta_1}^{p_1} \ldots x_{\beta_n}^{p_n}.$$

If any $p_i \neq 0$, then $uv = 0$. On the other hand, v is an eigenvector of each h_i. Thus an element of V_λ is a linear combination of vectors

$w = y_{\beta_1}^{r_1} \dots y_{\beta_n}^{r_n} v$. Now w is an eigenvector of each $h \in$ **h** with the eigenvalue

$$\mu(h) = \lambda(h) - \sum_{k=1}^{n} r_k \beta_k(h).$$

Since each β_k can be written as a linear combination of the simple roots α_i with non-negative coefficients, it follows that $\mu = \lambda$ only if $r_k = 0 \, \forall k$. This proves (1),(2), and the first part of (3). We leave the second part of (3) as an exercise to the reader.

THEOREM 1.5.5. *Let $\lambda \in \Lambda^+$. The subspace M_λ of V_λ consisting of the vectors*

$$\sum_{i=1}^{l} u_i (x_{\alpha_i})^{n_{\alpha_i}+1} \cdot (1 + I_\lambda), \qquad n_\alpha = \langle \lambda, \alpha \rangle,$$

where $u_i \in \mathcal{U}(\mathbf{g})$, is invariant under the action of **g** *and the quotient space $L_\lambda = V_\lambda / M_\lambda$ carries an irreducible finite-dimensional representation of* **g** *with highest weight λ.*

CHAPTER 2 REPRESENTATIONS OF AFFINE KAC-MOODY ALGEBRAS

2.1. Affine Kac-Moody algebras from generalized Cartan matrices

In the first chapter we explained how simple finite-dimensional Lie algebras can be completely characterized in terms of their Cartan matrices or Dynkin diagrams. The same holds for an arbitrary semisimple finite-dimensional Lie algebra. A semisimple Lie algebra is a direct sum of simple ideals which are pairwise orthogonal with respect to the Killing form. It follows that the Cartan matrix of a semisimple Lie algebra decomposes to a block diagonal form, each block representing a simple ideal. Similarly, the Dynkin diagram is a disconnected union of Dynkin diagrams of simple Lie algebras. Next we shall study certain infinite-dimensional Lie algebras which have many similarities with the simple finite-dimensional Lie algebras. In particular, they can be described in terms of generalized Cartan matrices. These algebras were independently introduced in Kac [1968] and Moody [1968].

A *generalized Cartan matrix* is a real $n \times n$ matrix $A = (a_{ij})$ such that

(C1) $a_{ii} = 2$ for $i = 1, 2, \ldots, n$
(C2) a_{ij} is a nonpositive integer for $i \neq j$
(C3) $a_{ij} = 0$ iff $a_{ji} = 0$.

To each generalized Cartan matrix one can associate a Lie algebra using the method of generators and relations as explained in Kac [1985]. However, we shall not take that road since we shall describe in the next section a simple method for constructing those algebras which we shall deal with in this book. The set of *indecomposable* matrices A, i.e., those which cannot be written in a block diagonal form by reordering the indices {1,2, ... ,n}, can be divided into three disjoint subsets:

(1) There is a vector $v \in \mathbf{N}_+^n$ such that also $Av \in \mathbf{N}_+^n$. In this case the Lie algebra $\mathbf{g}(A)$ corresponding to A is a simple finite-dimensional Lie algebra.
(2) There is $v \in \mathbf{N}_+^n$ such that $Av = 0$. The algebra $\mathbf{g}(A)$ is an *affine Lie algebra* and $\dim \mathbf{g}(A) = \infty$.
(3) There is $v \in \mathbf{N}_+^n$ such that $(Av)_i < 0 \, \forall i$.

In this book we shall concentrate to the theory of affine Kac-Moody algebras, which is much better understood than the Kac-Moody algebras of class (3). However, the class (3) contains the subclass of the so-called *hyperbolic* Lie algebras which seem to have interesting mathematical structures; see the discussion in Feingold and Frenkel [1983], where the hyperbolic algebra corresponding to the Cartan matrix

$$A = \begin{pmatrix} 2 & -2 & 0 \\ -2 & 2 & -1 \\ 0 & -1 & 2 \end{pmatrix}$$

has been studied in detail. We shall now give a list of the Dynkin diagrams of the affine Lie algebras. For the proofs see Kac [1985]. The diagrams with the upper index 1 correspond to the *untwisted affine Lie algebras* and the rest describe the *twisted affine Lie algebras* . The reason for this division will become apparent in the next section. Note that each of the Dynkin diagrams is obtained by adjoining the node labeled by 0 to a Dynkin diagram of a simple finite-dimensional algebra.

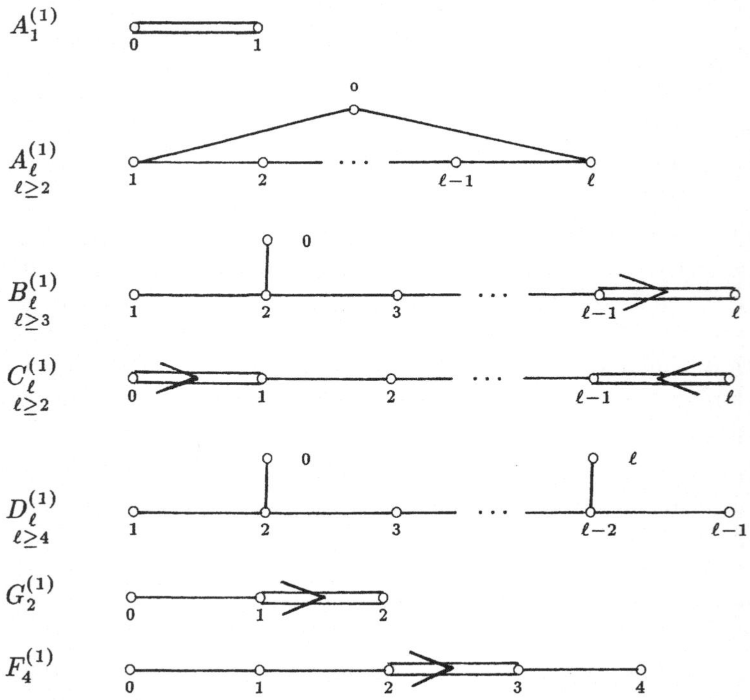

Figure 2a Dynkin diagrams for affine Lie algebras.

2.2. Affine Lie algebras as central extensions of loop algebras: the untwisted case

Let **g** be an arbitrary finite-dimensional complex Lie algebra and denote by $S^1\mathbf{g}$ the space of smooth maps (loops) $f : S^1 \to \mathbf{g}$, where S^1 is the unit circle. Consider $S^1\mathbf{g}$ as a vector space by pointwise addition of the loops and the natural multiplication of functions by complex numbers. Furthermore, $S^1\mathbf{g}$ is naturally an infinite-dimensional Lie algebra through the commutator $[\cdot,\cdot]_{(0)}$,

$$[f,g]_{(0)} = [f(z), g(z)], \ z \in S^1.$$

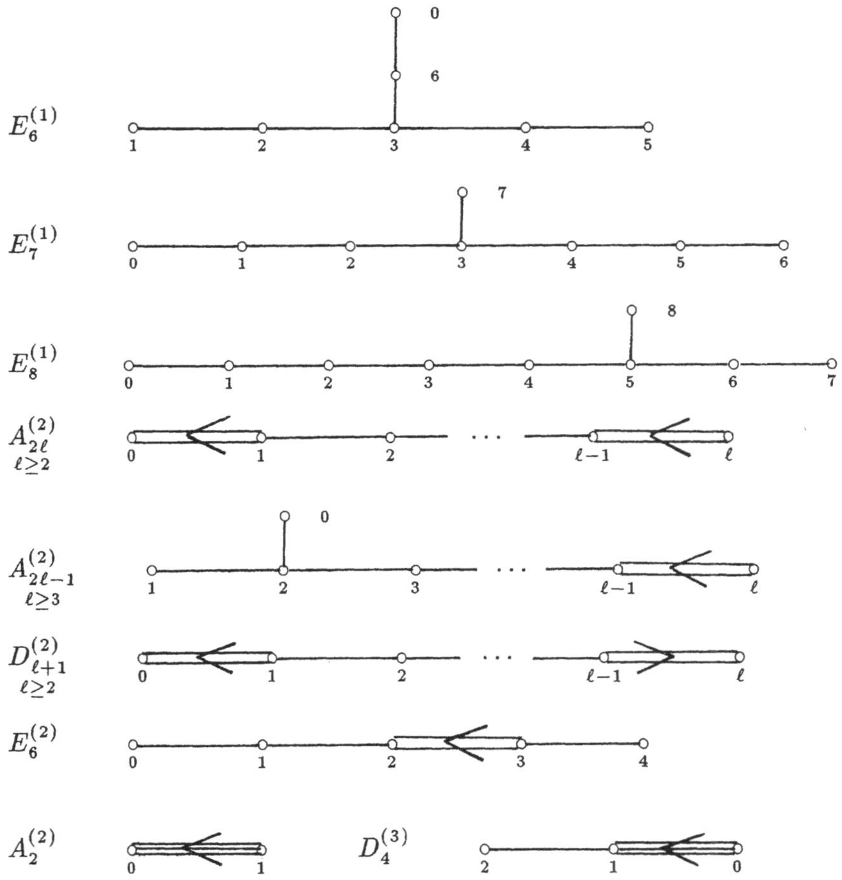

Figure 2b Dynkin diagrams for affine Lie algebras.

A smooth function on S^1 is always square-integrable and a basis for square-integrable functions is given by the Fourier modes. Let $\{T_1, T_2, \ldots, T_r\}$ be a basis of $\mathbf{g}$ and denote

$$T_a^n = e^{in\phi} T_a,$$

where $0 \leq \phi \leq 2\pi$ parametrizes the circle and $n \in \mathbf{Z}$. Define the structure constants of $\mathbf{g}$ by

$$[T_a, T_b] = \sum_{c=1}^{r} \lambda_{ab}^c T_c.$$

Then

$$[T_a^n, T_b^m]_{(0)} = \sum_c \lambda_{ab}^c T_c^{n+m}.$$

Let $(\cdot,\cdot) : \mathbf{g} \times \mathbf{g} \to \mathbf{C}$ be any *invariant* bilinear symmetric form, that means

$$([x,y],z) = (y,[z,x]) \,\forall x,y,z \in \mathbf{g}.$$

Let $\widehat{\mathbf{g}}$ denote the vector space $S^1\mathbf{g} \oplus \mathbf{C}$. We define in $\widehat{\mathbf{g}}$ the following commutator:

$$\text{(A)} \qquad [(f,\alpha),(g,\beta)] = \left([f,g]_{(o)}, \frac{k}{2\pi}\int_0^{2\pi} (f(\phi), g'(\phi))d\phi\right).$$

Here $0 \neq k \in \mathbf{C}$ is an arbitrary constant. For brevity, we shall denote the pair $(f,0)$ by f. For the Fourier modes the equation (A) gives

$$\text{(B)} \qquad [T_a^n, T_b^m] = \sum \lambda_{ab}^c T_c^{n+m} + km\delta_{n,-m}(T_a, T_b).$$

Next let $\mathbf{g}$ be a *simple* Lie algebra. We shall show that the commutation relations (B) define an untwisted affine Lie algebra. Choose a Cartan subalgebra $\mathbf{h} \subset \mathbf{g}$. We shall identify $\mathbf{g}$ with a subalgebra of $\widehat{\mathbf{g}}$ by $\mathbf{g} \ni x \mapsto$ (the constant function $x : S^1 \to \mathbf{g}$ taking the value x). We can write

$$\widehat{\mathbf{g}} = \mathbf{C} \oplus \sum_{n \in \mathbf{Z}} \mathbf{g}^{(n)},$$

where $\mathbf{g}^{(n)}$ is spanned by the vectors T_a^n, $1 \leq a \leq r$, and $\mathbf{C}$ stands for the *center* of $\widehat{\mathbf{g}}$ spanned by the vector $k = (0,k)$. In particular, $\mathbf{g}^{(0)} = \mathbf{g}$.

Let Φ be the system of roots for $(\mathbf{g}, \mathbf{h})$ and $\Delta \subset \Phi^+$ a system of simple roots. Choose $0 \neq x_\alpha \in \mathbf{g}_\alpha, 0 \neq y_\alpha \in \mathbf{g}_{-\alpha} \forall \alpha \in \Phi^+$. From (B) we get

$$\begin{aligned}
[h, x_\alpha^n] &= \alpha(h)x_\alpha^n, \\
[h, y_\alpha^n] &= -\alpha(h)y_\alpha^n, \\
[k, x_\alpha^n] &= [k, y_\alpha^n] = 0,
\end{aligned}$$

where we have also used Lemma 1.5.1 (1). We notice that if we define $\mathbf{h} \oplus \mathbf{C}k$ to be the Cartan subalgebra of $\widehat{\mathbf{g}}$, then each of the roots α has an infinite multiplicity. For this reason we extend the algebra $\widehat{\mathbf{g}}$ by an element d (and to add confusion we shall denote the new algebra also by $\widehat{\mathbf{g}}$) which has the following commutation relations:

$$\text{(C)} \qquad [d, T_a^n] = nT_a^n \qquad [d, k] = 0.$$

A concrete realization for the new element is $d = -i\frac{d}{d\phi}$. We then define the Cartan subalgebra of $\widehat{\mathbf{g}}$ as

$$\widehat{\mathbf{h}} = \mathbf{h} \oplus \mathbf{C}k \oplus \mathbf{C}d.$$

Correspondingly, we write a root of $(\widehat{\mathbf{g}}, \widehat{\mathbf{h}})$ in the component form $(\alpha, 0, n)$; this root corresponds to the root vector x_α^n. Thus the set of nonzero roots for $(\widehat{\mathbf{g}}, \widehat{\mathbf{h}})$ is

$$\widehat{\Phi} = \{(\pm\alpha, 0, n) \mid \alpha \in \Phi^+, n \in \mathbf{Z}\} \cup \{(0, 0, n) \mid 0 \neq n \in \mathbf{Z}\}.$$

The root subspace of the root $(0, 0, n)$, $(n \neq 0)$ is spanned by the vectors h_i^n, where $\{h_1, \ldots, h_l\}$ is an orthonormal basis of $\mathbf{h}$. Each of the roots $(\pm\alpha, 0, n)$ has multiplicity $=1$ and each of the nonzero roots $(0, 0, n)$ has multiplicity $=l$. We define a system of simple roots

$$\widehat{\Delta} = \{(\alpha, 0, 0) \mid \alpha \in \Delta\} \cup \{(-\psi, 0, 1)\},$$

where ψ is the *highest root* of $(\mathbf{g}, \mathbf{h})$, that is, ψ is the highest weight of the adjoint representation $ad(x)y = [x, y]$ of $\mathbf{g}$. The set of positive roots is then

$$\widehat{\Phi}^+ = \{(\alpha, 0, n) \mid \alpha \in \Phi, n > 0\} \cup \{(\alpha, 0, 0) \mid \alpha \in \Phi^+\}$$

and $\widehat{\Phi}^- = -\widehat{\Phi}^+$ as in the case of finite-dimensional semisimple Lie algebras.

Example 2.2.1. Let $\mathbf{g} = A_l$. We use the notation of Section 1.2. The highest weight vector in the adjoint representation is $e_{1,l+1}$ since $[x_{\alpha_{ij}}, e_{1,l+1}] = [e_{ij}, e_{1,l+1}] = 0$ for $i < j$. The highest root is thus $\alpha_{1,l+1} = \alpha_{12} + \alpha_{23} + \cdots + \alpha_{l,l+1} = \alpha_1 + \alpha_2 + \cdots + \alpha_l$.

Exercise 2.2.2. Determine the highest roots for the Lie algebras B_l, C_l, D_l.

There is an *invariant symmetric bilinear form* on $\widehat{\mathbf{g}}$ given by

$$(f,g) = \frac{1}{2\pi}\int_0^{2\pi} (f(\phi), g(\phi))d\phi \tag{B1}$$

$$(k,f) = (d,f) = 0 \qquad f \in S^1\mathbf{g} \tag{B2}$$

$$(k,k) = (d,d) = 0 \tag{B3}$$

$$(k,d) = 1 \tag{B4}$$

where the form under the integral sign is the Killing form of $\mathbf{g}$.

PROPOSITION 2.2.3. *Up to a multiplicative constant any invariant symmetric bilinear form on $\widehat{\mathbf{g}}$ is obtained from the form above by replacing $(d,d) = 0$ by $(d,d) = s$, where $s \in \mathbf{C}$ is an arbitrary constant.*

PROOF: If $(\cdot,\cdot)$ is invariant we have

$$\begin{aligned}([d, T_a^n], T_b^m) &= n(T_a^n, T_b^m)\\ &= (T_a^n, [T_b^m, d]) = -m(T_a^n, T_b^m)\end{aligned}$$

and so $(T_a^n, T_b^m) = 0$ if $n \neq -m$. For a fixed n, write $\eta_{ab} = (T_a^n, T_b^{-n})$. Using the invariance of the Killing form of $\mathbf{g}$ we get $\lambda_{ab}^c = -\lambda_{ac}^b$ in the orthonormal basis $\{T_a\}$. Comparing with

$$\begin{aligned}([T_c, T_a^n], T_b^{-n}) &= \sum_e \lambda_{ca}^e (T_e^n, T_b^{-n}) = \sum \lambda_{ca}^e \eta_{eb}\\ &= (T_a^n, [T_b^{-n}, T_c]) = -\sum \lambda_{cb}^e (T_a^n, T_e^{-n})\\ &= -\sum \lambda_{cb}^e \eta_{ae}\end{aligned}$$

we conclude that the matrix η commutes with each of the matrices $\lambda_a = (\lambda_{a,bc})$, $\lambda_{a,bc} = -\lambda_{a,cb} = \lambda_{ac}^e g_{eb}$, and $g_{ab} = (T_a, T_b)$. [The antisymmetry of λ_a follows from the invariance of the form $(\cdot,\cdot)$ on $\mathbf{g}$.] The adjoint representation is irreducible for any simple Lie algebra and thus by Schur's lemma the matrix η has to be proportional to the identity,

$$(T_a^n, T_b^{-n}) = \xi(n)\delta_{ab}.$$

From

$$([T_a^1, T_b^n], T_c^{-n-1}) = (T_b^n, [T_c^{-n-1}, T_a^1])$$

we conclude that $\xi(n) = \xi(n+1)\,\forall n$. On the other hand,

$$\frac{1}{2\pi}\int (T_a^n, T_b^m)d\phi = \delta_{ab}\delta_{n,-m},$$

so after a renormalization the inner product takes the form (B1). We leave as an exercise for the reader to complete the proof by showing that with this normalization (B2),(B4) and $(k,k) = 0$ holds.

Let $\{h_1, h_2, \ldots, h_l\}$ be an orthonormal basis of $\mathbf{h}$. Then

$$\{h_1, \ldots, h_l, k, d\}$$

is a basis of $\widehat{\mathbf{h}}$ and the restriction of the invariant form (B) to $\widehat{\mathbf{h}}$ is described by the matrix

$$\begin{pmatrix} 1 & 0 & \ldots & 0 & 0 \\ 0 & 1 & \ldots & 0 & 0 \\ \vdots & \vdots & \ddots & \vdots & \vdots \\ 0 & 0 & \ldots & 0 & 1 \\ 0 & 0 & \ldots & 1 & 0 \end{pmatrix}.$$

It is nondegenerate but *Lorentzian* in signature: In the basis

$$\left\{h_1, \ldots, h_l, \frac{1}{\sqrt{2}}(k+d), \frac{1}{\sqrt{2}}(k-d)\right\}$$

it takes the form diag$(+1, \ldots, +1, -1)$. If $\mu, \mu' \in \widehat{\mathbf{h}}^*$ are arbitrary linear forms, then the dual of the inner product (B) on $\widehat{\mathbf{h}}^* \times \widehat{\mathbf{h}}^*$ is

$$(\mu, \mu') = \sum_{i=1}^{l} \mu(h_i)\mu'(h_i) + \mu(d)\mu'(k) + \mu(k)\mu'(d).$$

We can now compute the scalar products between the simple roots $\widehat{\Delta}$. We shall work only through the case $\mathbf{g} = A_l$; the other cases are handled in the same way. (All we need to know is the highest root ψ as a linear combination of the simple roots and the Dynkin diagram or the Cartan matrix of $\mathbf{g}$.)

Example 2.2.4. $\mathbf{g} = A_l$. Now $\psi = \alpha_1 + \alpha_2 + \cdots + \alpha_l$, where the α_i's are the simple roots. If $2 \leq i \leq l-1$, then

$$(\psi, \alpha_i) = (\alpha_{i-1} + \alpha_i + \alpha_{i+1}, \alpha_i) = -1 + 2 + (-1) = 0.$$

The only nonzero products involving ψ are $(\psi, \alpha_1) = (\alpha_1 + \alpha_2, \alpha_1) = 1$ and $(\psi, \alpha_l) = (\alpha_{l-1} + \alpha_l, \alpha_l) = 1$. Denoting the simple root $(-\psi, 0, 1)$ of $\widehat{\mathbf{g}}$ by α_0 we obtain the Dynkin diagram of $\widehat{\mathbf{g}}$ from that of $\mathbf{g}$ by adjoining the node labeled by 0 and connecting the new node to the nodes 1 and l.

Exercise 2.2.5. Show that the Dynkin diagram of $\widehat{\mathbf{g}}$ is

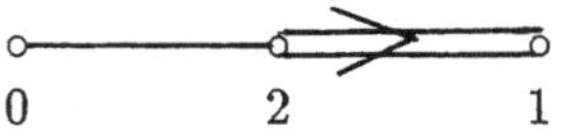

when $\mathbf{g} = G_2$.

2.3. Affine Lie algebras as central extensions of loop algebras: the twisted case

Let again $\mathbf{g}$ be a simple complex finite-dimensional Lie algebra and let $\sigma : \mathbf{g} \to \mathbf{g}$ be an automorphism such that $\sigma^N = 1$ for some integer $N > 0$; let N be the smallest positive integer for which this holds. Set $\epsilon = e^{2\pi i/N}$. Let $S^1_\sigma \mathbf{g}$ consist of the loops $f : S^1 \to \mathbf{g}$ such that

$$f(\epsilon^{-1} z) = \sigma f(z).$$

Clearly $S^1_\sigma \mathbf{g} \subset S^1 \mathbf{g}$ is a linear subspace and

$$\begin{aligned}[f, g](\epsilon_{-1} z) &= [f(\epsilon^{-1} z), g(\epsilon^{-1} z)] = [\sigma f(z), \sigma g(z)] \\ &= \sigma[f(z), g(z)] = \sigma[f, g](z),\end{aligned}$$

so $S^1_\sigma \mathbf{g}$ is closed under commutation. We define

$$\widehat{\mathbf{g}}(\sigma) = S^1_\sigma \mathbf{g} \oplus \mathbf{C}k \oplus \mathbf{C}d$$

as a vector space and we define the commutator by (A) and (C) as before. When $\sigma = 1$ we have $\widehat{\mathbf{g}}(\sigma) = \widehat{\mathbf{g}}$.

Example 2.3.1. Let $\mathbf{g} = A_2$. Define $\sigma : \mathbf{g} \to \mathbf{g}$ by

$$\sigma(e_{12}) = e_{23}, \ \sigma(e_{23}) = e_{12}, \ \sigma(e_{31}) = -e_{31}.$$

From the commutation relations follows then that $\sigma(e_{13}) = -e_{13}$, $\sigma(e_{32}) = e_{21}$, $\sigma(e_{21}) = e_{32}$, $\sigma(e_{11} - e_{22}) = e_{22} - e_{33}$, and $\sigma(e_{22} - e_{33}) = e_{11} - e_{22}$. Now $\sigma^2 = 1$ and $\epsilon = -1$. A basis for the polynomial loops in $S^1_\sigma \mathbf{g}$ is defined by

$$(e_{11} - e_{33}) z^{2n}, \ (e_{12} + e_{23}) z^{2n}; \ (e_{21} + e_{32}) z^{2n}, \ e_{13} z^{2n+1}$$
$$e_{31} z^{2n+1}, \ (e_{12} - e_{23}) z^{2n+1}, \ (e_{21} - e_{32}) z^{2n+1}, \ (e_{11} + e_{33} - 2e_{22}) z^{2n+1},$$

where $n \in \mathbf{Z}$. The coefficients of z^{2n} span the eigenspace $\mathbf{g}(1) \subset \mathbf{g}$ corresponding to the eigenvalue $+1$ of σ and the coefficients of z^{2n+1} correspond to the eigenvalue -1. A Cartan subalgebra of $\widehat{\mathbf{g}}(\sigma)$ is spanned by the vectors k, d, and $h = e_{11} - e_{33}$. In the ordered basis $\{h, k, d\}$ the nonzero roots are

$$\{(0,0,n) \mid 0 \neq n \in \mathbf{Z}\} \cup \{(\pm 1, 0, n) \mid n \in \mathbf{Z}\} \cup \{\pm 2, 0, 2n+1) \mid n \in \mathbf{Z}\}.$$

A system of simple roots is then

$$\widehat{\Delta}(\sigma) = \{(1,0,0), (-2,0,1)\} = \{\alpha_1, \alpha_0\}$$

and the positive roots are $\{(1,0,n-1), (-2,0,2n-1), (0,0,n) \mid n > 0\}$. Now $\langle \alpha_0, \alpha_1 \rangle = -1$ and $\langle \alpha_1, \alpha_0 \rangle = -4$ so that the Dynkin diagram is

$A_2^{(2)}$

In general, given an automorphism $\sigma : \mathbf{g} \to \mathbf{g}$ with $\sigma^N = 1$ (N minimal) one can write $\mathbf{g}$ as direct sum of eigenspaces

$$\mathbf{g} = \bigoplus_{j=0}^{N-1} \mathbf{g}(\epsilon^j).$$

Since $[\mathbf{g}(\epsilon^j), \mathbf{g}(\epsilon^i)] \subset \mathbf{g}(\epsilon^{i+j})$, only the subspace $\mathbf{g}(1)$ is a subalgebra. In the above example, $\mathbf{g}(1) \cong A_1$. One has a grading for $S^1_\sigma \mathbf{g}$,

$$S^1_\sigma \mathbf{g} = \bigoplus_{j=0}^{N-1} (\mathbf{g}(\epsilon^j) \otimes V_j(z)),$$

where $V_j(z)$ consists of linear combinations of the monomials z^{nN-j}, $n \in \mathbf{Z}$. The Cartan subalgebra of $\widehat{\mathbf{g}}(\sigma)$ consists of the Cartan subalgebra of $\mathbf{g}(1)$ and the elements k and d. One can show that with respect to this Cartan subalgebra a system of simple roots of $\widehat{\mathbf{g}}(\sigma)$ consists of the roots $(\alpha, 0, 0)$, where α goes through the simple roots of $\mathbf{g}(1)$, and the root $(-\psi, 0, 1)$, where ψ is a certain root of $\mathbf{g}(1)$. We are not going to study the twisted algebras in detail; see [Kac, 1985] for more information.

Exercise 2.3.2. The Dynkin diagram of D_4 is

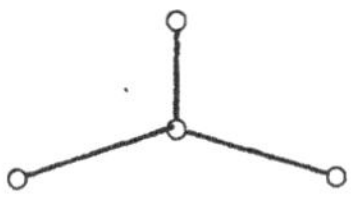

Rotations by the angles $k\dot{2}\pi/3$ are symmetries of the diagram. Corresponding to the rotation $2\pi/3$ construct an automorphism of D_4 which permutes the root subspaces $\mathbf{g}_{\alpha_1}, \mathbf{g}_{\alpha_1}$, and $\mathbf{g}_{\alpha_3}$. Construct the affine Lie algebra $D_4^{(3)}$ using this automorphism (of order 3). Show that the Dynkin diagram is

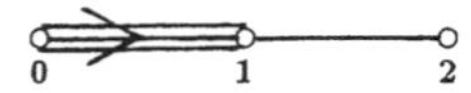

2.4. The highest weight representations of affine Lie algebras

Let $\mathbf{a}$ be an affine Lie algebra, $\mathbf{h} \subset \mathbf{a}$ a Cartan subalgebra, $\Delta \subset \mathbf{h}^*$ a system of simple roots, and $\Phi^+ \supset \Delta$ the set of positive roots. There is a splitting

$$\mathbf{a} = \mathbf{n}_- \oplus \mathbf{h} \oplus \mathbf{n}_+,$$

where the subalgebra $\mathbf{n}_+$ (respectively, $\mathbf{n}_-$) is spanned by the root subspaces $\mathbf{a}_\alpha$ corresponding to positive (respectively, negative) roots. Let $\lambda \in \mathbf{h}^*$ be arbitrary and define the *Verma module* as in the finite-dimensional case,

$$V_\lambda = \mathcal{U}(\mathbf{a})/I_\lambda,$$

where the left ideal is generated by $\mathbf{n}_+$ and the elements $h - \lambda(h)$, $h \in \mathbf{h}$. As in Section 1.5, the space V_λ is a direct sum of its weight subspaces $V_\lambda(\mu)$; this and the other assertions of Theorem 1.5.4 are proved exactly in the same way as for a finite-dimensional semisimple Lie algebra.

THEOREM 2.4.1. *The Verma module V_λ contains a unique maximal proper submodule M_λ (i.e., a proper invariant subspace $M \subset V_\lambda$ such that if $M' \subset V_\lambda$ is an invariant subspace containing M, then $M' = M$ or $M' = V_\lambda$) and $L_\lambda = V_\lambda/M_\lambda$ carries an irreducible highest weight representation of $\mathbf{a}$ with the highest weight $= \lambda$.*

PROOF: Let $M \underset{\neq}{\subset} V_\lambda$ be an invariant subspace. We can write

$$M = \underset{\mu}{\oplus} M(\mu),$$

by Theorem 1.5.4 (3). Now $M(\lambda) = 0$; otherwise, choosing $0 \neq v \in M(\lambda)$ and taking account of $dim V_\lambda(\lambda) = 1$, we would have $M \supset \mathcal{U}(\mathbf{a})v = V_\lambda$. Thus defining M_λ as the sum of all invariant proper subspaces we get a maximal invariant subspace not containing the vector $v_\lambda = 1 + I_\lambda$ in V_λ. The highest weight vector in L_λ is $v_\lambda + M_\lambda \neq 0$.

Before studying the irreducible modules L_λ in more detail, we need some more information about the structure of affine Lie algebras. Let

A be a linear operator in a vector space V. We say that A is *locally nilpotent* if for any $x \in V$ there is an integer $n = n(x) \in \mathbf{N}$ such that $A^n x = 0$. Let $0 \neq e_i \in a_{\alpha_i}$ and $0 \neq f_i \in a_{-\alpha_i}$ for $i = 0, 1, \ldots, l$, where $\{\alpha_0, \alpha_1, \ldots, \alpha_l\}$ is a set of simple roots; we shall normalize the vectors such that $[e_i, f_i] = h_{\alpha_i}$, $(e_i, f_i) = 1$. In the case of a finite-dimensional semisimple Lie algebra it is obvious that the operators $\text{ad}e_i$ and $\text{ad}f_i$ are locally nilpotent. By inspecting the root systems of the untwisted affine Lie algebras one can see that if β is a root then $\beta + n\alpha_i$ is a root only for *finitely* many values of $n \in \mathbf{Z}$. We state without proof that the same remains true for the twisted algebras. In conclusion:

THEOREM 2.4.2. *The operators $\text{ad}e_i$ and $\text{ad}f_i$ are locally nilpotent in any affine Lie algebra.*

In general, we call an **a**-module V *integrable*, if $\text{ad}e_i$ and $\text{ad}f_i$ are locally nilpotent for $0 \leq i \leq l$ and if V is a direct sum of weight subspaces. In particular by 2.4.2 the space **a** considered as an **a**-module through the adjoint action is an integrable **a**-module.

THEOREM 2.4.3. *Let V be an integrable* **a**-*module. If λ is a weight of V and if $\lambda + \alpha_i$ (respectively, $\lambda - \alpha_i$) is not a weight of V, then $(\lambda, \alpha_i) \geq 0$ [respectively, $(\lambda, \alpha_i) \leq 0$]. If λ is any weight of V, then $\lambda' = \lambda - \langle \lambda, \alpha_i \rangle \alpha_i$ is also a weight and $\dim V(\lambda) = \dim V(\lambda')$.*

PROOF: In the finite-dimensional case we proved that if α is any root, then the vectors x_α, y_α, and h_α span a subalgebra isomorphic to A_1. From our construction of the root systems in the untwisted case it is not difficult to see that the same holds for the simple roots of an affine Lie algebra. [It is *not* true for the nonsimple roots $(0, 0, n)$.] One can show that this result is valid also for the twisted affine algebras. For any fixed i, let A_1 be the Lie algebra spanned by e_i, f_i, and h_{α_i}. Let $0 \neq v$ be a vector of weight λ in V. Because V is integrable, $\mathcal{U}(A_1)v$ is a finite-dimensional A_1-module. If $\lambda + \alpha_i$ is not a weight, then $e_i v = 0$ and so $\langle \lambda, \alpha_i \rangle$ is a non-negative integer by our earlier analysis of A_1-modules in Section 1.4. If $\lambda - \alpha_i$ is not a weight, then $f_i v = 0$ and so v is the lowest weight vector for a finite-dimensional A_1-module. The lowest weight of an A_1-module is minus the highest weight; thus in this case $\langle \lambda, \alpha_i \rangle \leq 0$ and $(\lambda, \alpha_i) \leq 0$. If $0 \neq v \in V(\lambda)$, then

$$h_{\alpha_i} v = \lambda(h_{\alpha_i}) v = (\lambda, \alpha_i) v$$

and similarly $h_{\alpha_i} v' = (\lambda', \alpha_i) v'$ if there is $0 \neq v' \in V(\lambda')$. But

$$(\lambda', \alpha_i) = (\lambda, \alpha_i) - \langle \lambda, \alpha_i \rangle (\alpha_i, \alpha_i) = -(\lambda, \alpha_i).$$

Since in a finite-dimensional A_1-module the weights appear symmetrically (i.e., μ is a weight iff $-\mu$ is a weight) we can conclude that also λ' is a weight.

For each $0 \leq i \leq l$ we define the linear map

$$\sigma_i : \mathbf{h}^* \to \mathbf{h}^*, \ \sigma_i(\lambda) = \lambda - \langle \lambda, \alpha_i \rangle \alpha_i.$$

Let $W = W(\mathbf{a}, \mathbf{h})$ be the group generated by the *fundamental reflections* σ_i; W is called the *Weyl group* of $(\mathbf{a}, \mathbf{h})$. Note that $\langle \lambda, \alpha_i \rangle = 2(\lambda, \alpha_i)/(\alpha_i, \alpha_i)$ is well-defined for the simple roots because of $(\alpha_i, \alpha_i) \neq 0$. In the case of a finite-dimensional semisimple algebra the Weyl group can equivalently be defined as the group generated by *all* reflections σ_α, corresponding to an arbitrary nonzero root, because in that case the inner product is positive definite. From the Theorem 2.4.3 follows immediately that the Weyl group maps in an integrable representation the weight system onto itself. In particular, the set of roots Φ is mapped onto itself by W as a consequence of the facr that the adjoint representation is integrable. As in the finite-dimensional case, we define for the affine algebras

$$\Lambda = \{\lambda \in \mathbf{h}^* \mid \langle \lambda, \alpha_i \rangle \in \mathbf{Z} \, \forall i\}$$
$$\Lambda_+ = \{\lambda \in \Lambda \mid \langle \lambda, \alpha_i \rangle \geq 0 \, \forall i\}.$$

Let $\lambda \in \Lambda_+$. Using the fact that $(\alpha_0, \alpha_0) = \psi^2$ we observe that $\lambda(k) = \frac{\psi^2}{2} x$, where x is a positive integer called the *level* of λ.

THEOREM 2.4.4. *The irreducible highest weight module L_λ is integrable if and only if $\lambda \in \Lambda_+$.*

PROOF: 1) Let L_λ be integrable and let $v \neq 0$ be the vector of highest weight. Then there exists a smallest non-negative integer n_i such that $(f_i)^{n_i+1} v = 0$. Consequently

$$0 = e_i (f_i)^{n_i+1} v = (n_i + 1)[\lambda(h_i) - \tfrac{1}{2} n_i \alpha_i(h_i)] f_i^{n_i} v,$$

where $h_i = [e_i, f_i] = \frac{2}{(\alpha_i, \alpha_i)} h_{\alpha_i}$. Thus

$$0 = \lambda(h_{\alpha_i}) - \frac{1}{2} n_i \alpha_i(h_{\alpha_i}) = (\lambda, \alpha_i) - \frac{1}{2} n_i (\alpha_i, \alpha_i),$$

so that $\langle \lambda, \alpha_i \rangle = n_i$ is a non-negative integer.

2) Let $\lambda \in \Lambda_+$. By the same formula as above,

$$(f_i)^{\langle \lambda, \alpha_i \rangle + 1} = 0, \ 0 \leq i \leq l.$$

Let U be the maximal subspace of L_λ where the action of $\mathbf{a}$ is locally nilpotent; $U \neq 0$ because of $v \in U$. We shall show that U is invariant under the action of $\mathbf{a}$. Let $u \in U$ and $x \in \mathbf{a}$. Now for any $y \in \mathbf{a}$,

$$y^n x u = \sum_{j=0}^{n} \binom{n}{j} [(\mathrm{ad}y)^j x] y^{n-j} u,$$

which is proven by induction on n. For large enough j, $(\mathrm{ad}y)^j x = 0$ for $y = e_i$ or $y = f_i$. On the other hand, $y^{n-j}u = 0$ for large enough $n - j$ when $y = e_i, f_i$. Thus it follows that $y^n x u = 0$ for some n, when $y = e_i$ or $y = f_i$. Because of the irreducibility of L_λ we must have $U = L_\lambda$.

Let $\lambda, \mu \in \mathbf{h}^*$ and $\lambda' = \lambda - \langle \lambda, \alpha \rangle \alpha$, $\mu' = \mu - \langle \mu, \alpha \rangle \alpha$, where α is any root. Then

$$\begin{aligned}(\lambda', \mu') =& (\lambda - \langle \lambda, \alpha \rangle \alpha, \mu - \langle \mu, \alpha \rangle \alpha = (\lambda, \mu) - (\lambda, \alpha)\langle \mu, \alpha \rangle \\ & - \langle \lambda, \alpha \rangle (\alpha, \mu) + \langle \lambda, \alpha \rangle \langle \mu, \alpha \rangle (\alpha, \alpha) = (\lambda, \mu)\end{aligned}$$

by using $\langle \lambda, \alpha \rangle (\alpha, \alpha) = 2(\lambda, \alpha)$. Thus the inner product in $\mathbf{h}^*$ is *invariant under the action of the Weyl group.*As a consequence, also the brackets $\langle \lambda, \alpha \rangle$ are invariant under W.

LEMMA 2.4.5. *Let $\lambda \in \Lambda_+$ and let μ be a weight of L_λ. Then $(\lambda, \lambda - \mu) \geq 0$ and the equality holds if and only if $\lambda = \mu$.*

PROOF: Let $\lambda \neq \mu$. Let $\mathbf{m}$ be the subalgebra of $\mathbf{n}_-$ generated by those elements e_i, f_i for which $\langle \lambda - \mu, \alpha_i \rangle \neq 0$ (denote this set of indices i by S). Now

$$L_\lambda(\mu) \subset \mathcal{U}(\mathbf{n}_-)\mathbf{m}v$$

where $v \neq 0$ is the highest weight vector. By 2.4.3, $(\lambda, \alpha_i) \neq 0$ at least for one index $i \in S$ (otherwise $\mathbf{m}v = 0$ and thus $L_\lambda(\mu) = 0$). We can write $\lambda - \mu = \sum n_j \alpha_j$, where the n_j's are non-negative integers. Now $(\lambda, \lambda - \mu) = \sum (\lambda, \alpha_j) n_j$. Each term is non-negative and in the case $\mu \neq \lambda$ at least one is positive.

LEMMA 2.4.6. *Let $\lambda \in \Lambda_+$ and let μ be a weight of L_λ. Then there is $w \in W$ such that $w \cdot \mu \in \Lambda_+$.*

PROOF: Writing $\mu = \sum k_i \alpha_i$ we set $\mathrm{ht}\mu = \sum k_i$. Choose $w \in W$ such that $\mathrm{ht}(\lambda - w \cdot \mu)$ is minimal. Now $\langle w \cdot \mu, \alpha_i \rangle \geq 0$; otherwise $\mathrm{ht}\lambda - \sigma_{\alpha_i} w \cdot \mu < \mathrm{ht}\lambda - w \cdot \mu$.

We define $\rho \in \mathbf{h}^*$ by $\rho(h_{\alpha_i}) = 2(\alpha_i, \alpha_i)$, $0 \leq i \leq l$ and $\rho(k) = 0$.

PROPOSITION 2.4.7. *Let $\lambda \in \Lambda_+$ and let μ, ν be weights of L_λ. Then*

(1) $(\lambda, \lambda) - (\mu, \nu) \geq 0$; *the equality holds if and only if* $\mu = \nu$ *and* $\mu \in W \cdot \lambda$.

(2) $|\lambda + \rho|^2 - |\mu + \rho|^2 \geq 0$; *the equality holds only if* $\mu = \lambda$.

PROOF: (1) Using the invariance of the inner product under the Weyl group action and Lemma 2.4.6 we can assume that $\mu \in \Lambda_+$. We can write $(\lambda, \lambda) - (\mu, \nu) = (\lambda, \lambda - \mu) + (\mu, \lambda - \nu)$, both terms being non-negative (see the proof of Lemma 2.4.5). If the equality holds then $(\lambda, \lambda - \mu) = 0 = (\mu, \lambda - \nu)$ and from 2.4.5 follows that $\lambda = \mu$ and thus also $\mu = \nu$.

(2) Write

$$(\lambda + \rho, \lambda + \rho) - (\mu + \rho, \mu + \rho) = [(\lambda, \lambda) - (\mu, \mu)] + 2(\rho, \lambda - \mu).$$

The first term is non-negative by (1) and the second by a computation like in 2.4.5. Since $(\rho, \alpha_i) = \frac{1}{2}(\alpha_i, \alpha_i) > 0 \, \forall i$, in the case of equality sign we must have $(\lambda - \mu, \alpha_i) = 0 \, \forall i$.

There is one more property of the Weyl group which we shall need in the next section but which we state without proof:

LEMMA 2.4.8. *Let $w \in W$ and $\lambda \in \Lambda_+$ such that $(\lambda, \alpha) > 0$ for all $\alpha \in \Delta$. Then $w\lambda = \lambda$ implies $w = 1$.*

We shall define *an antilinear antiautomorphism* θ of $\mathbf{a}$ by

$$\theta(e_i) = f_i, \ \theta(f_i) = e_i, \ \theta(h_{\alpha_i}) = h_{\alpha_i}, \ \theta(d) = d, \ \theta(k) = k$$

for all $0 \leq i \leq l$. These relations determine θ uniquely, since all the vectors corresponding to positive (respectively, negative) roots are obtained by taking commutators of the elements e_i (respectively, f_i), and the vectors h_{α_i} form a basis of $\mathbf{h}$. Antilinearity means that $\theta(ax + by) = \bar{a}\theta(x) + \bar{b}\theta(y)$ for $x, y \in \mathbf{a}$ and $a, b \in \mathbf{C}$ and the antiautomorphism property is $\theta([x, y]) = -[\theta(x), \theta(y)]$. The antiautomorphism θ can be extended to an antilinear antiautomorphism of the enveloping algebra of $\mathbf{a}$ by setting $\theta(x_1 x_2 \ldots x_n) = \theta(x_n) \ldots \theta(x_2)\theta(x_1)$, $x_i \in \mathbf{a}$. It satisfies $\theta(uv) = \theta(v)\theta(u)$ for $u, v \in \mathcal{U}(\mathbf{a})$.

Exercise 2.4.9. Show that θ is really an antiautomorphism.

Example 2.4.10. Let $\mathbf{a} = A_l^{(1)}$. A basis for $\mathbf{a}$ is given by the following elements: $e_{ij} z^n$, $n \in \mathbf{Z}$ and $1 \leq i \neq j \leq l+1$; $h_i z^n$, $n \in \mathbf{Z}$ and $1 \leq i \leq l$ with $h_i = e_{ii} - e_{i+1,i+1}$; the elements d, k. Now

$$\theta(e_{ij} z^n) = -\frac{1}{2(l+1)} e_{ji} z^{-n}, \ i < j$$

$$\theta(e_{ji} z^n) = -2(l+1) e_{ij} z^{-n}, \ i < j$$

$$\theta(h_i z^n) = -h_i z^{-n}, \ \theta(d) = d, \ \theta(k) = k.$$

Note that the restriction of θ to root subspaces gives a linear isomorphism $\theta : \mathbf{g}_\alpha \to \mathbf{g}_{-\alpha}$. On the other hand, from the defining formula (B1) we observe that the restriction $\mathbf{g}_\alpha \times \mathbf{g}_{-\alpha} \to \mathbf{C}$ of the invariant bilinear form is nondegenerate; it is also positive definite in the sense that

$$(x, \theta(x)) \geq 0 \ \forall x \in \mathbf{g}_\alpha.$$

It follows that we can define a basis $\{x_\alpha^{(i)}\}$ in $\mathbf{g}_\alpha$ for each $\alpha \in \Phi^-$ such that $(x_\alpha^{(i)}, \theta(x_\alpha^{(j)})) = \delta_{ij}$. The multiplicity label i is really necessary only for the roots $(0,0,n)$; see Section 2.2. We set $x_\alpha^{(i)} = \theta(x_{-\alpha}^{(i)})$, $\alpha \in \Phi^+$. Fix also a basis $\{h^i\}$ of $\mathbf{h}$ dual to the basis $\{h^i\}$, $(h_i, h^j) = \delta_{ij}$.

In the case of a finite-dimensional semisimple Lie algebra one defines a *Casimir operator* Ω' by

$$\Omega' = \sum h_i h^i + \sum_{\alpha \in \Phi +} (x_\alpha x_{-\alpha} + x_{-\alpha} x_\alpha)$$

(no multiplicity label is needed here because the root subspaces are one-dimensional.) In the infinite-dimensional case we cannot use this formula because the sum will in general diverge. However, we can apply a "normal ordering" prescription to make the sum finite. We set

$$\Omega = \sum h_i h^i + 2 \sum_{\alpha \in \Phi +} \sum_i x_{-\alpha}^{(i)} x_\alpha^{(i)} + 2h_\rho.$$

Ω is a well-defined linear operator in any highest weight representation of $\mathbf{a}$. Namely, any vector in the representation space can be written as a polynomial in the generators of $\mathbf{n}_-$ acting on the highest weight vector. It follows that the action of the second term in Ω reduces to a finite polynomial.

PROPOSITION 2.4.11. *The element $\Omega \in \mathcal{U}(\mathbf{a})$ commutes with $\mathbf{a}$, and thus the action in a highest weight representation reduces to a multiplication with a scalar. The value of the scalar is $|\lambda + \rho|^2 - |\rho|^2$, where λ is the highest weight.*

PROOF: Denote by Ω_0 the part of Ω involving the x's. Let α, β be roots and $z \in \mathbf{a}_\beta$. Then $\mathrm{ad} z$ maps $\mathbf{a}_\alpha$ into $\mathbf{a}_{\alpha+\beta}$ and $\mathbf{a}_{-\beta-\alpha}$ into $\mathbf{a}_{-\alpha}$. By the invariance of the bilinear form, $([z,x],y) = -(x,[z,y])$, the former map is (-1) times the transpose of the latter. Let now β be a simple root. We obtain

$$\begin{aligned}[z, \Omega_0] &= 2 \sum_{\alpha \in \Phi +, i} ([z, x_{-\alpha}^{(i)}] x_\alpha^{(i)} + x_{-\alpha}^{(i)} [z, x_\alpha^{(i)}]) \\ &= 2[z, x_{-\beta}] x_\beta + 2 \sum_{\beta \neq \alpha \in \Phi +, i} ([z, x_{-\alpha}^{(i)}] x_\alpha^{(i)} + x_{-\alpha+\beta}^{(i)} [z, x_{\alpha-\beta}^{(i)}]),\end{aligned}$$

where we have done a simple renaming of the summation index in the last term. We have dropped the multiplicity index in the first term, since the simple roots have multiplicity =1. By the remark above, the second and the third term cancel on the right-hand side. Thus we get

$$[z, \Omega_0] = 2[z, x_{-\beta}]x_\beta = 2(z, x_{-\beta})h_\beta x_\beta = 2h_\beta z.$$

On the other hand,

$$\begin{aligned}[z, \sum h_i h^i] &= -\sum \beta(h_i) z h^i - \sum h_i \beta(h^i) z \\ &= -2\sum \beta(h_i) h^i z + \sum \beta(h_i)\beta(h^i) z = -2h_\beta z + (\beta, \beta) z.\end{aligned}$$

Finally $[z, 2h_\rho] = -2\beta(h_\rho)z = -2(\beta, \rho)z = -(\beta, \beta)z$ and combining this with the results above we get $[z, \Omega] = 0$. In the same way one can show that $[z, \Omega] = 0$ when β is *minus* a simple root. Taking commutators of vectors belonging to simple roots or to minus simple roots one can generate the whole algebra **a**. Thus $[z, \Omega] = 0$ for all $z \in \mathbf{a}$. Next we evaluate Ω by applying it to the highest weight vector v in a highest weight representation. We get

$$\Omega v = (\sum h_i h^i + 2h_\rho)v = [\lambda(h_i)\lambda(h^i) + 2\lambda(h_\rho)]v = [(\lambda, \lambda) + 2(\lambda, \rho)]v.$$

The coefficient in front of v is easily seen to be equal to $|\lambda + \rho|^2 - |\rho|^2$.

A Hermitian form H on a **a**-module V is *contravariant* if

$$H(xu, v) = H(u, \theta(x)v), \ \forall u, v \in V, \ x \in \mathbf{a}.$$

We use the convention that a Hermitian form is linear in the first and antilinear in the second argument. If V is a highest weight module, we define a contravariant Hermitian form in V as follows. Let v be a highest weight vector (unique up to a multiplicative constant) and set $H(v, v) = 1$. If $v_1, v_2 \in V$ are arbitrary, we can write $v_i = u_i \cdot v$, where $u_i \in \mathcal{U}(\mathbf{n}_-)$. Define

$$H(v_1, v_2) = H(u_1 v, u_2 v) = H(v, \theta(u_1)u_2 v).$$

Next we can write $\theta(u_1)u_2 v = uv$ for some $u \in \mathcal{U}(\mathbf{n}_-)$. Now we have

$$H(v_1, v_2) = H(v, uv) = \overline{H(uv, v)} = \overline{H(v, \theta(u)v)} = H(\theta(u)v, v).$$

Since $\theta(u) \in \mathcal{U}(\mathbf{n}_+)$, we obtain $\theta(u)v = a \cdot v$ for some $a \in \mathbf{C}$. Thus the value $H(v_1, v_2) = a$ has been uniquely determined by the contravariantness of the Hermitian form and by the normalization $H(v, v) = 1$.

THEOREM 2.4.12. *The Hermitian form H is positive definite in all integrable irreducible highest weight modules.*

PROOF: From the definitions follows at once that the different weight subspaces $L_\lambda(\mu)$ in L_λ are pairwise orthogonal. Thus it is sufficient to show that the restriction of H to any of these subspaces is positive definite. We prove it by induction on $n = \text{ht}(\lambda - \mu)$. The case $n = 0$ is clear by $H(v,v) = 1$. Using Theorem 2.4.11 we get

$$\begin{aligned}(|\lambda + \rho|^2 - |\rho|^2)&H(w,w)\\ &= H(\Omega w, w)\\ &= (|\mu|^2 + 2\mu(h_\rho))H(w,w) + \sum_{\alpha\in\Phi^+,i} H(x_\alpha^{(i)} w, x_\alpha^{(i)}),\end{aligned}$$

where we have also used $\theta(x_{-\alpha}^{(i)}) = x_\alpha^{(i)}$. If we subtract the first term on the right from the left-hand side we get $(|\lambda + \rho|^2 - |\mu + \rho|^2)H(w,w)$. The factor multiplying $H(w,w)$ is positive by (2.4.7) when $\mu \neq \lambda$. On the other hand, the height of the weight of $x_\alpha^{(i)} w$ is smaller than n and so by induction assumption each term in the sum on the right-hand side is also non-negative. To complete the proof we still have to show that the form H is nondegenerate. Because the representation is irreducible, we can choose $u_w \in \mathcal{U}(\mathbf{n}_+)$ such that $u_w \cdot w = v$. Now $H(w, \theta^{-1}(u_w)v) = H(u_w \cdot w, v) = H(v,v) \neq 0$ and thus H is non-degenerate.

2.5. The character formula

If V carries a finite-dimensional representation T of a semisimple Lie group G one can define the *character* of the representation by

$$ch(g) = \text{tr}T(g).$$

Thus the character is a complex valued function on G. Let H be a Cartan subgroup, $\mathbf{h}$ the corresponding Cartan subalgebra and denote by $V(\mu)$ the weight subspace belonging to the weight $\mu \in \mathbf{h}^*$. Then for $h = e^x \in H$,

$$ch(h) = \sum_{\mu\in\Lambda} e^{\mu(x)} \text{dim}V(\mu) \tag{2.5.1}$$

where the sum is over the set Λ of weights.

In an infinite-dimensional case one has to proceed in a more formal way since the sum (2.5.1) does not converge in general. We can still define the *formal character* by

$$ch\, V = \sum_{\mu\in\Lambda} e_\mu \dim V(\mu), \tag{2.5.2}$$

where the symbols e_μ are now formal exponentials; they are the generators of a commutative algebra subject to the defining relations $e_\mu \cdot e_\nu = e_{\mu+\nu}$. The element $e(0)$ is the neutral element with respect to multiplication and we write $e(0) = 1$. In this section we shall compute the formal characters of the highest weight representations of affine Lie algebras.

The formal characters of the Verma modules V_λ are easely computed. Let $x_{-\beta_i,p_i}$ be a basis of the root subspace $\mathbf{g}_{-\beta_i}$, where $1 \le p_i \le m(i) = mult\beta_i$ and $\{\beta_1, \beta_2, \ldots, \beta_\ell\} = \Phi^+$ is the set of positive roots. Then the vectors

$$(x_{-\beta_1,1})^{n(1,1)}(x_{-\beta_1,2})^{n(1,2)} \ldots$$
$$\ldots (x_{-\beta_1,m(1)})^{n(1,m(1))} \ldots (x_{-\beta_\ell,m(\ell)})^{n(\ell,m(\ell))} v$$

form a basis of the subspace $V_\lambda(\mu)$, where $[n(1,1) + \ldots n(1,m(1))]\beta_1 + [n(2,1) + \ldots n(2,m(2))]\beta_2 + \cdots + [n(\ell,1) + \cdots + n(\ell,m(\ell))]\beta_\ell = \lambda - \mu$ and $n(i,j)$'s are non-negative integers. Thus

$$ch\, V_\lambda = e(\lambda) \prod_{\beta\in\Phi^+} [1 + e(-\beta) + e(-2\beta) + \ldots]^{mult\beta}$$
$$= e(\lambda) \prod_{\beta\in\Phi^+} (1 - e(-\beta))^{-mult\beta}. \tag{2.5.3}$$

If V and V' are a pair of modules for a given Lie algebra and W is a submodule of V then

$$ch\, V/W = ch\, V - ch\, W, \quad ch(V \oplus V') = ch\, V + ch\, V'.$$

A highest weight module can always be thought as a quotient of a Verma module by a submodule; for this reason it is natural that the character of a highest weight module can be expanded as

$$ch\, V = \sum_\lambda c(\lambda) V_\lambda \tag{2.5.4}$$

where the $c(\lambda)$'s are integers. The proof is not completely trivial but we shall skip it here because it is not very illuminating [Kac, 1985].

Taking account that the value of the Casimir operator in a highest weight module with highest weight λ is equal to $|\lambda+\rho|^2 - |\rho|^2$ one can show that the only possible nonzero terms in the sum above are those which satisfy $|\lambda+\rho|^2 = |\Lambda+\rho|^2$ and $\lambda \leq \Lambda$ where Λ is the highest weight of V.

Next we let the Weyl group W act on the formal exponentials by $w[e(\lambda)] = e(w(\lambda))$. From Theorem 2.4.3 follows that

$$ch\, L_\lambda = w(ch\, L_\lambda)\ \forall w \in W. \tag{2.5.5}$$

For any $w \in W$ we can write $w = \sigma_1\sigma_2\ldots\sigma_s$ where σ_i is the fundamental reflection in the plane orthogonal to the simple root α_i, $1 \leq i \leq \ell$. Clearly the determinant of the linear transformation $\sigma_i : \mathbf{h}^* \to \mathbf{h}^*$ is equal to -1 and therefore the determinant of w is $\epsilon(w) = (-1)^s$. Define the formal character

$$R = \prod_{\alpha\in\Phi^+} [1-(-\alpha)]^{mult\alpha}.$$

We shall need the following fact: The action of a fundamental reflection σ_i in $\Phi^+ \setminus \{\alpha_i\}$ permutes the elements among themselves. This is a consequence of the fact that any positive root is a sum of simple roots and that $\langle\alpha_j,\alpha_i\rangle \leq 0$.

LEMMA 2.5.6. *$w[e(\rho)R] = \epsilon(w)e(\rho)R$ for all $w \in W$.*

PROOF: It is sufficient to prove the lemma in the case $w = \sigma_i$ for some i. Now $mult\alpha = mult\, w(\alpha)$ for any $\alpha \in \Phi^+$ and $\Phi^+ \setminus \{\alpha_i\}$ is invariant under w. Therefore,

$$\begin{aligned} w[e(\lambda)R] &= e(\rho-\alpha_i)[1-e(\alpha_i)]\sigma_i \prod_{\alpha\in\Phi^+\setminus\{\alpha_i\}} [1-e(-\alpha)]^{mult\alpha} \\ &= e(\rho)e(-\alpha_i)[1-e(\alpha_i)] \prod_{\alpha\in\Phi^+\setminus\{\alpha_i\}} [1-e(-\alpha)]^{mult\alpha} \\ &= -e(\rho)R = \epsilon(w)e(\rho)R. \end{aligned}$$

THEOREM 2.5.7. *Let $\lambda \in \Lambda_+$ and L_λ the irreducible module for an affine Lie algebra with highest weight λ. Then*

$$ch\, L_\lambda = \frac{\sum_{w\in W}\epsilon(w)e[w(\lambda+\rho)-\rho]}{\prod_{\alpha\in\Phi^+}[1-e(-\alpha)]^{mult\alpha}}.$$

PROOF: From (2.5.3) and (2.5.4) we obtain

$$e(\rho)\, Rch\, L_\lambda = \sum_{\mu\in B} c(\mu)e(\mu+\rho)$$

where B is the set of weights consisting of those $\mu \in \Lambda$ for which $\mu \leq \lambda$ and $|\mu+\rho|^2 = |\lambda+\rho|^2$. From (2.5.5) and (2.5.6) follows that

$$c(\mu) = \epsilon(w)c(\nu) \,\text{if}\, w(\mu+\rho) = \nu+\rho$$

for some $w \in W$. It follows that $c(\mu) \neq 0$ if and only if $c(w(\mu+\rho)-\rho) \neq 0$ and so $w(\mu+\rho) \leq \lambda+\rho$ if $c(\mu) \neq 0$. Assuming $c(\mu) \neq 0$ choose a weight $\nu \in \{w(\mu+\rho)-\rho \mid w \in W\}$ such that $\mathrm{ht}\lambda - \nu$ is minimal. Then $\nu+\rho \in \Lambda_+$ and $|\nu+\rho|^2 = |\lambda+\rho|^2$. Applying 2.4.7 we conclude that $\nu = \lambda$ and therefore $w(\mu+\rho) = \lambda+\rho$. Thus $c_\mu = \epsilon(w^{-1}) = \epsilon(w)$.

Since $(\lambda+\rho, \alpha) > 0$ for all $\alpha \in \Delta$ we get from 2.4.8 that $w(\lambda+\rho) = \lambda+\rho$ only if $w = 1$. Clearly $c(\lambda) = 1$ and therefore we have

$$e(\rho) R \, ch\, L_\lambda = \sum_{w \in W} \epsilon(w) e(w(\lambda+\rho)),$$

which gives the asserted formula for $ch\, L_\lambda$.

If $\lambda = 0$ then L_λ is the trivial one-dimensional representation and so $ch\, L_0 = e(0) = 1$. From the character formula we obtain the identity

$$\prod_{\alpha \in \Phi+} [1 - e(-\alpha)]^{mult\alpha} = \sum_{w \in W} \epsilon(w) e(w(\rho) - \rho). \tag{2.5.8}$$

We can now write 2.5.7 alternatively as

$$ch\, L_\lambda = \frac{\sum_{w \in W} \epsilon(w) e[w(\lambda+\rho) - \rho]}{\sum_{w \in W} \epsilon(w) e(w(\rho))}. \tag{2.5.9}$$

In the case of a finite-dimensional semisimple Lie algebra this is the classical Weyl character formula.

In the finite-dimensional case the multiplicities of the weights can be also obtained from the *Kostant multiplicity formula*

$$\dim L_\lambda(\mu) = \sum_{w \in W} \epsilon(w) K[w(\lambda+\rho) - (\lambda+\rho)] \tag{2.5.10}$$

where K is the Kostant partition function obtained from the expansion

$$\prod_{\alpha \in \Phi+} [1 - e(\alpha)]^{-mult\alpha} = \sum_{\beta \in \mathbf{h}^*} K(\beta) e(\beta). \tag{2.5.11}$$

Expanding $[1-e(\alpha)]^{-1}$ as a power series we can write the left-hand side of (2.5.11) also as

$$\prod_{\alpha \in \Phi+} [1 + e(\alpha) + e(2\alpha) + \ldots]^{mult\alpha}$$

and therefore $K(\beta)$ is equal to the number of partitions of β into a sum of positive roots, where each root is counted as many times as is its multiplicity. Clearly $K(0) = 1$ and in general, $K(\beta) = \dim V_\lambda(\beta)$ according to (2.5.3).

Exercise 2.5.12. Prove the formula (2.5.10) in the case of an affine Lie algebra starting from 2.5.7 and the definition (2.5.11).

We define a homomorphism F from the polynomial algebra generated by the formal exponentials $e(-\alpha)$, $\alpha \in \Delta$, to the polynomial algebra in one variable q by

$$F(e(-\alpha)) = q, \ \forall \alpha \in \Delta.$$

Since all weights of L_λ are of the form λ minus a sum of simple roots we can define the formal power series $\dim_q L_\lambda = F(e(-\lambda) ch\, L_\lambda)$. The coefficient of the monomial q^n is equal to the sum of the dimensions $\dim L_\lambda(\mu)$ where $\mathrm{ht}(\lambda - \mu) = n$, where ht is defined as in the proof of 2.4.6.

Let $a_{ij} = 2\frac{(\alpha_i,\alpha_j)}{(\alpha_j,\alpha_j)}$ be the Cartan matrix of an affine Lie algebra. The transposed matrix $b_{ij} = a_{ji}$ defines also an affine Lie algebra. The simple roots of the transposed algebra $\mathbf{g}^t$ are $\beta_i = 2\alpha_i/(\alpha_i,\alpha_i)$. Let $\rho^* \in \mathbf{h}^*$ be the weight such that $(\alpha_i, \rho^*) = 1$ for all simple roots. Then ρ^*, considered as a weight for $\mathbf{g}^t$, corresponds to the weight ρ of $\mathbf{g}$. Let Φ^* be the set of weights for the transpose Lie algebra.

THEOREM 2.5.13.

$$dim_q L_\lambda = \prod_{\alpha \in \Phi^* +} \left(\frac{1 - q^{(\lambda+\rho,\alpha)}}{1 - q^{(\rho,\alpha)}} \right)^{mult \alpha}$$

PROOF: For any positive dominant weight μ define

$$N(\mu) = \sum_{w \in W} \epsilon(w) e(w(\mu) - \mu).$$

Now $\mathrm{ht}(\mu - w(\mu)) = (\mu - w(\mu), \rho^*)$ and so

$$F(e(w(\mu) - \mu)) = q^{(\mu - w(\mu),\rho^*)}. \tag{2.5.14}$$

Applying the homomorphism F to both sides of (2.5.8) we get

$$\prod_{\alpha \in \Phi +} (1 - q^{(\alpha,\rho^*)})^{mult \alpha} = \sum_{w \in W} \epsilon(w) q^{(\rho - w(\rho),\rho^*)}$$

and combining this with (2.5.14) we get

$$\begin{aligned} F(N(\mu)) &= \sum_{w\in W} \epsilon(w) q^{(\mu-w(\mu),\rho^*)} \\ &= \sum_{w\in W} \epsilon(w) q^{(\mu,\rho^*-w(\rho^*))} \\ &= F'\left(\sum_{w\in W} \epsilon(w) e(w(\rho^*)-\rho^*)\right), \end{aligned}$$

where the homomorphism F' is defined by the relations $F'(e(-\alpha)) = q^{(\mu,\alpha^*)}$ for $\alpha \in \Delta$ with $\alpha^* = 2\alpha/(\alpha,\alpha)$. Applying the identity (2.5.8) to the transposed Lie algebra we get

$$F(N(\mu)) = F'\left(\prod_{\alpha\in\Phi^*_+} (1-e(-\alpha))^{mult\,\alpha}\right)$$

where Φ^* is the root system of the transposed algebra. Thus

$$F(N(\mu)) = \prod_{\alpha\in\Phi^*_+} (1-q^{(\mu,\alpha)})^{mult\,\alpha}.$$

Combining this with (2.5.9) we obtain

$$F(e(-\lambda)ch\,L(\lambda)) = \prod_{\alpha\in\Phi^*_+} \left(\frac{1-q^{(\lambda+\rho,\alpha)}}{1-q^{(\rho,\alpha)}}\right)^{mult\,\alpha}$$

which implies the theorem.

CHAPTER 3 PRINCIPAL BUNDLES

3.0. A short introduction to calculus of differential forms

Algebra of differential forms

A smooth manifold M is a Hausdorff space which has a complete set (atlas) of coordinate charts (U_α, ϕ_α). Each U_α is an open set of M and $\phi_\alpha : U_\alpha \to \mathbf{R}^n$ is a homeomorphism onto a subset of the standard Euclidean space such that the coordinate transformations $\phi_\alpha \circ \phi_b^{-1}$ are smooth functions. The space M is covered by the sets U_α. n is the dimension of the manifold M.

A smooth function f from a manifold M to a manifold N is a map such that $\psi_\alpha \circ f \circ \phi_\beta^{-1}$ is a smooth function (in its domain of definition) from an open set in $\mathbf{R}^n$ to an open set in $\mathbf{R}^m$; here m, n are the dimensions of M, N, respectively, and ϕ_α is any local coordinate on M and ψ_β any local coordinate on N.

The tangent space of M at a point $p \in M$ consists of equivalence classes of smooth curves through p. By a smooth curve through p we mean a smooth map h from an open interval of $\mathbf{R}$ (containing 0) such that $h(0) = p$. Two curves h and k are equivalent if the derivative of $\phi \circ h$ at $t = 0$ is equal to the derivative of $\phi \circ k$ at that point; ϕ is any local coordinate defined in a neighborhood of p. Since the equivalence classes are completely characterized by the vectors $\frac{d}{dt}\phi \circ h|_{t=0}$ in $\mathbf{R}^n$, we can use the linear structure of $\mathbf{R}^n$ to define addition and multiplication by scalars of the tangent vectors. The tangent space at p is denoted by T_pM. All tangent spaces are isomorphic to $\mathbf{R}^n$, but the isomorphism depends on the choice of coordinates.

For each $p \in M$ we denote by $\Omega_p^m M$ the space of *alternating m-linear forms* in the tangent space T_pM. A form $\omega(p; v_1, \dots, v_m)$, linear separately in each of the arguments $v_k \in T_pM$, is said to be alternating if a permutation π of its arguments only changes the value of ω_m by the sign of the permutation. The space $\Omega_p^m M$ is a linear space of dimension $\binom{n}{m}$ if $m \leq n = dim M$ and zero otherwise. We define $\Omega_p^0 M$ as the one-dimensional vector space $\mathbf{R}$.

Choose a coordinate chart (U, ϕ) at $p \in M$. We write $\phi(q) = (x_1(q), \dots, x_n(q))$, where the x_k's are real valued functions on U. The coordinate axis in $\mathbf{R}^n$ defines local curves in U via the inverse of the map ϕ. We denote the tangent vectors so defined by $\partial_k = \frac{\partial}{\partial x_k}$; these are defined at each point $q \in U$. In fact, the vectors $\partial_1, \ldots, \partial_n$ form a basis of the tangent space $T_q M$ for each $q \in U$.

Once we have fixed the system of local coordinates $x_1, \ldots, x_n$ we can construct a basis of $\Omega_p^m M$ as follows. Let $1 \le i_1 < i_2 < \cdots < i_m \le n$ be a sequence of integers. The alternating form

$$dx_{i_1} \wedge dx_{i_2} \wedge \cdots \wedge dx_{i_m} \tag{3.0.1}$$

is the m linear map which sends m tuplet $(v_1, \ldots, v_m)$ of tangent vectors at the point p to the real number $\det(v_{ji_\mu})$, where the $m \times m$ matrix (v_{ji_μ}) is formed from the components of the tangent vectors in the coordinate basis, $v_j = \sum v_{ji}\partial_i$. The collection of the alternating forms (3.0.1) is a basis of $\Omega_p^m M$. In particular, $dx_1, \ldots, dx_n$ is the dual basis of the coordinate basis $\partial_1, \ldots, \partial_n$, that is, they form a basis of the dual vector space $T_p^* M$ such that $dx_i(\partial_j) = \delta_{ij}$.

A differential form of degree m on the manifold M is a smooth distribution of alternating m linear forms $\omega(p; \cdot)$. The smoothness property means that when the form ω is written in terms of local coordinates then all the coefficients are smooth functions. In an open coordinate neighborhood U we may write

$$\omega(p; \cdot) = \sum_{1 \le i_1 < \cdots < i_m \le n} \omega_{i_1 i_2 \ldots i_m} dx_{i_1} \wedge dx_{i_2} \wedge \ldots dx_{i_m}, \tag{3.0.2}$$

where the $\omega_{i_1 i_2 \ldots i_m}$'s are smooth real valued functions on U. We shall denote by $\Omega^m M$ the space of all differential forms of degree m on M. It is naturally a real vector space under pointwise addition and multiplication by scalars. We define $\Omega^0 M$ as the space $C^\infty(M)$ of smooth functions.

The alternating forms (3.0.1) are not only defined for the increasing sequences but for all sequences of integers $i_1, \ldots, i_m$ between 1 and n. However, a permutation of the integers in a given sequence merely multiplies the form by -1 if the permutation is odd and by $+1$ if the permutation is even. We can define the exterior product of the m form (3.0.1) and a k form $dx_{j_1} \wedge \cdots \wedge dx_{j_k}$ simply by writing the factors in a row; this can be extended by linearity to give a product $\Omega_p^m M \times \Omega_p^k M \to \Omega_p^{m+k} M$. The product is associative but noncommutative. We have

$$\omega \wedge \theta = (-1)^{(deg\, \omega)(deg\, \theta)} \theta \wedge \omega. \tag{3.0.3}$$

The exterior product of two differential forms is defined pointwise on the manifold M. Denote by ΩM the direct sum of the vector spaces $\Omega^m M$ $(m = 0, 1, \ldots, dim\, M)$. ΩM is an associative algebra, the algebra of differential forms, with respect to the exterior product. The space ΩM is a module for the commutative algebra $C^\infty(M)$: a differential form can be multiplied pointwise by a smooth function to give a smooth differential form.

The exterior derivative

Next we define the *exterior derivative* $d\omega$ of a differential form ω. Using local coordinates the exterior differentiation d is uniquely defined by the properties

(1) $d(dx_{i_1} \wedge \cdots \wedge dx_{i_m}) = 0$ for all sequences of integers i_μ
(2) If $f \in C^\infty(M)$ then $df = \sum_{i=1}^n \frac{\partial f}{\partial x_i} dx_i$
(3) $d(f dx_{i_1} \wedge \cdots \wedge dx_{i_m}) = df \wedge dx_{i_1} \wedge \cdots \wedge dx_{i_m}$.

The exterior derivative of an arbitrary form is obtained by linearity from the above properties. The operator d has been defined using local coordinates but one can show by employing the chain rule of differentiation that d is independent from the choice of coordinates. By definition the operator d increases the degree of a differential form by 1; i.e., it is a linear operator from $\Omega^m M$ to $\Omega^{m+1} M$. The fundamental property of d is the following:

THEOREM 3.0.4. $d^2 = 0$.

PROOF: By linearity it is sufficient to show that the action of the operator d^2 to the form $\omega = f dx_{i_1} \wedge \cdots \wedge dx_{i_m}$ is zero.

$$\begin{aligned} d(d\omega) &= d \sum_{j=1}^n \frac{\partial f}{\partial x_j} dx_j \wedge dx_{i_1} \wedge \cdots \wedge dx_{i_m} \\ &= \sum_{j,k=1}^n \frac{\partial^2 f}{\partial x_k \partial x_j} dx_k \wedge dx_j \wedge \cdots \wedge dx_{i_m}. \end{aligned}$$

The claim follows now from the symmetry of the second derivative and from the antisymmetry $dx_j \wedge dx_k = -dx_k \wedge dx_j$.

A differential form ω is said to be *closed* if $d\omega = 0$. Denote by Z^m the subspace of closed forms of degree m on M. By the theorem, the space Z^m contains the space $B^m = \{d\omega \mid \omega \in \Omega^{m-1} M\}$ of *exact* forms of degree m. The quotient space $H^m(M, \mathbf{R}) = Z^m / B^m$ is called the mth *de Rham cohomoly group* of the manifold M. The zeroth cohomology is equal to $\mathbf{R}^k$, where k is the number of connected components of M. If $M \subset \mathbf{R}^n$ is any contractible domain then $H^m(M, \mathbf{R}) = 0$ for $m > 0$:

LEMMA 3.0.5. *(Poincaré's lemma) Let ω be a closed form of degree $m > 0$ on the contractible domain M. Then there is a form θ of degree $m-1$ such that $\omega = d\theta$.*

PROOF: Fix a point $p \in M$ such that for any $q \in M$ the line connecting p to q is contained in M. Set

$$\theta(q; v_1, \ldots, v_{m-1}) = \int_0^1 t^{m-1}\omega(p + t(q-p); q-p, v_1, \ldots, v_{m-1})dt$$

for tangent vectors $v_1, \ldots, v_{m-1}$. (Since $M \subset \mathbf{R}^n$ we may think of the tangent vectors as vectors in the background space $\mathbf{R}^n$.) The exterior derivative of θ is ω.

Let $\phi : M \to N$ be a smooth map. If $h(t)$ is a smooth curve through a point $p \in M$ then $\phi(h(t))$ is a smooth curve through $\phi(p) \in N$. Equivalent curves are mapped to equivalent curves and therefore we get a map $T_p\phi : T_pM \to T_{\phi(p)}N$. Using local coordinates and the linearity of the derivative it is shown that this map is linear. $T_p\phi$ is called the derivative of ϕ at the point p. Again using local coordinates and the chain rule of the ordinary differentiation one gets the chain rule $T_p(\phi \circ \psi) = T_{\psi(p)}\phi \circ T_p\psi$.

Let ω be a differential form of degree m on N. We define the *pull-back* $\phi^*\omega$ as the m form on M such that

$$(\phi^*\omega)(p; v_1, \ldots, v_m) = \omega(\phi(p); T_p\phi \cdot v_1, \ldots, T_p\phi \cdot v_m)$$

for $v_1, \ldots, v_m \in T_pM$. If $(x_1, \ldots, x_n)$ are local coordinates at $p \in M$, $(y_1, \ldots, y_k)$ local coordinates at $\phi(p)$ and the mapping ϕ in terms of coordinates is written as $y_i = y_i(x)$ then the pull-back of a form $f(y)dy_{i_1} \wedge \cdots \wedge dy_{i_m}$ is

$$\phi^*\omega = \sum_{(j)} f(y(x))\frac{\partial y_{i_1}}{\partial x_{j_1}} \cdots \frac{\partial y_{i_m}}{\partial x_{j_m}} dx_{j_1} \wedge \cdots \wedge dx_{j_m}. \tag{3.0.6}$$

The pull-back operation is contravariant, that means $(\phi \circ \psi)^* = \psi^* \circ \phi^*$.

THEOREM 3.0.7.

(1) *The pull-back operator ϕ^* commutes with the exterior derivative d.*
(2) *$\phi^*(\omega \wedge \theta) = (\phi^*\omega) \wedge (\phi^*\theta)$ for all forms ω, θ.*
(3) *$d(\omega \wedge \theta) = d\omega \wedge \theta + (-1)^m\omega \wedge d\theta$, where m is the degree of the form ω.*

PROOF: The proofs of all the three items above are completely straightforward computations when using local coordinates. As an example we prove (3). Because of linearity of d we may assume that $\omega = f(x)dx_{i_1} \wedge \cdots \wedge dx_{i_m}$ and $\theta = g(x)dx_{j_1} \wedge \cdots \wedge dx_{j_k}$. Now

$$\begin{aligned} d(\omega \wedge \theta) &= \sum_l \frac{\partial(fg)}{\partial x_l} dx_l \wedge dx_{i_1} \wedge \dots dx_{i_m} \wedge dx_{j_1} \cdots \wedge dx_{j_k} \\ &= \sum_l \frac{\partial f}{\partial x_l} dx_l \wedge dx_{i_1} \wedge \cdots \wedge dx_{i_m} \wedge (g dx_{j_1} \wedge \dots dx_{j_k}) \\ &+ \sum_l f \frac{\partial g}{\partial x_l} dx_l \wedge dx_{i_1} \wedge \cdots \wedge dx_{j_k}. \end{aligned}$$

The first term in the last expression is equal to $d\omega \wedge \theta$, whereas by (3.0.3), permuting the factor dx_l to the right of the dx_{i_μ}'s, the second term is $(-1)^m d\omega \wedge d\theta$.

Vector fields and differential forms

A vector field on a manifold M is a smooth distribution X of tangent vectors $X(p) \in T_pM$. Smooth means that when written in local coordinates as

$$X = \sum X_j(x)\partial_j$$

all the coefficients X_j are smooth real valued functions on M. A *diffeomorphism* is a one-to-one map ϕ between two manifolds such that both ϕ and its inverse ϕ^{-1} are a smooth map. A diffeomorphism $\phi : M \to N$ sends a vector field X on M to a vector field $Y = \phi_* X$ on N such that $Y(q) = T_p\phi \cdot X(p)$, where $q = \phi(p)$. Let us denote by $D(M)$ the vector space of all smooth vector fields on M. $D(M)$ is a left $C^\infty(M)$ module: Given a smooth function f and a vector field X we can construct a new vector field fX as $(fX)(p) = f(p)X(p)$.

A vector field acts as a differentiation of the algebra of smooth functions on M. Given a vector field X and a smooth function f we define the function $X \cdot f$ by $(X \cdot f)(p) = T_pf \cdot X(p)$. In local coordinates this operation is written as $f \mapsto \sum X_j \partial_j f$ for $X = \sum X_j \partial_j$. Using the usual Leibnitz rule one proves that $X(fg) = (Xf)g + (Xg)f$ for a pair of smooth functions f, g. One can also prove the converse: Any derivation of the commutative algebra $C^\infty(M)$ can be written as $f \mapsto Xf$ for some vector field X. A commutator of two derivations X, Y of the algebra $C^\infty(M)$ is again a derivation; this can be taken as the definition of the commutator $[X,Y] \in D(M)$ of a pair of vector fields. In local coordinates,

$$[X,Y] = \sum_i (X \cdot Y_i - Y \cdot X_i)\partial_i.$$

The commutator is antisymmetric and satisfies $[X, fY] = f[X,Y] + (Xf)Y$ for any smooth function f. The commutator and the action of a vector field on real functions are examples of *Lie derivations*; sometimes one writes $Xf = \mathcal{L}_X f$ and $[X,Y] = \mathcal{L}_X Y$. A diffeomorphism $\phi : M \to N$ induces an isomorphism between the algebras of vector fields: The mapping $\phi_* : D(M) \to D(N)$ has the property

$$\phi_*[X,Y] = [\phi_* X, \phi_* Y]. \tag{3.0.8}$$

A differential form ω of degree m on M can be viewed as an alternating map from $D(M) \times \ldots D(M)$ (m factors) to C^∞. Given the vector fields $X_1, \ldots, X_m$ we can define the smooth function $\omega(X_1, X_2, \ldots, X_m)$ as

$$\omega(X_1, \ldots, X_m)(p) = \omega(p; X_1(p), \ldots, X_m(p)).$$

The map ω is in fact $C^\infty(M)$ linear in its arguments X_i. A multiplication of any of the vector fields X_i multiplies $\omega(X_1, \ldots, X_m)$ by the function f.

Conversely, given an alternating $C^\infty(M)$ linear function $T : D(M) \times \cdots \times D(M) \to C^\infty$ there is a uniquely defined differential form ω such that $\omega(X_1, \ldots, X_m) = T(X_1, \ldots, X_m)$ for all vector fields $X_1, \ldots, X_m$ on M. The differential form ω can be reconstructed from T using local coordinates. Local coordinates define a local basis $\partial_1, \ldots, \partial_n$ of vector fields. Choose a smooth real valued function ρ on M such that $\rho = 1$ in a neighborhood $U' \subset U$ of the point p and zero outside U, where U is the domain of the local coordinates. Write

$$T(\rho\partial_{i_1}, \ldots, \rho\partial_{i_m}) = f_{i_1 \ldots i_m}.$$

The form ω in the neighborhood U' can be written as

$$\omega = \sum_{1 \le i_1 < \cdots < i_m \le n} f_{i_1 \ldots i_m} dx_{i_1} \wedge \cdots \wedge dx_{i_m}.$$

The exterior derivative $d\omega$ can now alternatively be defined as the $m+1$ linear alternating map

$$\begin{aligned} d\omega(X_1, \ldots, X_{m+1}) = {} & \sum_{i=1}^{m+1} (-1)^{i+1} X \cdot \omega(X_1, \ldots, \hat{X}_i, \ldots, X_{m+1}) \\ & + \sum_{i<j} (-1)^{i+j} \omega([X_i, X_j], X_1, \ldots, \hat{X}_i, \ldots, \hat{X}_j, \ldots, X_{m+1}) \end{aligned} \tag{3.0.9}$$

where the caret means that the variable has been deleted. By choosing local coordinates it is rather straightforward to show that this definition of $d\omega$ agrees with the previous one.

A vector field X defines a *contraction* of the differential forms. Given a m form ω we can define a $m-1$ form $\imath(X)\omega$ by

$$\imath(X)\omega(X_1, \ldots, X_{m-1}) = \omega(X, X_1, \ldots, X_{m-1}).$$

By the antisymmetry of a differential form we have

$$(\imath_X)^2 = 0$$

for all vector fields X.

Differential forms on Lie groups

A *Lie group* G is a group which is also a smooth manifold such that the product $(a, b) \mapsto ab$ is smooth in both arguments and the inverse a^{-1} is smooth in a. We can think of a finite-dimensional Lie group for the most part as a subgroup of the general linear group $GL(n, \mathbf{F})$, where $\mathbf{F} = \mathbf{R}, \mathbf{C}$. Any closed subgroup G of $GL(n, \mathbf{F})$ inherits the structure of a smooth manifold from the natural smooth structure of the linear group, defined by coordinatization by matrix entries [Helgason, 1978]. Examples of Lie groups are $SU(n)$, $O(n)$, $Sp(2n)$, and $U(n)$, which were already discussed in Chapter 1.

A vector field X on a Lie group G is said to be *left (right) invariant* if it is invariant under all left (right) translations ℓ_g (r_g), $g \in G$. The left translation is the diffeomorphism of G onto itself defined by $\ell_g(a) = ga$. The right translation r_g is defined by $r_g(a) = ag$. Invariance under left translations means that $(\ell_g)_* X = X$ for all $g \in G$. Let $\mathbf{g}$ be the space of left invariant vector fields on G. A left invariant vector field is uniquely defined by giving its value at the identity $e \in G$. Thus we may identify the space $\mathbf{g}$ as the tangent space $T_e G$. The space of left invariant vector fields is a Lie algebra with respect to the commutator of vector fields: If X, Y are left invariant then $(\ell_g)_*[X, Y] = [(\ell_g)_* X, (\ell_g)_* Y] = [X, Y]$ and so $[X, Y]$ is left invariant. The dimension of the Lie algebra $\mathbf{g}$ of the Lie group G is equal to $\dim G$.

For each $X \in \mathbf{g}$ there is a uniquely defined one-parameter subgroup $h(t)$ of G such that the tangent vector of the curve $h(t)$ at $h(0) = e$ is equal to X [Helgason, 1978]. In the case of a matrix Lie group the one-parameter subgroup is given by the matrix valued exponential function, $h(t) = \exp(tX)$. In general, one can define the exponential function $\exp : \mathbf{g} \to G$ by $\exp(X) = h(1)$.

A *representation* of a Lie group G is a homomorphism ϕ to the group of invertible linear operators of some vector space V. If V is finite dimensional a representation of G induces a representation ρ of its Lie algebra $\mathbf{g}$ by

$$\rho(X) = \frac{d}{dt}\phi(\exp(tX))|_{t=0}, \qquad X \in \mathbf{g}.$$

If V is infinite-dimensional one must worry about domains of definition of the operators $\rho(X)$. In case of a unitary representation of a finite-dimensional Lie group in a Hilbert space one can show that there is an invariant dense domain for the operators $\rho(X)$ [Warner, 1972; see also Flato *et al.*, 1972].

Let $X_1, \ldots, X_n$ be a basis of left invariant vector fields. Any vector field X on G can then be written as

$$X = f_1 X_1 + \cdots + f_n X_n$$

where the f_i's are smooth real valued functions on G. Thinking of differential form as alternating $C^\infty(G)$ valued functions we conclude that a m form ω on G is completely determined by giving its values

$$\omega(X_{i_1}, X_{i_2}, \ldots, X_{i_m}), \qquad 1 \le i_1 < \cdots < i_m \le n$$

for the left invariant vector fields.

Stokes's theorem

Let M be a smooth manifold of dimension n. *An orientation* on M is a nowhere vanishing differential form ω of degree n. If ω and ω' are two orientations then there is a smooth function f on M such that $\omega' = f\omega$. The function f is either everywhere positive or everywhere negative. Thus the orientations (if they exist) decompose to two classes depending on the sign of f. In fact, we shall need only the equivalence class of ω with respect to multiplication by positive functions, and so we shall use the word "orientation" also for the equivalence class. An orientation does not need to exist: An example of this is the real two-dimensional projective space $P\mathbf{R}^2$, points of which are the lines through origin in $\mathbf{R}^3$. A manifold is said to be *orientable* if it has an orientation. A group manifold is always orientable: we can define a nowhere vanishing form of degree n on G by setting $\omega(X_1, \ldots, X_n) = 1$, where $X_1, \ldots, X_n$ is a basis of $\mathbf{g}$.

Let M be an orientable manifold with a chosen orientation ω. Let θ be a differential form on M of degree $n = \mathit{dim}\, M$. To begin with, suppose that θ is nonzero only in an open coordinate neighborhood U where we have chosen coordinates $x_1, \ldots, x_n$. In U we can write $\omega = f dx_1 \wedge \cdots \wedge dx_n$ for some function f. If $f > 0$ we say that the coordinate system is compatible with the orientation. If $f < 0$ we write $\omega = -fd(-x_1) \wedge dx_2 \wedge \cdots \wedge dx_n$ and so the system $-x_1, x_2 \ldots, x_n$ is compatible with the orientation. Thus a compatible coordinate system always exists and we assume that $x_1, \ldots, x_n$ is already compatible with the orientation.

In U we can write $\theta = h\, dx_1 \wedge \cdots \wedge dx_n$ for some real function h. The integral of the n form θ over the manifold M is now defined as

$$\int_M \theta = \int_U h\, dx_1 dx_2 \ldots dx_n$$

where the integral on the right-hand-side is the ordinary Riemann integral over an open subset of $\mathbf{R}^n$.

In the general case the integral of a n form over M can be defined by chopping the manifold into pieces such that in each piece one can define local coordinates, integrate the form over each piece, and then add up the integrals. A more formal definition is given as follows. Choose a coordinate atlas (U_α, ϕ_α), $M = \bigcup_\alpha U_\alpha$. One can show that there exists *a partition of unity* subject to the given covering by coordinate neighborhoods. That means that there are smooth functions $\rho_\alpha : M \to \mathbf{R}$ such that

(1) $\sum \rho_\alpha(p) = 1$ for all $p \in M$
(2) The support *supp*ρ_α of ρ_α is contained in U_α.

[The support of a function f is the closure of the set of points where $f(x) \neq 0$.] The integral of θ is defined then as

$$\int_M \theta = \sum_\alpha \int_{U_\alpha} \rho_\alpha h_\alpha(x)\, dx_1 dx_2 \ldots dx_n,$$

where $h_\alpha\, dx_1 \wedge \cdots \wedge dx_n$ is the local presentation of θ in U_α.

In case of a compact manifold M the integral converges for any smooth form θ and defines a linear map from $\Omega^n M$ to the real numbers. A change of the (equivalence class of) orientation changes the sign of the integral.

Up to now we have considered only manifolds without boundary. A manifold with a boundary is a space M such that $M = N \cup \partial M$, where N (the set of interior points of M) is an open manifold of dimension n and ∂M (the boundary of M) is a manifold without boundary, of dimension $n - 1$. In addition, it is assumed that near the boundary M is like a product of a piece of the boundary and the half-closed interval $[0, 1)$. We shall not give a more precise treatment of manifolds with boundary here because in this book we shall deal with very simple cases where it is evident what one means by the boundary. Example: The unit ball $B(0,1) \subset \mathbf{R}^n$ consisting of points x with $||x|| \leq 1$ is a manifold with boundary $\partial B(0,1) = S^{n-1}$.

THEOREM 3.0.10. *Let M be a compact oriented manifold with boundary and θ a differential form on M of degree $\dim M - 1$. Then*

$$\int_M d\theta = \int_{\partial M} \theta.$$

In particular, the integral of a closed form in M over the boundary vanishes.

The proof can be found in Spivak [1979], for example.

3.1. Definition of a principal bundle and examples

Let G be a Lie group and M a smooth manifold. *A principal G bundle over M* is a manifold which locally looks like $M \times G$.

DEFINITION 3.1.1. *A smooth manifold P is a principal G bundle over the manifold M, if a smooth right action of G on P is given, i. e., a map $P \times G \to P$, $(p,g) \mapsto pg$, such that $p(gg') = (pg)g' \forall p \in P$ and g, g' in G, and if there is given a smooth map $\pi : P \to M$ such that*

1. *$\pi(pg) = \pi(p)$ for all g in G.*
2. *$\forall x \in M \, \exists$ an open neighborhood U of x and a diffeomorphism (local trivialization) $f : \pi^{-1}(U) \to U \times G$ of the form $f(p) = (\pi(p), \phi(p))$ such that $\phi(pg) = \phi(p)g \; \forall p \in \pi^{-1}(U), g \in G$.*

The manifold P is the *total space* of the bundle, M is the *the base space,* and π is the bundle projection. The *trivial bundle* $P = M \times G$ is defined by the projection $\pi(x,g) = x$ and by the natural right action of G on itself.

Consider two bundles $P_i = (P_i, \pi_i, M_i; G)$ with the same structure group G. A smooth map $\phi : P_1 \to P_2$ is a *G bundle map* , if $\phi(pg) = \phi(p)g$ for all p and g. Two bundles P_1 and P_2 are *isomorphic* if there is a bijective bundle map $P_1 \to P_2$. An isomorphism of a bundle onto itself is an *automorphism* .

If $H \subset G$ is a closed subgroup then G is a principal H bundle over the homogeneous space G/H. The right action of H on G is just the right multiplication in G and the projection is the canonical projection on the quotient.

Example 3.1.2. Take $G = SU(2)$ and $H = U(1)$

$$H : \begin{pmatrix} e^{i\varphi} & 0 \\ 0 & e^{-i\varphi} \end{pmatrix}, \varphi \in \mathbf{R}.$$

A general element g of G is

$$g = \begin{pmatrix} z_1 & -\bar{z}_2 \\ z_2 & \bar{z}_1 \end{pmatrix},$$

with $|z_1|^2 + |z_2|^2 = 1$. Writing z_1 and z_2 in terms of their real and imaginary parts we see that the group G can be identified with the unit sphere S^3 in $\mathbf{R}^4$. We can define a map $\pi : G \to S^2$ by $\pi(g) = g\sigma_3 g^{-1}$, where σ_3 is the matrix $diag(1, -1)$; elements of $\mathbf{R}^3$ are represented by Hermitian traceless 2×2 matrices. The Euclidean metric is given by $\|x\|^2 = -\det x$. The kernel of the map π is precisely $U(1)$; thus we have a $U(1)$ fibration over $S^2 = SU(2)/U(1)$ in S^3.

Exercise 3.1.3. Let $S_+ = \{x \in S^2 | x_3 \neq -1\}$ and $S_- = \{x \in S^2 | x_3 \neq +1\}$. Construct local trivializations $f_\pm : \pi^{-1}(S_\pm) \to S_\pm \times U(1)$.

The bundle $S^3 \to S^2$ is nontrivial; it is not isomorphic to $S^2 \times S^1$ for topological reasons. Namely, S^3 is a simply connected manifold whereas the fundamental group of $S^2 \times S^1$ is equal to $\pi_1(S^1) = \mathbf{Z}$ [Greenberg, 1966].

Let $\{U_\alpha\}_{\alpha \in \Lambda}$ be an open cover of the base space M of a principal bundle P and let $p \mapsto (\pi(p), \phi_\alpha(p)) \in U_\alpha \times G$ be a set of local trivializations. If $p \in \pi^{-1}(U_\alpha \cap U_\beta)$, we can write

$$\phi_\alpha(p) = \xi_{\alpha\beta}(p)\phi_\beta(p),$$

where $\xi_{\alpha\beta}(p) \in G$. Now $\phi_\alpha(pg) = \phi_\alpha(p)g$ and $\phi_\beta(pg) = \phi_\beta(p)g$ from which follows that $\xi_{\alpha\beta}(pg) = \xi_{\alpha\beta}(p)$ and thus $\xi_{\alpha\beta}$ can be thought of as a function on the base space $U_\alpha \cap U_\beta$. If $p \in \pi^{-1}(U_\alpha \cap U_\beta \cap U_\gamma)$ and $x = \pi(p)$, then $\phi_\alpha(p) = \xi_{\alpha\beta}(x)\phi_\beta(p) = \xi_{\alpha\beta}(x)\xi_{\beta\gamma}(x)\phi_\gamma(p)$ so that

$$\xi_{\alpha\beta}(x)\xi_{\beta\gamma}(x) = \xi_{\alpha\gamma}(x).$$

In general, a collection of G-valued functions $\{\xi_{\alpha\beta}\}$ for the covering $\{U_\alpha\}$ is a *one-cocycle (with values in G)* if the above equation holds for all x in $U_\alpha \cap U_\beta \cap U_\gamma$ and for all triples of indices.

If we make the transformations $\phi'_\alpha = \eta_\alpha \phi_\alpha$ for some functions $\eta_\alpha : U_\alpha \to G$, then

$$\xi_{\alpha\beta} \mapsto \xi'_{\alpha\beta} = \eta_\alpha^{-1} \xi_{\alpha\beta} \eta_\beta.$$

If we can find the maps η_α such that $\xi'_{\alpha\beta} = 1 \forall \alpha, \beta$, then $\xi_{\alpha\beta} = \eta_\alpha \eta_\beta^{-1}$ and we say that the one-cocycle ξ is a *coboundary*.

Let $(P, \pi, M), (P', \pi', M')$ be a pair of principal G bundles and let $f : P \to P'$ be a bundle map. We define the induced map $\hat{f} : M \to M'$ by $\hat{f}(x) = \pi'(f(p))$, where p is an arbitrary element in the fiber $\pi^{-1}(x)$.

THEOREM 3.1.4. *Let P and P' be a pair of principal G bundles over M. Let $\{U_\alpha, \phi_\alpha\}_{\alpha\in\Lambda}$ (respectively, $\{U_\alpha, \phi'_\alpha\}_{\alpha\in\Lambda}$) be a system of local trivializations for P (respectively, for P'). Let $\xi_{\alpha\beta}$ and $\xi'_{\alpha\beta}$ be the corresponding transition functions. Then there exists an isomorphism $f: P \to P'$ such that $\hat{f} = id_M$ if and only if the transition functions differ by a coboundary, that is, $\xi'_{\alpha\beta}(x) = \eta_\alpha(x)^{-1}\xi_{\alpha\beta}(x)\eta_\beta(x)$ in $U_\alpha \cap U_\beta$ for some functions $\eta_\alpha : U_\alpha \to G$.*

PROOF: 1) Suppose first that $\xi'_{\alpha\beta} = \eta_\alpha^{-1}\xi_{\alpha\beta}\eta_\beta$ for all $\alpha, \beta \in \Lambda$. Define $f: P \to P'$ as follows. Let $p \in P$ and $x = \pi(p)$. Choose $\alpha \in \Lambda$ such that $x \in U_\alpha$. Using a local trivialization (U_α, ϕ'_α) at x we set $f(p) = (x, f_\alpha(p))$, where $f_\alpha(p) = \eta_\alpha(x)^{-1}\phi_\alpha(p)$. We have to show that the map is well-defined: If $x \in U_\alpha \cap U_\beta$ then $\phi_\beta(p) = \xi_{\beta\alpha}(x)\phi_\alpha(p)$ and thus

$$\begin{aligned} f_\beta(p) &= \eta_\beta(x)^{-1}\phi_\beta(p) = \eta_\beta(x)^{-1}\xi_{\beta\alpha}(x)\phi_\alpha(p) \\ &= \xi'_{\beta\alpha}(x)[\eta_\alpha(x)^{-1}\phi_\alpha(p)] = \xi'_{\beta\alpha}(x)f_\alpha(p). \end{aligned}$$

We conclude that $(x, f_\alpha(p))$ and $(x, f_\beta(p))$ represent the same element in P'. The equation $f(pg) = f(p)g$ follows from $\phi_\alpha(pg) = \phi_\alpha(p)g$.

2) Let $f: P \to P'$ be an isomorphism. We can define

$$\eta_\alpha(x) = \phi_\alpha(p)\phi'_\alpha(f(p))^{-1},$$

where $p \in \pi^{-1}(x)$ is arbitrary. It follows at once from the definition of the transition functions that the collection $\{\eta_\alpha\}_{\alpha\in\Lambda}$ satisfies the requirements.

Let $\{\xi_{\alpha\beta}\}_{\alpha,\beta\in\Lambda}$ be a one-cocycle with values in G, subordinate to an open cover $\{U_\alpha\}$ on a manifold M. We can construct a principal G bundle P from this data. Let $C = \amalg(\alpha, U_\alpha \times G)$ be the disjoint union of all the sets $U_\alpha \times G$. Define an equivalence relation in C by $(\alpha, x, g) \sim (\alpha', x', g')$ if and only if $x = x'$ and $g' = \xi_{\alpha'\alpha}(x)g$. Set $P = C/\sim$. The action of G in P is given by $(\alpha, x, g)g_0 = (\alpha, x, gg_0)$. The smooth structure on P is defined such that the sets $U_\alpha \times G$ are smooth coordinate charts for P.

Exercise 3.1.5. Complete the construction of P above.

Let (P, π, M) be a principal G bundle. A *(global) section* of P is a map $\psi: M \to P$ such that $\pi \circ \psi = id_M$.

Exercise 3.1.6. Show that a principal bundle is trivial if and only if it has a global section.

A local section consists of an open set $U \subset M$ and a map $\psi: U \to P$ such that $\pi \circ \psi = id_U$. If $f: \pi^{-1}(U) \to U \times G$ is a local trivialization we can define a local section by $\psi(x) = f^{-1}(x, h(x))$, where $h : U \rightarrow G$ is an arbitrary (smooth) function.

Let $H \subset G$ be a closed subgroup. We say that the bundle P has been *reduced* to a principal H *subbundle* Q, if $Q \subset P$ is a submanifold such that $qh \in Q$ for all $q \in Q, h \in H, \pi(Q) = M$ and H acts transitively in each fiber $Q_x = \pi^{-1}(x) \cap Q$.

Any manifold M of dimension n carries a natural principal $GL(n, \mathbf{R})$ bundle, namely, the bundle FM of linear frames. The fiber F_xM at a point $x \in M$ consists of all frames (ordered basis) of the tangent space T_xM. The group $GL(n,\mathbf{R})$ acts in F_xM by $(f_1, f_2, ..., f_n)A = (\sum_{i=1}^n A_{i1} f_i, \sum_{i=1}^n A_{i2} f_i, ..., \sum_{i=1}^n A_{in} f_i)$, where the f_i's are tangent vectors at x and $A = (A_{ij}) \in GL(n,\mathbf{R})$. One can construct a local trivialization by choosing a local coordinate system $(x_1, x_2, ..., x_n)$ in M. In local coordinates the vectors of a frame f can be written as $f_i = \sum f_{ij}\partial_j$. This defines a mapping $f \mapsto (f_{ij}) \in GL(n, \mathbf{R})$. The collection $(\partial_1, ..., \partial_n)$ of vector fields defines a local section of FM.

If the manifold M has some additional structure the bundle FM can generally be reduced to a subbundle. For example, if M is a Riemannian manifold with metric g, then we can define the subbundle $OFM \subset FM$ consisting of *orthonormal* frames with respect to the metric g. If in addition M is oriented, then it makes sense to speak of the bundle *SOFM of oriented orthonormal* frames: A frame $(f_1, \ldots, f_n)$ at a point x is oriented if $\omega(x; f_1, \ldots, f_n)$ is positive, where ω is a n form defining the orientation. The structure group of OFM is the orthogonal group $O(n)$ and of $SOFM$ the special orthogonal group $SO(n)$ consisting of orthogonal matrices with determinant=1.

Let $\mathbf{g}$ be the Lie algebra of the Lie group G. To any $A \in \mathbf{g}$ there corresponds canonically a one-parameter subgroup $h_A(t) = \exp tA$. We define a vector field $\hat{A}$ on the G bundle P such that the tangent vector $\hat{A}(p)$ at $p \in P$ is equal to $\frac{d}{dt}[p \cdot h_A(t)]|_{t=0}$. Let $g \in G$ be any fixed element. The right translation $r_g(p) = pg$ on P determines canonically a transformation $X \mapsto r_g^* X$ on vector fields: The tangent vector of the transformed field at a point p is simply obtained by applying the derivative of the mapping r_g to the tangent vector $X(pg^{-1})$.

We denote by ad_g the adjoint action of an element $g \in G$ on the Lie algebra $\mathbf{g}$. When G is represented as a group of matrices, this action is simply $ad_g(x) = g^{-1}xg$, where $x \in \mathbf{g}$.

PROPOSITION 3.1.7. *For any $A \in \mathbf{g}$ the vector field $\hat{A}$ is equivariant, that is,* $r_g^* \hat{A} = \widehat{ad_g A} \forall g \in G$.

PROOF: Using a local trivialization,

$$\hat{A}(p) = \frac{d}{dt}(\pi(p), \phi(pe^{tA}))\Big|_{t=0}$$

and therefore

$$\begin{aligned}(r_g^*\hat{A})(p) &= T_{pg^{-1}}r_g \cdot \frac{d}{dt}(\pi(pg^{-1}, \phi(pg^{-1}e^{tA}))\Big|_{t=0} \\ &= \frac{d}{dt}(\pi(pg^{-1}), \phi(pg^{-1}e^{tA}g))\Big|_{t=0} \\ &= \frac{d}{dt}(0, \phi(pe^{tad_g A})\Big|_{t=0} \\ &= \widehat{ad_g A}(p).\end{aligned}$$

3.2. Connection and curvature in a principal bundle

Let E and M be a pair of manifolds, V a vector space and $\pi : E \to M$ a smooth surjective map.

DEFINITION 3.2.1. *The manifold E is a vector bundle over M with fiber V, if*

(1) *$E_x = \pi^{-1}(x)$ is isomorphic with the vector space V for each $x \in M$*
(2) *$\pi : E \to M$ is locally trivial: Any $x \in M$ has an open neighborhood U with a diffeomorphism $\phi : \pi^{-1}(U) \to U \times V$, $\phi(z) = (\pi(z), \xi(z))$, where the restriction of ξ to a fiber E_x is a linear isomorphism onto V.*

The product $M \times V$ is the *trivial vector bundle* over M, with fiber V. In this case the projection map $M \times V \to M$ is simply the projection onto the first factor.

A direct sum of two vector bundles E and F over the same manifold M is the bundle $E \oplus F$ with fiber $E_x \oplus F_x$ at a point $x \in M$. The *tensor product bundle* $E \otimes F$ is the vector bundle with fiber $E_x \otimes F_x$ at $x \in M$.

Example 3.2.2. The tangent bundle TM of a manifold M is a vector bundle over M with fiber $T_xM \simeq \mathbf{R}^n$, where $n = dim M$. The local trivializations are given by local coordinates: If $(x_1, x_2, ..., x_n)$ are local coordinates on $U \subset M$, then the value of ξ for a tangent vector $w \in T_xM$, $x \in M$, is obtained by expanding w in the basis defined by the vector fields $(\partial_1, ..., \partial_n)$.

A section of a vector bundle E is a map $\psi : M \to E$ such that $\pi \circ \psi = id_M$. The space $\Gamma(E)$ of sections of E is a linear vector space; the addition and multiplication by scalars is defined pointwise. A principal bundle may or may not have global sections but a vector bundle always has nonzero sections. A section can be multiplied by a smooth function $f \in C^\infty(M)$ pointwise, $(f\psi)(x) = f(x)\psi(x)$.

Let (P, π, M) be a principal G bundle. *The space V of vertical vectors* in the tangent bundle TP is the subbundle of TP with fiber $\{v \in T_pP | \pi(v) = 0\}$ at $p \in P$. If P is trivial, $P = M \times G$, then the vertical subspace at $p = (x, g)$ consists of vectors tangential to G at g. In general, the dimension of the fiber V_p is equal to $\dim G$.

DEFINITION 3.2.3. *A connection in the principal bundle P is a smooth distribution $p \mapsto H_p$ of subspaces of T_p such that*

(1) *The tangent space T_p is a direct sum of V_p and H_p $\forall p \in P$*
(2) *The distribution is equivariant, i.e., $r_g H_p = H_{pg}$ $\forall p \in P, g \in G$.*

Smoothness means that the distribution can be locally spanned by smooth vector fields. We shall denote by pr_h (respectively, pr_v) the projection in T_p to the *horizontal* subspace H_p (respectively, vertical subspace V_p).

Let $A \in \mathbf{g}$ and let $\hat{A}$ be the corresponding equivariant vector field on P. The field $\hat{A}$ is vertical at each point. Since the group G acts freely and transitively on P, the mapping $A \mapsto \hat{A}(p)$ is a linear isomorphism onto V_p for all $p \in P$. Thus for each $X \in T_pP$ there is a uniquely defined element $\omega_p(X) \in \mathbf{g}$ such that

$$\widehat{\omega_p(X)} = pr_v X$$

at p. The mapping $\omega_p : T_pP \to \mathbf{g}$ is linear, thus defining a differential form of degree one on P, with values in the Lie algebra $\mathbf{g}$. The form ω is the *connection form* of the connection H.

PROPOSITION 3.2.4. *The connection form satifies*

(1) $\omega_p(\hat{A}(p)) = A \, \forall A \in \mathbf{g}$,
(2) $r_a^* \omega = ad_a \omega \, \forall a \in G$.

Furthermore, each $\mathbf{g}$-valued differential form on P which satisfies the above conditions is a connection form of a uniquely defined connection in P.

PROOF: The first equation follows immediately from the definition of ω. To prove the second, we first note that

$$(\widehat{ad_a A})(p) = \frac{d}{dt} p e^{t ad_a A}|_{t=0} = \frac{d}{dt} p a^{-1} e^{tA} a|_{t=0} = r_a \hat{A}(pa^{-1}).$$

By the equivariantness property of the distribution H_p, the right translations r_a commute with the horizontal and vertical projection operators. Thus [we write $(A)^\frown$ for $\hat{A}$ in case of long expressions]

$$\begin{aligned}(ad_a \omega_p(X))^\frown(pa) &= r_a \cdot \widehat{\omega_p(X)}(p)\\ &= r_a(pr_v X) = pr_v(r_a X)\\ &= (\omega_p(r_a X))^\frown(pa).\end{aligned}$$

Taking account that $(r_a^* \omega)_p(X) = \omega_{pa}(r_a X)$ we get the second relation.

Let then ω be any form satisfying both equations. We define the horizontal subspaces $H_p = \{X \in T_p | \omega_p(X) = 0\}$. If $X \in H_p \cap V_p$, then $X = \hat{A}(p)$ for some $A \in \mathbf{g}$ and $\omega_p(\hat{A}(p)) = A = \omega_p(X) = 0$, from which follows $H_p \cap V_p = 0$. By (1) and a simple dimensional argument we get $T_p = H_p + V_p$. For $X \in H_p$ and $a \in G$ we obtain

$$\omega_{pa}(r_a X) = (r_a^* \omega)_p(X) = ad_a^{-1} \omega_p(X) = 0,$$

and therefore $r_a X \in H_{pa}$, which shows that the distribution H_p is equivariant and indeed defines a connection in P.

Let ω be a connection form in (P, π, M). Let $U \subset M$ be open and $\psi : U \to P$ a local section. The pull-back $A = \psi^* \omega$ is a one-form on U. Consider another local section $\phi : V \to P$ and set $A' = \phi^* \omega$. We can write $\psi(x) = \phi(x) g(x)$ for $g \in U \cap V$, where $g(x)$ is a smooth G valued function. We want to relate A to A'. Noting that

$$T_x \psi = r_{g(x)} T_x \phi + (g^{-1} T_x g)^\frown(\psi(x))$$

by the Leibnitz rule, we get

$$\begin{aligned}A_x(u) &= \omega_{\psi(x)}(T_x \psi \cdot u) = \omega_{\psi(x)}(r_{g(x)} T_x \phi \cdot u + (g^{-1} T_x g \cdot u)^\frown(\psi(x)))\\ &= ad_{g(x)} \omega_{\phi(x)}(T_x \phi \cdot u) + g^{-1} T_x g \cdot u.\end{aligned}$$

For a matrix group G we can simply write

$$A = g^{-1} A' g + g^{-1} dg.$$

The transformation relating A to A' is called *a gauge transformation* . Next we define the two-form

$$F = dA + \frac{1}{2}[A, A] \tag{3.2.5}$$

on U. The commutator of Lie algebra valued one-forms is defined by

$$[A, B](u, v) = [A(u), B(v)] - [A(v), B(u)]$$

for a pair u, v of tangent vectors. We shall compute the effect of a gauge transformation $(U, \psi) \to (V, \phi)$ on F:

$$\begin{aligned} F &= dA + \tfrac{1}{2}[A, A] \\ &= g^{-1} dA' g + [g^{-1}dg, g^{-1}A'g] - \tfrac{1}{2}[g^{-1}dg, g^{-1}dg] \\ &\quad + \tfrac{1}{2}[g^{-1}A'g + g^{-1}dg, g^{-1}A'g + g^{-1}dg] \\ &= g^{-1}(dA' + \tfrac{1}{2}[A', A'])g = g^{-1}F'g. \end{aligned}$$

The *curvature form* F is a pull-back under ψ of a gobally defined two-form Ω on P. The latter is defined by

$$\Omega_p(u, v) = a^{-1} F_x(\pi u, \pi v) a,$$

where $p \in \pi^{-1}(x)$, u, v tangent vectors at p and $a \in G$ is an element such that $p = \psi(x)a$. The left-hand side does not depend on the local section. Writing $p = \phi(x)a' = \psi(x)g(x)a'$ we get

$$a'^{-1} F'_x(\pi u, \pi v) a' = a'^{-1} g(x)^{-1} F_x(\pi u, \pi v) g(x) a' = a^{-1} F_x(\pi u, \pi v) a.$$

Since A is the pull-back of ω and F is the pull-back of Ω we obtain from 3.2.5

$$\Omega = d\omega + \frac{1}{2}[\omega, \omega] \tag{3.2.6}$$

Exercise 3.2.7. Prove the *Bianchi identity* $dF + [A, F] = 0$. (The 3-form $[A, F]$ is defined by an antisymmetrization of $[A(u), F(v, w)]$ with respect to the triplet (u, v, w) of tangent vectors.)

Let (P, π, M) be a principal G bundle and $\rho : G \to Aut V$ a linear representation of G in a vector space V. We define the manifold $P \times_G V$ to be the set of equivalence classes $P \times V / \sim$, where the equivalence relation is defined by $(p, v) \sim (pg^{-1}, \rho(g)v)$, for $g \in G$. There is a natural projection $\theta : P \times_G V \to M$, $[(p, v)] \mapsto \pi(p)$. The inverse image $\theta^{-1}(x) \cong V$, since G acts freely and transitively in the fibers of P. The linear structure in a fiber $\theta^{-1}(x)$ is defined by $[(p, v)] + [(p, w)] = [(p, v + w)]$, $\lambda[(p, v)] = [(p, \lambda v)]$. Local trivializations of $P \times_G V$ are obtained from local trivializations $p \mapsto (\pi(p), \phi(p)) \in M \times G$ of P by $[(p, v)] \mapsto (\pi(p), \rho(\phi(p))v)$. Thus $P \times_G V$ is a vector bundle over M, *the vector bundle associated to P* via the representation ρ of G.

Example 3.2.8. Let $P = SU(2), M = S^2 = SU(2)/U(1), G = U(1), V = \mathbf{C}$ and $\rho(\lambda) = \lambda^2$ for $\lambda \in U(1)$. The associated vector bundle $E = SU(2) \underset{U(1)}{\times} \mathbf{C}$ is in fact the tangent bundle of the sphere S^2. The isomorphism is obtained as follows. Fix a linear isomorphism of $\mathbf{C} \cong \mathbf{R}^2$ with the tangent space of S^2 at the point x, which has as its isotropy group the given $U(1)$. The map $E \to TS^2$ is defined by $(g, v) \mapsto D(g)v$, where $D(g)$ is the 2-1 representation of $SU(2)$ in $\mathbf{R}^3$. The tangent vectors of S^2 are represented by vectors in $\mathbf{R}^3$ by the natural embedding $S^2 \subset \mathbf{R}^3$.

3.3. Parallel transport

Let H be a connection in a principal G bundle (P, π, M). *A horizontal lift* of a smooth curve $\gamma(t)$ on the base manifold M is a smooth curve $\gamma^*(t)$ on P such that the tangent vector $\dot\gamma^*(t)$ is *horizontal* at each point on the curve and $\pi(\gamma^*(t)) = \gamma(t)$.

LEMMA 3.3.1. *Let $X(t)$ be a smooth curve on the Lie algebra $\mathbf{g}$ of G, defined on an interval $[t_0, t_1]$. Then there exists a unique smooth curve $a(t)$ on G such that $\dot a(t)a(t)^{-1} = X(t) \forall t \in [t_0, t_1]$ and such that $a(t_0) = e$.*

PROOF: See Kobayashi and Nomizu, vol. I, p. 69.

PROPOSITION 3.3.2. *Let $\gamma(t)$ be a smooth curve on M and p an element in the fiber over $\gamma(t_0)$. Then there exists a unique horizontal lift $\gamma^*(t)$ of $\gamma(t)$ such that $\gamma^*(t_0) = p$.*

PROOF: Choose first any (smooth) curve $\phi(t)$ on P such that $\pi(\phi) = \gamma$ and $\phi(t_0) = p$. We are looking for the solution in the form $\gamma^*(t) = \phi(t)g(t)$, where $g(t)$ is a curve on G such that $g(t_0) = e$. Now $\gamma^*(t)$ is a solution if

$$\dot\gamma^*(t) = r_{g(t)} \cdot \dot\phi(t) + (g(t)^{-1}\dot g(t))^\frown[\phi(t)g(t)]$$

is horizontal. Let ω be the connection form of the connection H. A tangent vector on P is horizontal if and only if it is in the kernel of ω. We get the differential equation

$$\begin{aligned} 0 = \omega(\dot\gamma^*(t)) &= \omega(r_{g(t)}\dot\phi(t)) + \omega([g(t)^{-1}\dot g(t)]^\frown[\phi(t)g(t)]) \\ &= ad_{g(t)}\omega(\dot\phi(t)) + g(t)^{-1}\dot g(t). \end{aligned}$$

Applying ad_g^{-1} to this equation we get

$$\dot g(t)g(t)^{-1} = -\omega(\dot\phi(t)).$$

The solution $g(t)$ exists and is unique by the previous lemma.

Example 3.3.3. Let $P = M \times U(1)$. A connection form ω can be written as $\omega_{(x,g)}(u,a) = -A_x(u) + g^{-1} \cdot a$, where u is a tangent vector at $x \in M$ and a is a tangent vector at $g \in U(1)$; the Lie algebra of $U(1)$ is identified by the set of purely imaginary complex numbers. Let $\gamma(t)$ be a curve on M. The horizontal lift of $\gamma(t)$ which goes through $(\gamma(t), g)$ at time $t = 0$ is $\gamma^*(t) = (\gamma(t), g(t))$ with

$$g(t) = g \cdot \exp\left(\int_0^t A_{\gamma(s)}(\dot\gamma(s))ds\right).$$

In particular, for a *closed curve*, $\gamma(0) = \gamma(1)$, we get by Stokes's theorem

$$g(1) = g \cdot \exp(\int_S F),$$

where $F = dA$ is the curvature two-form and the integration is taken over any surface on M bounded by the closed curve γ.

We define *the parallel transport* along a curve $\gamma(t)$ on M as a mapping $\tau : \pi^{-1}(x_0) \to \pi^{-1}(x_1)$ ($x_0 = \gamma(t_0), x_1 = \gamma(t_1)$ points on the curve). The value $\tau(p_0)$ for $p_0 \in \pi^{-1}(x_0)$ is given as follows: Let $\gamma^*(t)$ be a horizontal lift of $\gamma(t)$ such that $\gamma^*(t_0) = p_0$. Then $\tau(p_0) = \gamma^*(t_1)$.

Exercise 3.3.4. Prove the following properties of the parallel transport.

(1) $\tau \circ r_g = r_g \circ \tau \forall g \in G$
(2) If γ_1 is a path from x_0 to x_1 and γ_2 is a path from x_1 to x_2 then the parallel transport along the composed path $\gamma_2 * \gamma_1$ is equal to the product of parallel transport along γ_1 followed by a parallel transport along γ_2.
(3) The parallel transport is a one-to-one mapping between the fibers $\pi^{-1}(x_0)$ and $\pi^{-1}(x_1)$.

3.4. Covariant differentiation in vector bundles

Let E be a vector bundle over a manifold M with fiber V, $\dim V = n$. The vector space V is defined over the field $\mathbf{K} = \mathbf{R}$ or $K = \mathbf{C}$. A vector bundle can always be thought of as an associated bundle to a principal bundle. Namely, let P_x denote the space of all linear frames in the fiber E_x for $x \in M$. Using the local trivializations of E it is not difficult to see that the spaces P_x fit together and form naturally a smooth manifold P. Fix a basis $w = \{w_1, \ldots, w_n\}$ in E_x. Then any other basis of E_x can be obtained from w by a linear tranformation $w'_i = \sum A_{ji} w_j$

and therefore P_x can be identified with the group $GL(n, \mathbf{K})$ of all linear transformations in $\mathbf{K}^n$; it should be stressed that this identification depends on the choice of w. However, we have a well-defined mapping $P \times GL(n, \mathbf{K}) \to P$ given by the basis transformations and this shows that P can be thought of as a principal $GL(n, \mathbf{K})$ bundle over M.

The vector bundle E is now isomorphic with the associated bundle $P \times_\rho \mathbf{K}^n$, where ρ is the natural representation of $GL(n, \mathbf{K})$ in $\mathbf{K}^n$. The isomorphism is defined as follows. Let $w \in P_x$ and $a \in \mathbf{K}^n$. We set $\phi(w, a) = \sum a_i w_i$. This gives a mapping from $P \times \mathbf{K}^n$ to E which is obviously linear in a. For a fixed w the mapping $a \mapsto \phi(w, a)$ gives an isomorphism between $\mathbf{K}^n$ and E_x. Let $w' = w \cdot g$ and $a' = \rho(g^{-1})a$ for some $g \in GL(n, \mathbf{K})$. We have to show that $\phi(w', a') = \phi(w, a)$; but this follows immediately from the definitions.

Often the bundle E can be thought of as an associated bundle to a principal bundle with a smaller structure group than the group $GL(n, \mathbf{K})$. This happens when there is some extra structure in E. For example, assume there is a *fiber metric* in E: This means that there is an inner product $< \cdot, \cdot >_x$ in each fiber E_x such that $x \mapsto < \psi(x), \psi(x) >_x$ is a smooth function for any (local) section ψ. We can then define the bundle of orthonormal frames in E with structure group $U(n)$ in the complex case and $O(n)$ in the real case. The vector bundle E is now an associated bundle to the bundle of orthonormal frames.

We shall now assume that E is given as an associated vector bundle $P \times_\rho V$ to some principal bundle P, with a connection H, over M. Let G be the structure group of P. For each vector field X on M we can define a linear map ∇_X of the space $\Gamma(E)$ of sections into itself such that

(1) $\nabla_{X+Y} = \nabla_X + \nabla_Y$
(2) $\nabla_{fX} = f\nabla_X$
(3) $\nabla_X(f\psi) = (Xf)\psi + f\nabla_X\psi$

for all vector fields X, Y, smooth functions f and sections ψ. We shall give the definition in terms of a local trivialization $\xi : U \to P$, where $U \subset M$ is open. Locally, a section $\psi : M \to E$ can be written as

$$\psi(x) = (\xi(x), \phi(x)),$$

where $\phi : U \to V$ is some smooth function. Let A denote the pull-back $\xi^*\omega$ of the connection form ω in P. The representation ρ of G in V defines also an action of the Lie algebra $\mathbf{g}$ in V. We set

$$\nabla_X\psi = (\xi, X\phi + A(X)\phi),$$

where $A(X)$ is the Lie algebra valued function giving the value of the one-form A in the direction of the vector field X.

We have to check that our definition does not depend on the choice of the local trivialization. So let $\xi'(x) = \xi(x) \cdot g(x)$ be another local trivialization, where $g : U \to G$ is a smooth function. The vector potential with respect to the trivialization ξ' is $A' = g^{-1}Ag + g^{-1}dg$. Now $(\xi, \phi) \sim (\xi', \phi')$, where $\phi' = g^{-1}\phi$ (we simplify the notation by dropping ρ) and therefore $(\xi', X\phi' + A'(X)\phi')$ is equal to

$$\begin{aligned}(\xi', -g^{-1}(Xg)g^{-1}\phi + g^{-1}X\phi + (g^{-1}Ag + g^{-1}Xg)g^{-1}\phi) \\ = (\xi', g^{-1}(X\phi + A(X)\phi)) \sim (\xi, X\phi + A(X)\phi)\end{aligned}$$

which shows that ∇_X is well-defined.

Exercise 3.4.1. Prove that ∇_X defined above satisfies (1)-(3).

The commutator of the *covariant derivatives* ∇_X is related to the curvature of the connection in the following way:

$$\begin{aligned}[\nabla_X, \nabla_Y]\psi &= (\xi, [X + A(X), Y + A(Y)]\phi) \\ &= (\xi, ([X, Y] + X \cdot A(Y) - Y \cdot A(X) + [A(X), A(Y)])\phi) \\ &= (\xi, (F(X, Y) + [X, Y] + A([X, Y]))\phi)\end{aligned}$$

where $F = dA + \frac{1}{2}[A, A]$. Thus we can write

$$[\nabla_X, \nabla_Y] - \nabla_{[X,Y]} = F(X, Y)$$

when acting on the functions ϕ.

A section ψ is *covariantly constant* if $\nabla_X\psi = 0$ for all vector fields. From the above commutator formula we conclude that one can find at each point in the base space a local basis of covariantly constant sections of the vector bundle if and only if the curvature vanishes.

3.5. An example: The monopole line bundle

Construction of the basic monopole bundle

Let G be a Lie group and $\mathbf{g}$ its Lie algebra. Let us denote by ℓ_g the left translation $\ell_g(a) = ga$ in G. The left invariant Maurer-Cartan form $\theta_L = g^{-1}dg$ is the $\mathbf{g}$-valued one form on G which sends a tangent vector X at $g \in G$ to the element $\ell_g^{-1}X \in T_eG$ in the Lie algebra. Similarly, we can define the right invariant Maurer-Cartan form $\theta_R = dgg^{-1}$, $\theta_R(g; X) = r_g^{-1}X$. By taking commutators, we can define higher order forms. For example, the form $[g^{-1}dg, g^{-1}dg]$ sends the pair (X, Y) of tangent vectors at g to $2[\ell_g^{-1}X, \ell_g^{-1}Y] \in \mathbf{g}$.

Taking projections to one dimensional subspaces of **g** we get real valued one-forms on G.

Let $< \cdot, \cdot >$ be a bilinear form on **g** and $\sigma \in \mathbf{g}$. Then $\alpha = < \sigma, g^{-1}dg >$ is a well-defined one form. Let us compute the exterior derivative of α. Let X, Y be a pair of left invariant vector fields on G. Now

$$\begin{aligned} d\alpha(g; X, Y) &= X \cdot \alpha(Y) - Y \cdot \alpha(X) - \alpha([X, Y]) \\ &= -\alpha([X, Y]) \end{aligned}$$

since $\alpha(Y)(g) = < \sigma, \ell_g^{-1} Y >$ is a constant function on G and similarly for $\alpha(X)$. Since the left invariant vector fields on a Lie group span the tangent space at each point, we conclude

$$d\alpha = - < \sigma, \tfrac{1}{2}[g^{-1}dg, g^{-1}dg] > .$$

We have not yet defined the exterior derivative of a Lie algebra valued differential form, but motivated by the computation above we set

$$d(g^{-1}dg) = -\frac{1}{2}[g^{-1}dg, g^{-1}dg].$$

A bilinear form $< \cdot, \cdot >$ on **g** is *invariant* if

$$< [X, Y], Z > = - < Y, [X, Z] >$$

for all X, Y, and Z. Given an invariant bilinear form, the group G has a natural **closed** three-form c_3 which is defined by

$$c_3(g; X, Y, Z) = < \ell_g^{-1} X, [\ell_g^{-1} Y, \ell_g^{-1} Z] > .$$

Thus

$$c_3 = < g^{-1}dg, \tfrac{1}{2}[g^{-1}dg, g^{-1}dg] > .$$

PROPOSITION 3.5.1. $dc_3 = 0$.

PROOF.: Recall the definition of the exterior differentiation d: If ω is a n-form and $V_1, ..., V_{n+1}$ are vector fields, then

$$\begin{aligned} d\omega(V_1, ..., V_{n+1}) &= \sum_{i=1}^{n+1} (-1)^{i+1} V_i \cdot \omega(V_1, ..., \hat{V}_i, ..., V_{n+1}) \\ &\quad + \sum_{i<j} (-1)^{i+j} \omega([V_i, V_j], V_1, ..., \hat{V}_i, ..., \hat{V}_j, ..., V_{n+1}), \end{aligned}$$

where the caret means that the corresponding variable has been dropped. Let us compute dc_3 for left invariant vector fields $X_1, ..., X_4$. Taking account that $c_3(X_i, X_j, X_k)$ is a constant function we get

$$\begin{aligned} dc_3(X_1, ...X_4) = & -2 < [X_1, X_2], [X_3, X_4] > +2 < [X_1, X_3], [X_2, X_4] > \\ & -2 < [X_1, X_4], [X_2, X_3] > \\ = & 2 < X_1, [[X_3, X_4], X_2] - [[X_2, X_4], X_3] + [[X_2, X_3], X_4] > \\ = & 0 \end{aligned}$$

by Jacobi's identity.

If G is a group of matrices we can define an invariant form on $\mathbf{g}$ by $< X, Y >= \text{tr}XY$. Then the form c_3 can be written as

$$c_3 = \text{tr}(g^{-1}dg)^3.$$

This will be important in the construction of Kac-Moody groups in Chapter 4.

As an example we shall consider in detail the case $G = SU(2)$. Let $\sigma_3 = \begin{pmatrix} i & 0 \\ 0 & -i \end{pmatrix}$ and define the one-form $\alpha = -\frac{1}{2}\text{tr}\sigma_3 g^{-1}dg$. Remember that $SU(2) \to SU(2)/U(1) = S^2$ is a principal $U(1)$ bundle. The form α is invariant with respect to right translations $g \mapsto gh$ by $h \in U(1)$. Thus α is a connection form in the bundle $SU(2)$ [the Lie algebra of the structure group $U(1)$ can be identified with $i\mathbf{R}$]. Let us compute the curvature. The exterior derivative of α is $\frac{1}{4}\text{tr}\sigma_3[g^{-1}dg, g^{-1}dg]$. A tangent vector at $x \in S^2$ can be represented by a tangent vector $\ell_g X$ at $g \in \pi^{-1}(x), X \in \mathbf{g}$, such that X is orthogonal to the $U(1)$ direction, $\text{tr}\sigma_3 X = 0$. The curvature in the base space S^2 is then $\Omega(X, Y) = \frac{1}{4}tr\sigma_3[X, Y]$. The form Ω is $\frac{1}{2}\times$ the volume form on S^2: If {X,Y} is an ortonormal system at $x \in S^2$, then $[X, Y] = \pm\sigma_3$ (exercise), the sign depending on the orientation. We obtain $\Omega(X, Y) = \pm\frac{1}{4}\text{tr}\sigma_3^2 = \pm\frac{1}{2}$.

The basic monopole line bundle is defined as the associated bundle to the bundle $SU(2) \to S^2$, constructed using the natural one dimensional representation of $U(1)$ in $\mathbf{C}$.

Embedding $S^2 \subset \mathbf{R}^3$ and using Cartesian coordinates $\{x_1, x_2, x_3\}$ we can write the curvature form as

$$\Omega = \frac{1}{4r^3}\varepsilon^{ijk}x_i dx_j \wedge dx_k,$$

where $r^2 = x_1^2 + x_2^2 + x_3^2$ is equal to 1 on S^2. However, we can extend Ω to the space $\mathbf{R}^3 \setminus \{0\}$ using the above formula. The coefficients of the linearly independent forms $dx_2 \wedge dx_3$, $dx_3 \wedge dx_1$ and $dx_1 \wedge dx_2$ form a vector $\vec{B} = \frac{1}{2r^3}(x_1, x_2, x_3) = \frac{\vec{x}}{r^3}$. The field $\vec{B}$ satisfies

$$\vec{\nabla} \cdot \vec{B} = 0 \tag{1}$$

$$\vec{\nabla} \times \vec{B} = 0, \tag{2}$$

i.e., it satisfies Maxwell's equations in vacuum. On the other hand,

$$\int_{S^2} \vec{B} \cdot d\vec{S} = 2\pi \tag{3}$$

for *any* sphere containing the origin. Because of these properties, the field $\vec{B}$ can be interpreted as the magnetic field of a magnetic monopole located at the origin. The integral (3) multiplied by the dimensional constant $1/e$ (e is the unit electric charge) is called the *monopole strength.*

The first Chern class

The magnetic field of the monopole is the curvature of a circle bundle over the unit sphere S^2. The circle bundle we have constructed is a "generator" for the set of all circle bundles over S^2. In general, a principal $U(1)$ bundle over S^2 can be constructed from the transition function $\xi : S_- \cap S_+ \to U(1)$ (cf. 3.1.3). The intersection of the coordinate neighborhoods $S_\pm$ is homeomorphic with the product of an open interval with the circle S^1. It follows that the set of maps ξ decomposes to connected components labelled by the winding number of a map $S^1 \to U(1)$. Let ξ_1 be the transition function of the bundle $SU(2) \to S^2$ with respect to some fixed local trivializations on $S_\pm$. The winding number of ξ_1 is equal to one. The winding number of $\xi_n = (\xi_1)^n$ is equal to n. Let $P(n)$ be the bundle constructed from ξ_n. Let $A_\pm$ be the vector potentials on $S_\pm$ corresponding to the chosen local trivializations and the connection in $SU(2)$ described above.

We have $A_+ = A_- + \xi^{-1} d\xi$ on $S_- \cap S_+$ and therefore $nA_+ = nA_- + \xi_n^{-1} d\xi_n$. Thus nA is a connection in the bundle $P(n)$ and the curvature of $P(n)$ is n times the curvature form Ω of the (basic) monopole bundle. The monopole strength of the bundle $P(n)$ is $2\pi n/e$.

The cohomology class $[\Omega] \in H^2(S^2, \mathbf{R})$ is the *first Chern class* of the bundle. It depends only on the equivalence class of the bundle and not on the chosen connection; we shall return to the proof of the topological invariance of the Chern classes in a more general context in

Section 4.5, but as an illustration of the general ideas we give a simple proof for the case at hand. Let $B_\pm$ be the vector potentials on $S_\pm$ of some connection in the bundle $P(n)$. We have $B_+ = B_- + n\xi^{-1}d\xi$ and therefore $A_+ - B_+ = A_- - B_-$ on $S_+ \cap S_-$. It follows that $A - B$ is a globally defined one-form on S^2; the difference of the curvatures corresponding to the connections A and B is equal to $d(A - B)$.

The first Chern class of a circle bundle (or an associated complex line bundle) over a manifold M can be evaluated from the knowledge of the $U(1)$ valued transition functions [Bott and Tu, 1982]. In the example above we needed only one transition function ξ. A representative Ω for the Chern class can be constructed from a vector potential (A_+, A_-) such that $A_- = 0$ for $x_3 < \frac{1}{2}$, A_+ is equal to $\xi^{-1}d\xi$ on the strip $-\frac{1}{2} < x_3 < \frac{1}{2}$, and A_+ is contracted smoothly to zero when approaching the north pole $x_3 = 1$. The first Chern class is always quantized in the sense that the integral of the two-form Ω over any two-dimensional compact surface is 2π times an integer.

Holomorphic sections

Let E^n be the complex line bundle associated to the principal $U(1)$ bundle $SU(2)$ via the representation $\lambda \mapsto \lambda^n$ of $U(1)$ in $\mathbf{C}$. Recall that the $n = 2$ gives the tangent bundle of S^2. Let Γ^n be the space of sections of E^n. We shall study in detail the space Γ^n, but we need first a general theorem:

PROPOSITION 3.5.2. *Let (P, π, M) be a principal G bundle and $\varrho : G \to AutV$ a linear representation of G in the complex vector space V. Denote by E the associated vector bundle over M. Then there is a natural linear isomorphism between the space $\Gamma(E)$ of E and the space of equivariant functions $f : P \to V$; f is equivariant if $f(pg) = \varrho(g)^{-1}f(p)$ for all p in P, g in G.*

PROOF: 1) Let $\psi : M \to E$ be a section. Let $p \in P$. We can write $\psi(\pi(p)) = [(p, v)]$, where $v \in V$. Define $f(p) = v$. This is equivariant.

2) Let $f : P \to V$ be equivariant. Let $x \in M$. Choose $p \in \pi^{-1}(x)$ and set $\psi(x) = [(p, f(p)]$. If $p' \in \pi^{-1}(x)$, then $p' = pg$ for some $g \in G$ and $(p', f(p')) = (pg, \varrho(g)^{-1}f(p)) \sim (p, f(p))$. Thus ψ is well-defined.

We shall think of a section $\psi \in \Gamma^n$ as a function $\psi : SU(2) \to \mathbf{C}$ with $\psi(gh) = h^{-n}\psi(g)$ for $h \in U(1)$. Any L^2 (and thus also any C^∞) function on $SU(2)$ can be expanded in the basis given by the matrix elements $D^j_{m_1m_2}(\phi_1, \theta, \phi_2) = D^j_{m_1m_2}(g)$ in irreducible finite dimensional representations of $SU(2)$; here $j = 0, 1/2, 1, ...$ is the spin and $m = -j, -j+1, ..., j$ is a label for the basis vectors in the representation D^j.

The basis is chosen such that the action of the diagonal subgroup $U(1)$ on the vector labelled by m is $\lambda \mapsto \lambda^m$. The condition $\psi(gh) = h^{-n}\psi(g)$ means that the label m_2 is fixed to be equal to $-n$. Thus the space Γ^n carries a direct sum of irreducible representations of $SU(2)$ labelled by spin $j = |n|, |n|+1, \ldots$.

The line bundle E^n has a *complex structure*. To describe this we write first $S^2 = SU(2)/U(1) = SL(2, \mathbf{C})/B$, where the subgroup B consists of matrices

$$b = \begin{pmatrix} \lambda & w \\ 0 & 1/\lambda \end{pmatrix}$$

with $\lambda, w \in \mathbf{C}$, $\lambda \neq 0$. The cosets gB corresponding to points in $S^2 \setminus (0,0,-1)$ can be parametrized by matrices

$$g = \begin{pmatrix} 1 & 0 \\ z & 1 \end{pmatrix}$$

giving the complex structure of the Riemann sphere.

Exercise 3.5.3. Find a complex coordinate z' in $SL(2, \mathbf{C})$, well-defined in $S^2 \setminus (0,0,1)$, such that the coordinate transformation $z' = z'(z)$ is holomorphic in $S^2 \setminus (0,0,\pm 1)$.

A *complex manifold* is a smooth manifold M of even real dimension such that it has a coordinate atlas $\{(U_\alpha, \phi_\alpha)\}$ consisting of coordinates $\phi_\alpha : U_\alpha \to \mathbf{C}^n$ with the property that all coordinate transformations are complex analytic functions. A holomorphic function on a complex manifold is a function such that when written in terms of local coordinates it is complex analytic.

PROPOSITION 3.5.4. *The restriction map to the subgroup $SU(2) \subset SL(2, \mathbf{C})$ gives a linear isomorphism from the space $\{\psi : SL(2, \mathbf{C}) \to \mathbf{C} | \psi(gb) = \lambda^{-n}\psi(g) \forall b \in B\}$ to the space Γ^n.*

PROOF: If $\psi : SL(2, \mathbf{C}) \to \mathbf{C}$ has the property above, then $\psi|_{SU(2)} \in \Gamma^n$ follows immediately from the definitions. On the other hand, each matrix $g \in SL(2, \mathbf{C})$ can be uniquely decomposed as $g = un$, where $u \in SU(2)$ and $n \in B$ such that λ is real and positive (this is the Cartan decomposition of complex matrices). If $\psi \in \Gamma^n$, we have thus a unique extension $\psi : SL(2, \mathbf{C}) \to \mathbf{C}$ characterized by $\psi(gn) = \lambda^{-n}\psi(g)$ for all matrices n as above. The extension satisfies automatically $\psi(gb) = \lambda^{-n}\psi(g)$ for all $b \in B$.

We shall now think of the Γ^n as the space of B-equivariant functions on $SL(2, \mathbf{C})$ as in Proposition 3.5.4. The group $SL(2, \mathbf{C})$ has a natural complex manifold structure given by the complex matrix entries. The space $\Gamma^n_a \subset \Gamma^n$ of *holomorphic sections* consists by definition of holomorphic equivariant functions ψ. There is a simple group theoretical characterization of the space of holomorphic sections as follows. First, by the equivariantness condition

$$0 = \frac{d}{dt}\psi(gn(t))|_{t=0},$$

where $n(t) = \exp t \begin{pmatrix} 0 & 1 \\ 0 & 0 \end{pmatrix}$. Denote the basis of the Lie algebra $\mathbf{su}(2)$ by

$$L_1 = \frac{1}{2}\begin{pmatrix} 0 & 1 \\ -1 & 0 \end{pmatrix},$$
$$L_2 = \frac{1}{2}\begin{pmatrix} 0 & i \\ i & 0 \end{pmatrix},$$
$$L_3 = \frac{1}{2}\begin{pmatrix} i & 0 \\ 0 & -i \end{pmatrix}$$

with $[L_1, L_2] = L_3$, and similarly for cyclic permutations. The Lie algebra acts through Lie derivatives on smooth functions in G. In the space of *holomorphic* functions it is clear that the linear combination $L_- = L_1 - iL_2$ is represented by the same operator as the element $\begin{pmatrix} 0 & 1 \\ 0 & 0 \end{pmatrix}$ in $SL(2, \mathbf{C})$. Thus

$$L^r_- \psi = 0$$

for a holomorphic section ψ; here the upper index r refers to the *right* action of group $SL(2, \mathbf{C})$ on itself; the same symbol has been used both for the group generators and the corresponding vector fields. Since $[L_3, L_-] = iL_-$, the element L_- lowers the eigenvalue of the Hermitian generator iL_3 by one unit. It follows that the subspace $\Gamma^n_a \subset \Gamma^n$ is spanned by the functions $D^j_{m_1 m_2}$ with $-m_2 = n = j$. Since $0 \leq j$, the space Γ^n_a is nonzero only if $0 \leq n$ and in that case $\dim \Gamma^n_a = 2n + 1$. In particular, Γ^n_a for $0 \leq n$ carries *an irreducible representation* of $SU(2)$ with spin $j = n$.

3.6. Invariant connections

Let P be a principal G-bundle over M and let K be a group of automorphisms acting on P. *A K-invariant connection on P* is a connection such that the pull-back $k^*\omega$ of the connection form is equal to ω for every $k \in K$. This is equivalent to the condition that $T_p k \cdot H_p = H_{kp}$ for all $k \in K$ and $p \in P$, where H_p is the horizontal subspace at p. In this section we shall discuss a method to generate invariant connections and compute the curvature.

LEMMA 3.6.1. *Let $t \mapsto \phi_t$ be a one parameter family of automorphism of the principal bundle P and let $p \mapsto X_p = \frac{d}{dt}\phi_t(p)|_{t=0}$ the associated vector field on P. Let ω be a ϕ_t- invariant connection. Let $p_0 \in P$ and let p_t be the horizontal lift of the curve $t \mapsto \pi(\phi_t(p_0))$ going through the given point p_0; π is the bundle projection. We can write $\phi_t(p_0) = p_t g_t$, where $g_t \in G$. Then g_t is the one-parameter subgroup of G generated by the Lie algebra element $\omega_{p_0}(X)$.*

PROOF: By the Leibnitz formula

$$\frac{d}{dt}\phi_t(p_0) = (\frac{d}{dt}p_t) \cdot g_t + \hat{A}(p_t)$$

where $A = A_t = g_t^{-1}\frac{d}{dt}g_t$ and $\hat{A}$ is the vector field defined as in Section 3.1. By Proposition 3.2.2 we have $\omega(\hat{A}) = A$ and therefore

$$\omega_{p_t}\left(\frac{d}{dt}\phi_t(p_0)\right) = ad_{g_t}\omega\left(\frac{d}{dt}p_t\right) + A.$$

Since p_t is horizontal the first term on the right vanishes and so $A = \omega_{p_t}(\frac{d}{dt}\phi_t(p_0)) = \omega_{p_t}(\phi_{t*}X_{p_0}) = \omega_{p_0}(X)$, where in the last step we have used the invariance of ω.

By a projection onto the base any automorphism of a principal bundle induces a diffeomorphism on the base manifold. Let $L \subset K$ be the stability subgroup at $x_0 = \pi(p_0)$ for some fixed $p_0 \in P$. We define a map $\lambda : L \to G$ by the equation

$$k \cdot p_0 = p_0 \cdot \lambda(k).$$

This map is actually a homomorphism: Let $k, k' \in L$. Then

$$\begin{aligned} p_0 \cdot \lambda(kk') &= (kk') \cdot p_0 = k(p_0\lambda(k')) = (kp_0)\lambda(k') \\ &= (p_0\lambda(k))\lambda(k') = p_0 \cdot (\lambda(k)\lambda(k')). \end{aligned}$$

The map λ depends on the reference point p_0 which we shall keep fixed throughout this section. We shall denote by the same letter λ the induced Lie algebra homomorphism $\mathbf{l} \to \mathbf{g}$. We define also a linear mapping $\Lambda : \mathbf{k} \to \mathbf{g}$ by $\Lambda(X) = \omega_{p_0}(\hat{X})$.

PROPOSITION 3.6.2. *Let ω be a K-invariant connection in P. Then*

(1) $\Lambda(X) = \lambda(X)$ *for* $X \in \mathbf{l}$;
(2) $\Lambda(ad_k(X)) = ad_{\lambda(k)}\Lambda(X)$ *for* $k \in L$ *and* $X \in \mathbf{k}$.

PROOF: (1) Let ϕ_t be the one-parameter subgroup generated by $X \in \mathbf{l}$. Now $t \mapsto \pi(\phi_t(p_0))$ is the constant path at x_0 and so $\phi_t(p_0) = p_0\lambda(\phi_t)$. Differentiating this equation at $t = 0$ and using Proposition 3.6.1 we obtain $\Lambda(X) = \lambda(X)$.

(2) Let $X \in \mathbf{k}$ and $k \in L$. Then the one-parameter subgroup of K generated by $Y = ad_k(X)$ is $t \mapsto k\phi_t k^{-1}$. Now

$$k\phi_t k^{-1} \cdot p_0 = k\phi_t(p_0\lambda(k^{-1})) = k(r_{\lambda(k^{-1})}\phi_t p_0)$$

and it follows that $\hat{Y}_{p_0} = k(r_{\lambda(k^{-1})}\hat{X}_{p_0})$. By the invariance of the connection,

$$\begin{aligned}\omega_{p_0}(\hat{Y}) &= \omega_{p_0}(k(r_{\lambda(k^{-1})}\hat{X}_{p_0})) = \omega_{k^{-1}p_0}(r_{\lambda(j^{-1})}\hat{X}_{p_0}) \\ &= ad_{\lambda(k)}\omega_{p_0}(\hat{X}_{p_0}) = ad_{\lambda(k)}\Lambda(X).\end{aligned}$$

PROPOSITION 3.6.3. *The curvature form Ω of a K-invariant connection ω satisfies*

$$\Omega(\hat{X}, \hat{Y}) = [\Lambda(X), \Lambda(Y)] + \Lambda([X, Y]) \qquad \textit{for } X, Y \in \mathbf{k}.$$

PROOF: From 3.2.4 we obtain

$$\begin{aligned}\Omega(\hat{X}, \hat{Y}) &= d\omega(\hat{X}, \hat{Y}) + [\omega(\hat{X}), \omega(\hat{Y})] \\ &= \hat{X} \cdot \omega(\hat{Y}) - \hat{Y} \cdot \omega(\hat{X}) - \omega([\hat{X}, \hat{Y}]) + [\omega(\hat{X}), \omega(\hat{Y})].\end{aligned}$$

By the invariance of ω we have $\hat{X} \cdot \omega(\hat{Y}) - \omega([\hat{X}, \hat{Y}]) = 0$ and similarly with X and Y interchanged. Thus

$$\begin{aligned}\Omega(\hat{X}, \hat{Y}) &= \omega([\hat{X}, \hat{Y}]) + [\omega(\hat{X}), \omega(\hat{Y})] \\ &= \Lambda([X, Y]) + [\Lambda(X), \Lambda(Y)].\end{aligned}$$

THEOREM 3.6.4. *Assume that the group K acts transitively on the base M and that $\mathbf{k}$ is a direct sum of $\mathbf{l}$ and a subspace $\mathbf{m}$ such that $ad_k(\mathbf{m}) = \mathbf{m}$ for all $k \in L$. Then*

(1) *There is a one-to-one correspondence between the set of K-invariant connections on P and the set of linear maps $\Lambda_m : \mathbf{m} \to \mathbf{g}$ such that*

$$\Lambda_m(ad_k X) = ad_{\lambda(k)}\Lambda_m(X) \qquad \textit{for } X \in \mathbf{m},\ k \in L.$$

The correspondence is given by

$$\Lambda(X) = \begin{cases} \lambda(X), & X \in \mathbf{l} \\ \Lambda_m(X), & X \in \mathbf{m}. \end{cases}$$

(2) *The curvature of the connection ω corresponding to the map Λ_m satisfies*

$$\Omega_{p_0}(\hat{X},\hat{Y}) = [\Lambda_m(X),\Lambda_m(Y)] + \Lambda_m([X,Y]_m) + \lambda([X,Y]_l),$$

where Z_m (respectively, Z_l) denotes the $\mathbf{m}$ (respectively, $\mathbf{l}$) - component of an element $Z \in \mathbf{k}$.

PROOF: If the K-invariant connection ω is given then from Proposition 3.6.2 follows that the map $\Lambda_m = \Lambda|_{\mathbf{m}}$ satisfies the condition (1) above. The curvature formula is a consequence of Proposition 3.6.3.

Since the group K acts transitively on M and the connection is invariant the values of ω at one point p_0 fix the form ω completely. From the transitivity it follows also that the vectors $\hat{X}_{p_0}$, $X \in \mathbf{k}$, together with the vertical vectors span the tangent space $T_{p_0}P$. The values of a connection form along the vertical directions are fixed by Proposition 3.2.2; thus an invariant connection is completely determined when the map Λ (Proposition 3.6.2) is given; but Λ is determined by Λ_m. It follows that the map $\omega \mapsto \Lambda_m$ is injective.

To complete the proof one needs to show that to each Λ_m there corresponds a K-invariant connection ω. Let Λ_m be given such that it satisfies the condition (1). Define the horizontal subspace H_{p_0} at the base point p_0 by

$$H_{p_0} = \{\tilde{X}_{p_0} - \Lambda_m(X)\hat{}_{p_0} | X \in \mathbf{m}\},$$

where $\tilde{X}$ is the vector field on P induced by the action of K on P. Since $M = K/L$ we observe by a simple dimensional argument that the linear map $\mathbf{m} \to T_{x_0}M$, $X \mapsto \pi(\tilde{X}_{p_0})$ is injective. Since $\pi(\hat{A}) = 0$ for all $A \in \mathbf{g}$ we reduce that the map $X \mapsto \tilde{X}_{p_0} - \Lambda_m(X)\hat{}_{p_0}$ is injective, too. Again by dimensional grounds we see that H_{p_0} is a complement of the subspace V_{p_0} in $T_{p_0}P$. Next we can define the horizontal subspace H_p at any point p in the fiber $\pi^{-1}(x_0)$ using the characteristic relation $r_g \cdot H_p = H_{pg}$ for $g \in G$. The distribution H_p can then be extended to other fibers simply by the action of the group K, $H_{kp_0} = k \cdot H_{p_0}$; we leave it as an exercise to the reader to show that this really defines an invariant connection and that it is related to Λ_m by (1).

Exercise 3.6.5. Complete the proof of 3.6.4.

3.7. The Levi-Cività connection

Let M be a Riemannian manifold with metric g. The metric defines a positive definite inner product $g_p(\cdot,\cdot)$ in each of the tangent spaces T_pM. The inner product defines a connection, the *Levi-Cività connection* , as follows. We shall define the connection by giving the operator ∇_X when acting in the space of sections of the tangent bundle TM for any vector field X on M. The Levi-Cività connection has the following characteristic features:

(1) The torsion vanishes: $\nabla_X Y - \nabla_Y X = [X,Y]$ for all vector fields X, Y;
(2) $Z \cdot g(X,Y) = g(\nabla_Z X, Y) + g(X, \nabla_Z X)$ for all X, Y, Z.

In fact, there is precisely one connection ∇ which satisfies these two conditions. We prove the uniqueness. First we observe that the right-hand-side of (2) is equal to

$$g(\nabla_X Z, Y) + g(X, \nabla_Z Y) + g([Z,X],Y)$$

where we have used (1). In the same way we can compute $X \cdot g(Y,Z)$ and $Y \cdot g(X,Z)$ and then we can eliminate from the resulting set of equations the terms involving ∇_X and ∇_Y and we obtain

$$\begin{aligned} 2g(X,\nabla_Z Y) =& Z \cdot g(X,Y) + Y \cdot g(X,Z) - X \cdot g(Y,Z) \\ &+ g(Z,[X,Y]) + g(Y,[X,Z]) - g(X,[Y,Z]). \end{aligned} \tag{3.7.1}$$

Since g is nondegenerate this formula fixes $\nabla_Z Y$ uniquely. Conversely, we can use equation (3.7.1) to *define* ∇_Z. Going backwards the steps leading to (3.7.1) one easily deduces that the operator ∇ so defined satisfies (1) and (2). The only thing left is to show that ∇ is indeed a connection, that is,

(1) $Y \mapsto \nabla_X Y$ and $Y \mapsto \nabla_Y X$ are linear
(2) $\nabla_{fX} Y = f\nabla_X Y$ for any smooth function f on M
(3) $\nabla_X(fY) = f\nabla_X Y + (Xf)Y$.

The property (1) is clear from the formula (3.7.1), (2) follows from the same formula by applying $[X, fZ] = f[X,Z] + (Xf)Z$ to the last to terms, and (3) is proven by a similar computation.

In local coordinates the Levi-Cività connection is conveniently expressed through the *Christoffel symbols* Γ^k_{ij}. Denoting $\nabla_i = \nabla_{\frac{\partial}{\partial x_i}}$ we have

$$\nabla_i \frac{\partial}{\partial x_j} = \sum_k \Gamma^k_{ij} \frac{\partial}{\partial x_k}.$$

Let X, Y be a pair of vector fields on M. The curvature operator $R(X, Y)$ is a $C^\infty(M)$ linear map of the space of vector fields into itself, defined by

$$R(X,Y) = \nabla_X\nabla_Y - \nabla_Y\nabla_X - \nabla_{[X,Y]}. \tag{3.7.2}$$

Defining the coefficients of the curvature R by

$$R(\partial_i, \partial_j)\partial_k = \sum_m R^m_{ijk}\partial_m$$

we obtain from (3.7.2)

$$R^m_{ijk} = \Gamma^m_{in}\Gamma^n_{jk} - \Gamma^m_{jn}\Gamma^n_{ik} + \partial_i\Gamma^m_{jk} - \partial_j\Gamma^m_{ik}. \tag{3.7.3}$$

This is of course a special case of the general formula (3.2.5) for the curvature in a principal bundle.

A change $x'_i = x'_i(x)$ of the local coordinates induces a transformation on the Christoffel symbols. Writing

$$\nabla'_i\partial'_j = \sum \Gamma'^m_{ij}\partial'_m$$

and taking into account $\partial'_i = \frac{\partial x_j}{\partial x'_i}\partial_j \equiv a_{ji}\partial_j$ we get

$$\Gamma'^k_{ij} = (a^{-1})_{km}a_{li}a_{nj}\Gamma^m_{ln} + (a^{-1})_{km}a_{li}\partial_l a_{mj}. \tag{3.7.4}$$

It is useful to think of the connection Γ^k_{ij} as a local one form on M taking values in the general linear Lie algebra $\mathbf{gl}(N)$ (N is the dimension of M) by considering k and j as matrix indices; we can then write (3.7.4) as

$$\Gamma'_i = a_{li}a^{-1}\Gamma_l a + a_{li}a^{-1}\partial_l a. \tag{3.7.5}$$

This is in accordance with the general transformation rule of a vector potential in a gauge transformation a except that in addition the transformation hits the differential form index i corresponding to the pull-back of a one-form with respect to the map $x \mapsto x'$.

CHAPTER 4 EXTENSIONS OF GROUPS OF GAUGE TRANSFORMATIONS

4.0. Introduction

In this chapter we shall discuss the structure of the infinite-dimensional Lie groups associated to the affine Kac-Moody algebras. We shall also construct the group of the current algebra of a gauge field theory in 3+1 space-time dimensions and we shall study the implications of the commutation relations for the spin-statistics relation in 3+1 dimensions.

To begin with, we shall explain in Section 4.1 the basic machinery of group cohomology theory. The second cohomology is important for the classification of group extensions. If a group G acts as a group of automorphisms of an Abelian group A then one can construct an extension of G by A as the group $\hat{G}$ which as a set is the product $G \times A$ but which has the group composition rule

$$(g, a)(g', a') = (gg', a + g(a') + c(g, g')), \tag{4.0.1}$$

where $c : G \times G \to A$ is a function which has to satisfy a *cocycle* property in order that the group multiplication is associative. Isomorphic group extensions correspond to equivalent cocycles and the equivalence classes of cocycles are elements of the second cohomology group $H^2(G, A)$. If $c \equiv 0$ then the extension is the semidirect product of the groups G and A. A more general extension is thus a "twisted semidirect product". In the case of a *central extension* (groups associated to affine algebras) the action of G on A is trivial.

The case of affine algebras and their generalizations is in fact slightly more complicated than described above. It happens that the extensions will be also *topologically twisted*; this means that the extension is a nontrivial principal bundle over the group G. For an illustration of this situation consider the following simple example. Take $G = SO(3)$ and $A = \mathbf{Z}_2$. The group $SU(2)$ may be viewed as an extension of $SO(3)$ by the center $\mathbf{Z}_2$ of $SU(2)$. One might try to describe the group multiplication table in $SU(2)$ by a cocycle $c : SO(3) \times SO(3) \to \mathbf{Z}_2$ but there is an obstruction to this (we require that the cocycle is a continuous function): The obstruction is precisely the fact that $SU(2)$ is a nontrivial $\mathbf{Z}_2$ bundle over $SO(3)$.

Since any principal bundle is locally a direct product of the base and the fiber one can still describe the group $SU(2)$ in terms of cocycles on

$SO(3)$, with values in $\mathbf{Z}_2$. However, to do that one must cover $SO(3)$ by open sets (at least by three sets) and use local trivializations and *local* cocycles in the different open sets. In fact, this can be done in the case of affine Lie groups [Mickelsson, 1985b; Reyman, 1985]. In Section 4.2 we shall explain a different, more direct method for constructing affine groups [Mickelsson, 1987a].

Without the central extension an affine Lie algebra would be the Lie algebra of the loop group LG consisting of smooth maps from the unit circle S^1 to a finite-dimensional simple compact group G. We are thus looking for central extensions of the group LG. As we said before, the extension is not the direct product of LG and a circle (corresponding to the one dimensional center of the affine Lie algebra), but it is a nontrivial circle bundle over LG. However, the group DG consisting of maps of the unit disk D to the group G has an interesting central extension $\widehat{DG}$ which is topologically (but not algebraically) trivial. It turns out that the extension $\widehat{LG}$ of LG can be obtained as a quotient of $\widehat{DG}$ by the "gauge group" $\mathcal{G}$ consisting of maps which are equal to the identity on the boundary S^1 of the disk.

There is also an algebraic construction by generators and relations of the affine groups which we shall not discuss in this book [Kac, 1985b]. One can also recover the group structure from an explicit linear realization [Pressley and Segal, 1986]; we shall return to this construction in Chapter 6. A variant of the geometric construction is explained in Murray [1988].

The loop group LG and its extensions appear in quantum field theory because an affine Lie algebra is the Lie algebra of non-Abelian current densities in the one-dimensional compactified space S^1. For a similar reason S^3G (or in general S^nG), the group of maps from the sphere S^3 to the group G, is important in quantum field theory in a 3+1 dimensional space-time. When going to dimension 3 (or higher) there is an important new phenomenon: The relevant group extensions are no longer central; one has to deal with extensions by infinite-dimensional Abelian groups. This is related to the so-called operator valued Schwinger terms. The Schwinger terms arise because of the *anomalies of chiral fermions*. We shall discuss the structure of anomalies and their relation to Schwinger terms in Chapter 5. The existence of the anomalies has been known for a long time; in case of fermions coupled to a $U(1)$ gauge field the form of the Schwinger terms was derived using perturbation theory by Jackiw and Johnson [1969]. The Schwinger terms for a non-Abelian gauge group were computed in Faddeev [1984] (see also Reyman, Semenov-Tyan-Shanskii, and Faddeev [1985]), Mickelsson [1985a], and Singer [1985] using cohomological arguments. Other derivations and consequences for the construction of quantum fields have been extensively discussed in the physics literature: Niemi and Semenoff [1985]; Bao and Nair [1985]; Jo [1985]; Hosono and Seo [1988]; Harada and Tsutsui [1987]; Yamagishi [1987]; Fujiwara, Hosono, and Kitakado [1988]; Kolokolov and Yelkhovsky [1987], among others. For recent discussions on the geometry of loop groups and generalizations, related to anomalies, see Blau [1988]; Carey and Palmer [1988]; Floreanini and Percacci [1988]; Takasaki [1988].

Anomalous commutation relations between the charge density and the space components of electric current operators were already observed in Schwinger [1959].

In Section 4.3 we shall explain an application of the extension $\widehat{S^3G}$ to physics. We show that fermions can be consistently described as kinks (solitons) in the field configuration space of a bosonic sigma model, obeying the usual connection between spin and statistics. The idea of kinks as fermions was first discussed in detail in Finkelstein and Rubinstein [1968] and it was later shown in Witten [1983] that the idea is realized in the sigma model in 3+1 dimensions. The proof of the spin statistic relation is from Mickelsson [1988a]; an earlier argument in this direction was given in Rajeev [1984].

The explicit form of the commutator anomaly in any odd space dimension can be derived by the so called transgression formula from characteristic classes of vector bundles. We complete this chapter by a discussion of the Chern classes in Section 4.5. In even space dimensions there are no commutator anomalies if the space has no boundary.

4.1. Some group cohomology

Cocycles and coboundaries

Let G be a group acting as a group of automorphisms of an Abelian group A. An *n-cochain* for the group action is any function $c^n : G \times G \times \ldots G \to A$ of n arguments. Let C^n denote the linear space of all n-cochains (pointwise addition of functions). We set $C^0 = A$. We define the *coboundary operator* $\delta^n : C^n \to C^{n+1}$ by

$$\begin{aligned}(\delta^n f)&(g_1, g_2, \ldots, g_{n+1}) \\ &= g_1 \cdot c^n(g_2, \ldots, g_{n+1}) + (-1)^{n+1} c^n(g_1, g_2, \ldots, g_n) \\ &\quad + \sum_{i=1}^{n} (-1)^i c^n(g_1, \ldots, g_{i-1}, g_i g_{i+1}, g_{i+2}, \ldots, g_{n+1}).\end{aligned} \tag{4.1.1}$$

If there is no danger of confusion we shall drop the index n in the coboundary operator. If $c \in C^1$ then $(\delta c)(g_1, g_2) = -c(g_1 g_2) + c(g_1) + g_1 \cdot c(g_2)$ and if $c \in C^2$ then $(\delta c)(g_1, g_2, g_3) = -c(g_1 g_2, g_3) + c(g_1, g_2 g_3) - c(g_1, g_2) + g_1 \cdot c(g_2, g_3)$.

PROPOSITION 4.1.2. $\delta^{n+1} \circ \delta^n = 0$.

PROOF: Let $c^n \in C^n$ and $c^{n+2} = \delta^{n+1} \circ \delta^n c^n$. Then

$$c^{n+2}(g_1, \ldots, g_{n+2}) = \sum_{i=1}^{n+1} (\delta^n c^n)(g_1, \ldots, g_i g_{i+1}, \ldots, g_{n+2})(-1)^i$$
$$+ (\delta^n c^n)(g_1, \ldots, g_{n+1})(-1)^n$$
$$+ g_1 \cdot (\delta^n c^n)(g_2, \ldots, g_{n+2}). \tag{4.1.3}$$

The first term on the right is equal to

$$\sum_{i=1}^{n+1} \sum_{j=i+2}^{n} c^n(g_1, \ldots, g_i g_{i+1}, \ldots, g_j g_{j+1}, \ldots, g_{n+2})(-1)^{i+j-1}$$
$$+ \sum_{i=1}^{n+1} \sum_{j=1}^{i-2} c^n(g_1, \ldots, g_j g_{j+1}, \ldots, g_i g_{i+1}, \ldots, g_{n+2})(-1)^{i+j}$$
$$\sum_{i=2}^{n+1} c^n(g_2, \ldots, g_i g_{i+1}, \ldots, g_{n+2})(-1)^i - (g_1 g_2) \cdot c^n(g_3, \ldots, g_{n+2})$$
$$+ \sum_{i=1}^{n} c^n(g_1, \ldots, g_i g_{i+1}, \ldots, g_{n+1})(-1)^{i+n+1} + c_n(g_1, \ldots, g_n). \tag{4.1.14}$$

The sum of the first two terms in (4.1.4) is zero and the rest is $(-1)\times$ the contribution from the last two terms in (4.1.3).

The space Z^n of *n-cocycles* consists of elements $c^n \in C^n$ with $\delta^n c^n = 0$ and the space B^n of *coboundaries* is $\delta^{n-1} C^{n-1}$. Because of Proposition 4.1.2, $B^n \subset Z^n$ and we can define the cohomology groups

$$H^n(G, A) = Z^n / B^n. \tag{4.1.5}$$

It is customary to define $B^0 = 0$ and so $H^0 = C^0$. Thus H^0 is the fixed point set of G in A.

The cohomology groups H^1 and H^2 will be important in the following discussion so let us look more closely at the meaning of the 1- and 2-cocycles. Since G acts as a group of automorphisms on A we can define the semidirect product $G \ltimes A$ with the multiplication table

$$(g, a) \cdot (g', a') = (gg', a + g(a')). \tag{4.1.6}$$

Let $c \in C^1(G, A)$. Define $\phi : G \ltimes A \to G \ltimes A$ by $\phi(g,a) = (g, a + c(g))$. Is this a homomorphism? By a simple computation we see that the condition for ϕ to be a homomorphism is

$$c(g_1 g_2) = c(g_1) + g_1 \cdot c(g_2).$$

In other words, c is a 1-cocycle. Let c' be another 1-cocycle of G. Define a second homomorphism ϕ' using the cocycle c'. We can ask wether the maps ϕ, ϕ' can be connected by an inner automorphism, $\phi'(g,a) = (1,b)\phi(g,a)(1,-b)$. The condition for this to be true is

$$c(g) - c'(g) = g(b) - b,$$

which means that the difference $c - c'$ is the coboundary of the 0-cocycle b. Thus the set of homomorphisms ϕ modulo the inner automorphisms by elements of A are parametrized by $H^1(G, A)$.

Next let $c \in C^2$. We can try to define a new group multiplication by setting

$$(g_1, a_1) \cdot (g_2, a_2) = (g_1 g_2, a_1 + g_1(a_2) + c(g_1, g_2)). \tag{4.1.7}$$

Now

$$(g_1, a_1)[(g_2, a_2)(g_3, a_3)]$$
$$= (g_1 g_2 g_3, a_1 + g_1(a_2) + g_1(g_2(a_3)) + g_1[c(g_2, g_3) + c(g_1, g_2 g_3)])$$

and

$$[(g_1, a_1)(g_2, a_2)](g_3, a_3)$$
$$= (g_1 g_2 g_3, a_1 + g_1(a_2) + c(g_1, g_2) + (g_1 g_2)(a_3) + c(g_1 g_2, g_3)).$$

Thus the associativity is equivalent to the condition

$$g_1 \cdot c(g_2, g_3) + c(g_1, g_2 g_3) = c(g_1, g_2) + c(g_1 g_2, g_3)$$

which is the same as the vanishing of the coboundary δc. Thus each 2-cocycle defines a deformation of the group $G \ltimes A$. We can now ask whether two deformations c, c' are equivalent by automorphisms of $G \ltimes A$ of the type $\phi(g,a) = (g, a + c^1(g))$, i.e., whether $\phi((g_1, a_1)(g_2, a_2))$ with respect to the primed product is the same as $\phi(g_1, a_1) \cdot \phi(g_2, a_2)$ with respect to the unprimed product. This property is equivalent to the formula

$$c'(g_1, g_2) - c(g_1, g_2) = -c^1(g_1 g_2) + c^1(g_1) + g_1 \cdot c^1(g_2).$$

Thus the deformations of the group law by 2-cocycles c modulo the automorphisms ϕ are parametrized by elements of $H^2(G, A)$.

The descent equations

We shall next apply the general formalism to the following situation. Let P be a principal bundle with structure group G over a manifold M. We assume that the total space P is contractible (remember that any principal bundle with a finite-dimensional structure group is a pull-back of a bundle over the classifying space, with contractible base space). Let $\mathcal{A}$ denote the space of $\mathbf{g}$ valued one-forms on P ($\mathbf{g}$ is the Lie algebra of G), let $\Omega_k(P)$ be the space of complex valued differential forms of degree k on P and V_k the space of polynomial functions on $\mathcal{A}$ with values in $\Omega_k(P)$. Let $\mathcal{G}$ be a subgroup of the group of automorphisms of P; it acts on V_k by $(g \cdot f)(A) = f((g^{-1})^* A)$.

We consider cochains on $\mathcal{G}$ with values in the Abelian group $V = \oplus V_k$. The space of cochains is now graded by the degree of a cochain and the degree of a differential form, $C = \oplus C^{n,k}$, where n is the cochain degree and k is the form degree. The pull-back operator on differential forms commutes with the exterior differentiation on P. It follows that the exterior derivative d commutes with the coboundary operator δ. Elements of C are called *local cochains*. The action by δ increases the degree n by one unit and action by d the degree k by one. The local cohomology groups are defined by $H^{n,k}(\mathcal{G}, V) = Z^n(\mathcal{G}, V_k)/B^n(\mathcal{G}, V_k)$.

We shall meet later certain cocycles derived by the so-called *descent equations* from the cohomology class

$$\begin{aligned} c^{0,2n} &= q_{2n}\, \mathrm{tr} F \wedge F \wedge \cdots \wedge F \\ &= q_{2n}\, \epsilon^{\mu_1 \nu_1 \mu_2 \nu_2 \ldots \mu_n \nu_n} \mathrm{tr} F_{\mu_1 \nu_1} F_{\mu_2 \nu_2} \ldots F_{\mu_n \nu_n} dx_1 \wedge \cdots \wedge dx_{2n}, \end{aligned} \tag{4.1.8}$$

where q_{2n} is a normalization coefficient and F is the curvature defined by the one-form A. The traces are computed in a fixed faithful representation of the Lie algebra $\mathbf{g}$. The normalization coefficients depend on the representation and are so defined that the integral of the differential form over an arbitrary compact oriented manifold is an integer. For example, if $G = SU(N)$ then in the defining representation the coefficients are

$$q_{2n} = \frac{1}{n}\left(\frac{i}{2\pi}\right)^n.$$

Let $\mathcal{G} \subset Aut\, P$ consisting of those automorphisms which project onto the identity transformation on the base space M. Suppose for a moment that the bundle P is trivial. Then the elements of $\mathcal{G}$ can be represented by maps $g : M \to G$ and the action of an automorphism g on the curvature form F is given by $F \mapsto gFg^{-1}$, so the forms $c^{0,2n}$ (which are related to the *Chern classes* to be discussed in Section 4.5) are invariant under the gauge transformations g. This means that $c^{0,2n}$ is a 0-cocycle for $(\mathcal{G}, V_n)$. The invariance of a differential form under the mappings g is a local property and does not depend on the global topology of the bundle; therefore $c^{0,2n}$ is a cocycle for any principal bundle P. Since P is contractible and $dc^{0,n} = 0$ there exists a differential form $c^{0,2n-1}$ of degree $2n-1$ on P such that $c^{0,2n} = dc^{0,2n-1}$. The form $c^{0,2n-1}$ is not invariant under the gauge transformations. Let us define

$$c^{1,2n-1}(g) = \delta c^{0,2n-1} = g \cdot c^{0,2n-1} - c^{0,2n-1}. \tag{4.1.9}$$

Now $dc^{1,2n-1} = d\delta c^{0,2n-1} = \delta dc^{0,2n-1} = \delta c^{0,2n} = 0$. It follows that there is a form $c^{1,2n-2}$ of degree $2n-2$ such that $dc^{1,2n-2}(g) = c^{1,2n-1}(g)$. Next we define a coboundary of degree 2 by

$$c^{2,2n-2} = \delta c^{1,2n-2}. \tag{4.1.10}$$

Again using the commutativity of the pair (d, δ) we conclude that $c^{2,2n-2}$ is the exterior differential of a form $c^{2,2n-3}$ of degree $2n-3$. We can continue this process until we end up at a differential form $c^{2n-1,0}$ of degree zero; this is a cocycle of cochain degree $2n-1$.

When $A^{(1)}, \ldots, A^{(k)}$ are one-forms with values in the space of $N \times N$ matrices we shall denote by $\mathrm{tr} A^{(1)} \ldots A^{(k)}$ the complex valued differential form of degree k, which in local coordinates is

$$\mathrm{tr} A^{(1)} \ldots A^{(k)} = \mathrm{tr} A^{(1)}_{i_1} \ldots A^{(k)}_{i_k}\, dx_{i_1} \wedge \cdots \wedge dx_{i_k}$$

with a summation over repeated indices. The trace of a product of higher order forms is defined similarly.

Example 4.1.11. Consider $c^{0,4} = q_4 \mathrm{tr} FF$ in the case when $P = M \times G$ is a trivial bundle and M is contractible. The connection form in the total space P, at a point (x, h), is now $h^{-1}A(x)h + h^{-1}dh$, where A is a Lie algebra valued one form on M. The curvature is $F = dA + \frac{1}{2}[A, A]$ and by simple computation $c^{0,4} = dc^{0,3}$ with $c^{0,3} = q_4 \mathrm{tr}(AdA + \frac{2}{3}A^3)$. Next

$$c^{1,3} = q_4 \mathrm{tr}(d(Adgg^{-1}) - \frac{1}{3}(dgg^{-1})^3) = dc^{1,2}(A; g),$$

where $c^{1,2} = q_4(\mathrm{tr} Adgg^{-1} + H(g))$. Here H is a two form constructed as follows [Cronström and Mickelsson, 1983]. Since M is contractible we can write $g(x) = \exp Z(x)$ for some Lie algebra valued function Z. A solution of $dH(g) = \mathrm{tr}(dgg^{-1})^3$ is then given by

$$H = \sum_{k\geq 0} \frac{1}{(2k+3)!} \mathrm{tr} dZ \wedge (ad\, Z)^{2k+1} dZ \tag{4.1.12}$$

The next form in the descent process is $c^{2,2} = \delta c^{1,2} = -q_4(-\mathrm{tr} g_1^{-1} dg_1 \wedge dg_2 g_2^{-1} + H(g_1) + H(g_2) - H(g_1 g_2)$. This is going to play an important role in the construction of Kac-Moody groups.

Lie algebra cohomology

Next we shall give the basic definitions of Lie algebra cohomology which we shall need later when discussing Lie algebra extensions. Let $\mathbf{g}$ be a Lie algebra and V a $\mathbf{g}$ module. *A cochain of degree n* with values in V is an antisymmetric multilinear map $c : \mathbf{g} \times \mathbf{g} \times \cdots \times \mathbf{g} \to V$ (n arguments). The vector space of all n cochains is denoted by $C^n = C^n(\mathbf{g}, V)$. The coboundary operator $\delta : C^n \to C^{n+1}$ is defined by

$$\begin{aligned}
&(\delta c^n)(x_1, x_2, \ldots, x_{n+1}) \\
&= \sum_{i<j} (-1)^{i+j} c^n([x_i, x_j], x_1, \ldots, \hat{x}_i, \ldots, \hat{x}_j, \ldots, x_{n+1}) \\
&+ \sum_{i=1}^{n+1} (-1)^{i+1} x_i \cdot c^n(x_1, \ldots, x_{i-1}, x_{i+1}, \ldots, x_{n+1}),
\end{aligned} \tag{4.1.13}$$

where the caret means the variable under it has been deleted. For example,

$$\begin{aligned}
(\delta c^0)(x_1) &= x_1 \cdot c^0 \\
(\delta c^1)(x_1, x_2) &= x_1 \cdot c^1(x_2) - x_2 \cdot c^1(x_1) - c^1([x_1, x_2]) \\
(\delta c^2)(x_1, x_2, x_3) &= x_1 \cdot c^2(x_2, x_3) - x_2 \cdot c^2(x_1, x_3) + x_3 \cdot c^2(x_1, x_2) \\
&\quad - c^2([x_1, x_2], x_3) + c^2([x_1, x_3], x_2) - c^2([x_2, x_3], x_1).
\end{aligned}$$

Exercise 4.1.14. Show that $\delta^2 = 0$.

The spaces of coboundaries $B^n(\mathbf{g}, V)$, cocycles $Z^n(\mathbf{g}, V)$ and the cohomology groups $H^n(\mathbf{g}, V) = Z^n(\mathbf{g}, V)/B^n(\mathbf{g}, V)$ are defined just as in the case of group cohomology. The second cohomology H^2 is related to Lie algebra extensions. Consider V as an Abelian Lie algebra with respect to the vector addition. The semidirect sum of $\mathbf{g}$ with V consists of pairs $(x, v) \in \mathbf{g} \oplus V$ with the Lie commutator $[(x, v), (x', v')] =$

$([x, x'], x \cdot v' - x' \cdot v)$. Let $c \in C^2(\mathbf{g}, V)$. We can try to define a modified commutator by

$$(4.1.15) \qquad [(x, v), (x', v')] = ([x, x'], x \cdot v' - x' \cdot v + c(x, x')).$$

The Jacobi identity for the modified commutator is easily seen to be equivalent to the cocycle condition $\delta c = 0$. Thus each $c \in Z^2$ defines a new Lie algebra. Let $c, c' \in Z^2$. We can ask whether the Lie algebras formed from the cocycles c, c' are isomorphic through a mapping of the type $\phi(x, v) = (x, v + c^1(x))$, where $c^1 : \mathbf{g} \to V$ is a linear mapping. The condition

$$[\phi(x_1, v_1), \phi(x_2, v_2)]_{c'} = \phi([(x_1, v_1), (x_2, v_2)]_c)$$

is the same as $c - c' = \delta c^1$. Thus up to an isomorphism of the above type the Lie algebra extensions are parametrized by elements of $H^2(\mathbf{g}, V)$.

Further reading on group and Lie algebra cohomology relevant in physics: Discussions on cohomology and applications to field theory can be found for example in Alvarez [1985], Jackiw [1985], Stasheff [1985], and Zumino [1984].

4.2. Groups associated to affine Kac-Moody algebras

The central extension $\widehat{LG}$

Let D denote the unit disc in $\mathbf{C}$ including the boundary circle S^1. Let G be a compact simply connected semisimple Lie group, and denote by DG the space of all smooth mappings $f : D \to G$ such that the radial derivatives of f approach zero (to all orders) at the boundary.

Let $\mathcal{G}$ be the subgroup of DG consisting of maps $f : D \to G$ such that $f = 1$ on the boundary. Clearly $\mathcal{G} \subset DG$ is a normal subgroup. Define

$$(4.2.1) \qquad \gamma(f_1, f_2) = \frac{\theta^2}{16\pi^2} \int_D (f_1^{-1} df_1, df_2 f_2^{-1}),$$

where θ is the length of the longest root of the Lie algebra $\mathbf{g}$, computed using the dual of the invariant inner product $(\cdot, \cdot)$ in $\mathbf{g}$. From the computations in Example 4.1.11 it follows that γ is a real valued 2-cocycle on DG,

$$\gamma(f_1, f_2) + \gamma(f_1 f_2, f_3) = \gamma(f_2, f_3) + \gamma(f_1, f_2 f_3).$$

This formula can also be obtained by a simple computation starting from the definition of γ. From the cocycle property it follows that we can define an extension of the group DG by S^1 using the multiplication rule

$$(f_1, \lambda_1)(f_2, \lambda_2) = (f_1 f_2, \lambda_1 \lambda_2 \exp[2\pi i \gamma(f_1, f_2)]). \tag{4.2.2}$$

Any $g \in \mathcal{G}$ can be thought of as a map $g : S^2 \to G$ by identifying the boundary S^1 of D with the north pole of S^2. There is a homomorphism $\phi : \mathcal{G} \to DG \times S^1$ defined by

$$\phi(g) = (g, \exp[2\pi i C(g)]), \tag{4.2.3}$$

where

$$\begin{aligned} C(g) &= \frac{\theta^2}{48\pi^2} \int_B (dgg^{-1}, \tfrac{1}{2}[dgg^{-1}, dgg^{-1}]) \\ &= \frac{\theta^2}{48\pi^2} \int_B \epsilon^{ijk}(g^{-1}\partial_i g, \tfrac{1}{2}[g^{-1}\partial_j g, g^{-1}\partial_k g]), \end{aligned} \tag{4.2.4}$$

and we have extended the map $g : S^2 \to G$ to a map $g : B \to G$, where B is the unit ball in $\mathbf{R}^3$ with boundary S^2. Owing to the arbitrariness in the choice of the extension $C(g)$ is not uniquely defined. Let g, g' be two extensions to B. Because they agree on the boundary we can glue them along the boundary to form a mapping $h : S^3 \to G$. The upper hemisphere of S^3 corresponds to one copy of B and the lower hemisphere to a second copy. If we can show that

$$\frac{\theta^2}{48\pi^2} \int_{S^3} (dhh^{-1}, \tfrac{1}{2}[dhh^{-1}, dhh^{-1}]) \tag{4.2.5}$$

is an integer we can conclude that the difference $C(g) - C(g') \in \mathbf{Z}$ and $\exp[2\pi i C(g)]$ is well-defined.

To prove that the integral (4.2.5) is an integer we need some information about the topology of the group G. Suppose first that G is simple. It is known that the third homotopy group $\pi_3 G$ of any compact simple Lie group is equal to the group of integers $\mathbf{Z}$ [Husemoller, 1975]. Furthermore, a generator of $\pi_3 G$ is constructed as follows. The sphere S^3 can be identified as the group manifold $SU(2)$. Choose a homomorphism from $SU(2)$ to G which is locally one-to-one [the Lie algebra of any simple G contains the Lie algebra of $SU(2)$ as a subalgebra]. The composite map $S^3 \to SU(2) \to G$ is a generator. For this particular mapping the value of the integral is equal to the value of the integral of the identity mapping $g \mapsto g$ from S^3 to $SU(2)$; we omit the straightforward but a bit tedious computation of the latter integral. The result is $= 1$. The integral of the mapping $h(g) = g^n$ is then equal to n for any integer n. The claim follows now from the homotopy invariance of the integral (4.2.5): If h' can be obtained by a deformation $t \mapsto h_t$ from h, $h_0 = h$ and $h_1 = h'$, then by Stokes's theorem the difference of the integrals for h and h' is equal to the integral over a four-dimensional manifold of the exterior derivative of the integrand; but by a simple computation one proves that the integrand is a closed 3-form and thus the homotopy invariance. The case of a nonsimple group reduces to the proof in the simple case since any semisimple group is a product of simple factors.

The condition $\phi(g_1 g_2) = \phi(g_1)\phi(g_2)$ is equivalent to

$$C(g_1 g_2) = C(g_1) + C(g_2) + \gamma(g_1, g_2) \; mod \, \mathbf{Z} \tag{4.2.6}$$

which can be verified by a direct substitution.

LEMMA 4.2.7. *The image $\phi(\mathcal{G})$ in $DG \times S^1$ is a normal subgroup.*

PROOF: This can be shown by a direct computation but we shall give a simpler proof which is due to V.Kac. Let $g \in \mathcal{G}$ and $f \in DG$. We have to show that $(f,1)\phi(g)(f,1)^{-1} \in \phi(\mathcal{G})$. This is equivalent to the condition

$$C(g) + \gamma(f,g) + \gamma(fg, f^{-1}) = C(fgf^{-1}) \; mod \, \mathbf{Z}. \tag{4.2.8}$$

Denote the left-hand-side of (4.2.8) by $C'(g)$. From (4.2.6) it follows that the difference $C'(g) - C(fgf^{-1}) = \Delta(g)$ satisfies $\Delta(g_1 g_2) = \Delta(g_1) + \Delta(g_2)$ and thus $g \mapsto e^{2\pi i \Delta(g)}$ is a homomorphism from the group $\mathcal{G}$ to S^1. The only homomorphism when G is semisimple is the trivial homomorphism $S^2 G \to 1$ and therefore (4.2.8) holds.

We define $\widehat{LG} = (DG \times S^1)/\phi(\mathcal{G})$. This group is a circle bundle over the base space $LG = \{f : S^1 \to G \mid f \, \text{smooth}\}$. The projection is $(f, \lambda) \mapsto f|_{S^1}$. The center of $\widehat{LG}$ is represented by the pairs $(1, \lambda) \in DG \times S^1$. The center acts transitively in each fiber and so $\widehat{LG}$ is *a central extension* of LG by the circle S^1.

So far we have not said anything about the differentiable structure of LG or $\widehat{LG}$. We shall use in LG the topology of uniform convergence of maps $f : S^1 \to G$ and of all their partial derivatives. Let U be an open neighborhood of 0 in $\mathbf{g}$ such that the exponential mapping $\mathbf{g} \to G$ is one-to-one on U. An open neighborhood U_f of an arbitrary point $f \in LG$ consists of all maps of the form $e^Z f$, where Z is any smooth map from S^1 to U. The mapping $\xi_f : e^Z f \mapsto Z$ defines a coordinate system on U_f.The topology of the function space $S^1\mathbf{g}$ is defined by the infinite system of seminorms

$$\|Z\|_{(n)} = \sqrt{\int_{S^1} \left(\frac{d^n}{d\phi^n} Z, \frac{d^n}{d\phi^n} Z \right)}.$$

The coordinate transformation $\xi_f \circ \xi_{f'}^{-1}(Z) = \log(e^Z ff'^{-1})$ is differentiable in the Frechet sense and by definition LG is a *Frechet manifold.*

We can now reconstruct the bundle $\widehat{LG}$ in terms of local sections and transition functions for the system of neighborhoods U_f. Fix f and choose an extension $f : D \to G$. Let $\rho : [0,1] \to [0,1]$ be a smooth function such that $\rho(1) = 1$ and ρ vanishes along with all its derivatives at the point 0. If $h \in U_f$ then $h = e^Z f$ and we can define an extension $h_D : D \to G$ by

$$h_D(\phi, r) = e^{\rho(r)Z(\phi)} f(\phi, r).$$

Thus we have a local section $U_f \to \widehat{LG}$ given by $\psi_f(h) = (h_D, 1) mod\, \mathcal{G}$. Let us compute the transition function $\chi_{ff'}$ on $U_f \cap U_{f'}$. Now $e^{\rho Z} fg = e^{\rho Z'} f'$, where $Z' = \log(e^Z ff'^{-1})$ and

$$g(\phi, r) = f(\phi, r)^{-1} e^{-\rho(r)Z(\phi)} e^{\rho(r)Z'(\phi)} f'(\phi, r)$$

depends smoothly on Z. We have

$$\psi_f(h) = (e^{\rho Z} f, 1) \equiv (e^{\rho Z'} f', e^{2\pi i \omega(h_D, g)}) \; mod\, \mathcal{G}, \tag{4.2.9}$$

where $\omega(f, g) = \gamma(f, g) + C(g)$ for $f \in DG$ and $g \in \mathcal{G}$. Thus the transition function is given by $\chi(h) = \exp[2\pi i \omega(h_D, g)]$ and it is smooth in the coordinate Z.

What we have constructed here could be called the *basic* central extension of LG. For any integer n we can construct a central extension of LG by replacing $\gamma(f_1, f_2)$ by $n\gamma(f_1, f_2)$ and $C(g)$ by $nC(g)$. The right-hand-side of the commutator formula 4.2.12 will be likewise multiplied by n. Any central extension of LG is isomorphic to one of these central extensions.

Connection in $\widehat{LG}$

Next we define a connection in the principal bundle $\widehat{LG}$. First we define a splitting of the tangent space $T_{(f,\lambda)}$ of $DG \times S^1$ at the point

(f,λ) to the vertical subspace $V_{(f,\lambda)}$ consisting of tangent vectors $(0,a)$, and to the horizontal subspace

$$H_{(f,\lambda)} = \{(Z,a) \mid a = \frac{\theta^2}{8\pi}\int_D (f^{-1}df, dZ)\},$$

where the tangent space $T_f DG$ has been identified as the space of smooth maps $Z : D \to \mathbf{g}$ through left translation by f^{-1}. (We shall use the convention that Lie algebra elements correspond to *left* invariant vector fields on the group manifold.) A horizontal distribution in the tangent bundle of $\widehat{LG}$ is then defined by the canonical projection $pr : DG \times S^1 \to (DG \times S^1)/\mathcal{G} = \widehat{LG}$. We have to show that $H_{[(f,\lambda)]} = prH_{(f,\lambda)}$ depends only on the class $[(f,\lambda)]$ and not of the representative (f,λ). Let $(Z,a) \in H_{(f,\lambda)}$. Then $(fg, \lambda e^{2\pi i\omega(f,g)})$ represents the same class as (f,λ) for any $g \in \mathcal{G}$ and $pr(Z,a) = pr(Z',a')$, where $Z' = g^{-1}Zg$ and

$$\begin{aligned} a' &= a + \frac{\theta^2}{8\pi}\int (dZ + [f^{-1}df, Z], dgg^{-1}) \\ &= \frac{\theta^2}{8\pi}\int ((fg)^{-1}d(fg), d(g^{-1}Zg)), \end{aligned}$$

so $(Z',a') \in H_{(f',\lambda')}$ and consequently $prH_{(f,\lambda)} \subset prH_{(f',\lambda')}$. The opposite inclusion $prH_{(f,\lambda)} \supset prH_{(f',\lambda')}$ is proven in the same way.

Using the local section ψ_f on $U_f \subset LG$ the local vector potential for the connection is

$$\alpha(h;Z) = \frac{\theta^2}{8\pi}\int_D (h^{-1}dh, dZ), \tag{4.2.10}$$

where $h : D \to G$ is the extension of a loop $h \in U_f$ as described earlier. Another choice of the extension effects a gauge transformation of the vector potential α. The curvature F is given by the 2-form

$$(d\alpha)(h; Z_1, Z_2) = \frac{\theta^2}{4\pi}\int_{S^1} (Z_1, dZ_2) \tag{4.2.11}$$

and is gauge invariant as expected. If we restrict F to the subspace ΩG of *based loops* (i.e., loops which at a fixed point, say $1 \in S^1$, obtain the value $1 \in G$) then the it is *nondegenerate* : Given a tangent vector X to ΩG [that means $X(1) = 0$] then $F(X,Y) = 0$ for all Y if and only if $X = 0$. By definition, a closed nondegenerate 2-form on a manifold is a *symplectic form.*

We shall compute the central extension $\widehat{\mathbf{g}}$ of the loop algebra $S^1\mathbf{g}$ corresponding to the central group extension $\widehat{LG}$. The exponential mapping $\exp : S^1\mathbf{g} \to \widehat{LG}$ is defined pointwise, $[\exp(Z)](\phi) = \exp[Z(\phi)]$, and similarly for the algebra $D\mathbf{g}$. The Lie algebra of $DG \times S^1$ as a vector space is $D\mathbf{g} \oplus i\mathbf{R}$. The commutator is

$$[(X,a),(Y,b)] = ([X,Y],c(X,Y)),$$

where the Lie algebra cocycle c is computed from

(4.2.12)

$$\begin{aligned}([X,Y],c(X,Y)) &= \frac{d^2}{dtds}(e^{tX},1)(e^{sY},1)(e^{-tX},1)(e^{-sY},1)|_{s=t=0} \\ &= 4\pi i \frac{d^2}{dtds}\gamma(e^{tX},e^{sY})|_{s=t=0} = i\frac{\theta^2}{4\pi}\int_{S^1}(X,dY).\end{aligned}$$

Thus the cocycle defining the central extension is the same (up to the factor i) as the curvature on LG. This is no surprise since the connection form on $\widehat{LG}$ is just the projection of the Maurer-Cartan form $g^{-1}dg$ of $\widehat{LG}$ onto the center $i\mathbf{R}$.

4.3. Extensions of $Map(S^3, G)$

The Abelian extension $\widehat{S^3}G$

In order to study the Abelian extensions of the group of smooth maps $S^3G = Map(S^3, G)$ we shall consider the descent chain starting from the form $c^{0,6} = q_6 \mathrm{tr} F \wedge F \wedge F$. The traces are computed in the adjoint representation of the compact semisimple Lie group G. Locally $F = dA + \frac{1}{2}[A, A]$ and $c^{0,6} = dc^{0,5}$, where

$$c^{0,5} = q_6 \mathrm{tr}\left(AdAdA - \frac{3}{2}A^3dA + \frac{3}{5}A^5\right). \tag{4.3.1}$$

Here the products of the differential forms are exterior multiplications; for example,

$$\mathrm{tr}A^5 = \epsilon^{\alpha\beta\gamma\delta\omega}A_\alpha A_\beta A_\gamma A_\delta A_\omega$$

where ϵ is totally antisymmetric and $\epsilon^{12345} = +1$. In the following we shall work with local formulas in the base space instead of the global formulas in the principal bundle P in 4.1; this simplifies our discussion without affecting our result (4.3.3) below, which is a differential form in a *contractible space.* The coboundary of $c^{5,0}$ is

$$\begin{aligned} c^{1,5}(A;g) =& q_6 d\,\mathrm{tr}\Big[-\frac{1}{2}dgg^{-1}(AdA + dA\,A + A^3) + \frac{1}{4}(dgg^{-1}A)^2 \\ &+ \frac{1}{2}(dgg^{-1})^3 A\Big] + q_6\frac{1}{10}\mathrm{tr}(dgg^{-1})^5, \end{aligned} \tag{4.3.2}$$

Denote by $c^{1,4}$ the term after the first trace operator. Define

$$c^{2,4}(A; g_1, g_2) = (\delta c^{1,4})(A; g_1, g_2).$$

By a somewhat tedious but completely straightforward computation one gets

$$\begin{aligned} c^{2,4}(A; g_1, g_2) =& -q_6 d\,\mathrm{tr}\left[(dg_2g_2^{-1})(g_1^{-1}dg_1)(g_1^{-1}Ag_1)\right. \\ &\left.-(dg_2g_2^{-1})(g_1^{-1}Ag_1)(g_1^{-1}dg_1)\right] + q_6\,\mathrm{tr}\left[(dg_2g_2^{-1})(g_1^{-1}dg_1)^3\right. \\ &\left.+\tfrac{1}{2}(dg_2g_2^{-1}g_1^{-1}dg_1)^2 + (dg_2g_2^{-1})^3(g_1^{-1}dg_1)\right]. \end{aligned} \tag{4.3.3}$$

Let $D^4 = \{x \in \mathbf{R}^4 \mid \|x\| \leq 1\}$ and D^4G be the group consisting of smooth maps $f : D^4 \to G$ such that the radial derivatives of f approach zero at the boundary S^3 of D^4. Let $\mathcal{A}_3$ denote the space of smooth $\mathfrak{g}$ valued one-forms on S^3. We define a cochain γ_3 with values in the Abelian group $Map(\mathcal{A}_3, \mathbf{R})$ by the formula

$$\gamma_3(A; g_1, g_2) = \int_{D^4} c^{4,2}(A; g_1, g_2). \tag{4.3.4}$$

Note that because the first term in $c^{4,2}$ is an exterior derivative of a 3-form and the second term does not depend on A, γ_3 really depends only on the boundary values of A on the sphere S^3. Since γ_3 is manifestly a coboundary we can define an extension of D^4G by the Abelian group of maps $Map(\mathcal{A}_3, S^1)$ using the product

$$(g_1, \lambda_1)(g_2, \lambda_2) = (g_1g_2, \lambda_1(g\cdot\lambda_2)e^{2\pi i\gamma_3(\cdot; g_1, g_2)}), \tag{4.3.5}$$

where $g \cdot \lambda$ is the function $(g \cdot \lambda)(A) = \lambda(g^{-1}Ag - g^{-1}dg)$.

We shall now make the assumption that the fourth homotopy group $\pi_4 G$ contains only the unit element. This excludes, for example, the group $SU(2)$ for which π_4 is equal to $\mathbf{Z}_2$. On the other hand, $\pi_4 SU(N) = 0$ for $N > 2$. We denote by $\mathcal{G}_4$ the group of smooth maps $f : D^4 \to G$ such that $f \equiv 1$ on the boundary S^3; this is a normal subgroup of D^4G. An element $g \in \mathcal{G}_4$ can be thought of as a map $S^4 \to G$ by identifying the boundary of D^4 with a point on the sphere S^4. Since $\pi_4 G = 0$ we can choose an extension $g : D^5 \to G$, where D^5 is the unit disc in $\mathbf{R}^5$ with boundary S^4. Define

$$C_5(g) = \frac{q_6}{10} \int_{D^5} \mathrm{tr}(dgg^{-1})^5. \tag{4.3.6}$$

[If $G = SU(N)$ and we compute the traces in the *defining representation* then $q_6 = \frac{i}{24\pi^3}$.] If g and g' are equal on the boundary we can glue them along the boundary to form a smooth map $h : S^5 \to G$.

It can be shown that the integral of $q_6 \mathrm{tr}(dgg^{-1})^5$ over S^5 is always an integer and we conclude that $C_5(g) - C_5(g') \in \mathbf{Z}$. Thus $e^{2\pi i C_5(g)}$ is well defined for any $g \in \mathcal{G}_4$.

LEMMA 4.3.7. *The set of elements $(g, \exp[2\pi i C_5(g)])$ with $g \in \mathcal{G}_4$ forms a normal subgroup $\phi(\mathcal{G}_4)$ in the extension $D^4G \ltimes Map(\mathcal{A}_3, S^1)$.*

PROOF: From the definitions we can reduce the formula

$$\begin{aligned} C_5(g_1 g_2) =& C_5(g_1) + C_5(g_2) \\ &+ 5q_6 \int_{D^5} \mathrm{tr}\left[(g_1^{-1}dg_1)(dg_2 g_2^{-1})^4 + (g_1^{-1}dg_1)^2(dg_2 g_2^{-1})^3\right] \\ &+ 5q_6 \int_{D^5} \mathrm{tr}\left[(g_1^{-1}dg_1)^3(dg_2 g_2^{-1})^2 + (g_1^{-1}dg_1)^4(dg_2 g_2^{-1})\right] \\ &+ 5q_6 \int_{D^5} \mathrm{tr}\left[(dg_2 g_2^{-1})^2(g_1^{-1}dg_1)(dg_2 g_2^{-1})(g_1^{-1}dg_1)\right. \\ &\left.+(g_1^{-1}dg_1)^2(dg_2 g_2^{-1})(g_1^{-1}dg_1)(dg_2 g_2^{-1})\right] \\ =& C_5(g_1) + C_5(g_2) + \gamma_3(A; g_1, g_2), \end{aligned} \tag{4.3.8}$$

where we have applied Stokes's theorem at the second step. Note that γ_3 does not depend on A when the arguments g_1 and g_2 are in the subgroup $\mathcal{G}_4$. From (4.3.8) follows at once the subgroup property $(g_1, e^{2\pi i C_5(g_1)}) \times (g_2, e^{2\pi i C_5(g_2)}) = (g_1 g_2, e^{2\pi i C_5(g_1 g_2)})$.

Using repeatedly Stokes's theorem one can show that

$$C_5(fgf^{-1}) = C_5(g) + \gamma_3(A; f, g) + \gamma_3(A; fg, f^{-1}) \tag{4.3.9}$$

for all $g \in \mathcal{G}_4$ and $f \in D^4G$. This means that

$$(f, \lambda)(g, e^{2\pi i C_5(g)})(f, \lambda)^{-1} = (fgf^{-1}, \exp[2\pi i C_5(fgf^{-1})]\lambda(g\lambda)^{-1});$$

but $g \cdot \lambda = \lambda$ since $g = 1$ on $S^3 = \partial D^4$ and thus the conjugated element is also in the subgroup $\phi(\mathcal{G}_4)$.

Exercise 4.3.10. Prove the relation (4.3.9).

Let us denote by $\widehat{S_0^3 G}$ the group $[D^4 G \ltimes Map(\mathcal{A}_3, S^1)]/\phi(\mathcal{G}_4)$. This is a principal bundle over the space of *contractible* maps $S^3 \to G$; the bundle projection is given by $(f, \lambda) \mapsto f|_{S^3}$. The structure group is $Map(\mathcal{A}_3, S^1)$. The right action of the structure group is given by the group multiplication, $(f,\lambda)\cdot\mu = (f, \lambda(f\cdot\mu))$. The group $Map(\mathcal{A}_3, S^1)$ can be embedded as a normal subgroup in $\widehat{S_0^3 G}$ by the map $\lambda \mapsto (1, \lambda)\, mod\ \phi(\mathcal{G}_4)$. Thus $\widehat{S_0^3 G}$ is an extension of $S_0^3 G = \{f : S^3 \to G \mid f\,\text{smooth, contractible}\}$ by the group $Map(\mathcal{A}_3, S^1)$. The Lie algebras of $S^3 G$ and $D^4 G$ consists of $\mathbf{g}$ valued smooth functions in the corresponding domains; the commutator of two functions is again defined pointwise. Let us compute the commutation relations for the Lie algebra extension $\mathbf{p} = S^3\mathbf{g} \oplus Map(\mathcal{A}_3, i\mathbf{R})$. We have to compute

$$\frac{d^2}{dtds}(e^{tX}, 1)(e^{sY}, 1)(e^{-tX}, 1)(e^{-sY}, 1)|_{t=s=0} = ([X,Y], c_3(A; X, Y)),$$

where $X, Y : D^4 \to \mathbf{g}$ are smooth functions. The result is

$$c_3(A; X, Y) = 2\pi q_6 \int_{S^3} \mathrm{tr} A(dXdY + dYdX). \tag{4.3.11}$$

Since the Lie algebra cocycle c_3 depends only on the boundary values of the functions X, Y it is really a cocycle for the Lie algebra $S^3\mathbf{g}$ as it should be. Taking account of the infinitesimal action of $S^3 G$ on $Map(\mathcal{A}_3, i\mathbf{R})$ we get the commutator formula

$$[(X,\lambda), (Y,\mu)] = ([X,Y], X\cdot\mu - Y\cdot\lambda + c_3(A; X, Y)). \tag{4.3.12}$$

The product $X \cdot \lambda$ is defined by the Lie derivative $\frac{d}{dt} e^{tX}\lambda|_{t=0}$.

Exercise 4.3.13. Show that c_3 vanishes identically when $\mathbf{g}$ is the Lie algebra of $SU(2)$.

The structure group $Map(\mathcal{A}_3, S^1)$ acts in the one-dimensional vector space $\mathbf{C}$ by the formula $f \cdot a = f(0)a$. Using this action we construct a complex line bundle E_0 over $S_0^3 G$ as the associated bundle to the principal bundle $\widehat{S_0^3 G}$. Deleting the zero section from E_0 defines a principal $\mathbf{C}^\times$ bundle; that is not a group unlike in the case of loop groups in Section 4.2.

Geometry of the extension $\widehat{S^3G}$

The mapping $(X,Y) \mapsto c_3(\cdot;X,Y)$ defines a closed 2-form on S^3_0 with values in the Abelian Lie algebra $Map(\mathcal{A}_3, i\mathbf{R})$. This form is the curvature form of a connection on the bundle $\widehat{S^3_0G}$. Namely, let α be the projection onto the ideal $Map(\mathcal{A}_3, i\mathbf{R})$ of the right invariant Maurer-Cartan 1-form dpp^{-1} on the group $\widehat{S^3_0G}$. The pull-back of α in $D^4G \ltimes Map(\mathcal{A}_3, S^1)$ at a point $(g,\lambda) \in D^4G \ltimes Map(\mathcal{A}_3, S^1)$ is evaluated as follows. Let (X,μ) be a tangent vector at (g,λ). Then

$$\begin{aligned}\alpha(g,\lambda;X,\mu) =& pr\frac{d}{dt}(e^{tX}g, e^{t\mu}\lambda)(g,\lambda)^{-1}\\ =&\mu - \lambda^{-1}X\cdot\lambda + 2\pi i\frac{d}{dt}\gamma_3(A;e^{tX}g,g^{-1})\\ =&-\pi i q_6\int_{D^4}\mathrm{tr}(dgg^{-1})^3 dX\\ &+\pi i q_6\int_{S^3}\mathrm{tr}(dgg^{-1})[A,dX] + \mu - \lambda^{-1}X\cdot\lambda.\end{aligned} \tag{4.3.14}$$

The curvature computed from (4.3.14) is

$$\begin{aligned}F(X,Y) =& 2\pi i\int_{S^3}\mathrm{tr}\,\{(dgg^{-1})^2(XdY+YdX) + A[dX,dY]\\ &+\tfrac{1}{2}(dgg^{-1})(dYAX + XAdY - dXAY - YAdX\\ &-AYdX - dXYA + AXdY + dYXA)\}.\end{aligned} \tag{4.3.15}$$

The induced curvature on the line bundle E_0 is obtained by the Lie algebra homomorphism $f(\cdot) \mapsto f(0)$ and so we have

$$F_E(X,Y) = 2\pi i q_6\int_{S^3}\mathrm{tr}(dgg^{-1})^2(XdY - Y\,dX). \tag{4.3.16}$$

Let $S \subset S^3G$ be a compact oriented two-dimensional surface without boundary parametrized by local coordinates (s,t). We want to compute the integral of the two-form F_E over the surface S. Thus we have to compute

$$\int_{S\times S^3}\mathrm{tr}(dgg^{-1})^2\left[(\partial_s gg^{-1})d(\partial_t gg^{-1}) - (\partial_t gg^{-1})d(\partial_s gg^{-1})\right]. \tag{4.3.17}$$

By a partial integration with respect to the variable t in the first term and variable s in the second term the left-hand-side is equal to

$$\int \mathrm{tr}\,\{(dgg^{-1})^3[\partial_s gg^{-1}, \partial_t gg^{-1}] + (\partial_s dgg^{-1})(dgg^{-1})(\partial_t gg^{-1})(dgg^{-1}) + (\partial_s dgg^{-1})(\partial_t gg^{-1})(dgg^{-1})^2 - (\partial_t dgg^{-1})(dgg^{-1})(\partial_s gg^{-1})(dgg^{-1}) - (\partial_t dgg^{-1})(\partial_s gg^{-1})(dgg^{-1})^2\}\,. \tag{4.3.18}$$

By a partial integration in the second, third, and fifth term one gets the expression

$$\int \mathrm{tr}\,\{(dgg^{-1})^3[\partial_t gg^{-1}, \partial_s gg^{-1}] - (\partial_s gg^{-1})(dgg^{-1})^2(\partial_t gg^{-1})(dgg^{-1}) + (\partial_t gg^{-1})(dgg^{-1})^2(\partial_s gg^{-1})(dgg^{-1})\} - \int \mathrm{tr}\,\{(dgg^{-1})^2[(\partial_s gg^{-1})d(\partial_t gg^{-1}) - (\partial_t gg^{-1})d(\partial_s gg^{-1})]\}\,. \tag{4.3.19}$$

Comparing (4.3.17) with (4.3.19) we get the simple formula

$$\int_S F_E = 2\pi i q_6 \int_{S\times S^3} \mathrm{tr}(dgg^{-1})^5. \tag{4.3.20}$$

Whether the bundle $\widehat{S_0^3G}$ over S_0^3G is nontrivial or not depends on the group G. If it happens that $C_5 \equiv 0$ then clearly $\widehat{S_0^3G} = S_0^3G \ltimes Map(\mathcal{A}_3, S^1)$. This is the case for example when $G = SO(N)$ is the real orthogonal group. Namely,

$$\mathrm{tr}[(dgg^{-1})^5]^t = \mathrm{tr}[(dgg^{-1})^t]^5 \tag{4.3.21}$$

by the cyclic symmetry of the trace and the antisymmetry properties of the wedge product. On the other hand $(dgg^{-1})^t = -dgg^{-1}$ since the Lie algebra of $SO(N)$ consists of antisymmetric matrices and therefore $(dgg^{-1})^5$ vanishes identically. However, in the case $G = SU(N)$, $N > 2$ we can show that the bundle is nontrivial. It is sufficient to find a sphere S^2 embedded in S_0^3G such that the integral of the curvature form F_E over S^2 is nonzero.

Next we construct the extension $\widehat{S^3G}$ of the the group S^3G in the case when there is a trivializing homomorphism $\rho : \pi_3 G = S^3G/S_0^3G \to S^3G$. This means that we can choose in each connected component of S^3G an element g_α such that the set $\{g_\alpha\}$ is closed under multiplication and the map $g(x) \equiv 1$ represents the unit element in $\pi_3 G$.

Next let K denote the group consisting of smooth maps $f : [0,1] \times S^3 \to G$ such that

(1) the derivatives of f with respect to the first variable $t \in [0,1]$ approach zero at the end points

(2) at the point $t = 0$ f is one of the functions g_α.

Let $\mathcal{G}$ be the subgroup of K consisting of maps f such that $f(0,x) = f(1,x) = 1$ for all $x \in S^3$. Any such a map can be thought of as a function on the sphere S^4 by identifying the boundary $S^3 \cup S^3$ with one point on S^4. We define an extension of K by the Abelian group $Map(\mathcal{A}_3, S^1)$ as in the case of D^4G. The product is defined by

$$(g_1, \lambda_1)(g_2, \lambda_2) = (g_1 g_2, \lambda_1\, g_1\lambda_2 \exp[2\pi i \gamma_3(\cdot; g_1, g_2)]),$$

where the action of an element $g \in K$ on a function λ is defined through a gauge transformation by $g(1, \cdot)$. Defining $\phi : \mathcal{G} \to K \ltimes Map(\mathcal{A}_3, S^1)$ by $\phi(g) = (g, \exp[2\pi i C_5(g)])$ we get an embedding of $\mathcal{G}$ as a normal subgroup in $K \ltimes Map(\mathcal{A}_3, S^1)$. Now

$$\widehat{S^3 G} = [K \ltimes Map(\mathcal{A}_3, S^1)]/\phi(\mathcal{G})$$

is the sought extension of S^3G. Using this extension we can define the complex line bundle E over S^3G analogously to the line bundle E_0 over $S^3_0 G$.

4.4. Spin and statistics from group extensions

The Lagrangian of the Wess-Zumino-Witten model

Keeping the notation of the previous section, let us consider a quantum system whose configuration space is the circle bundle SE obtained from E by restricting the structure group $\mathbf{C}^\times$ to S^1; remember that E is an associated line bundle to the principal bundle P defined by an *unitary* representation of $Map(\mathcal{A}_3, S^1)$ in $\mathbf{C}$, so there is a natural reduction of E to SE. The infinite-dimensional quantum system we shall describe behaves in many ways like the simple Abelian monopole, so let us first recall the construction of the Lagrangian for the latter. Consider a magnetic monopole situated at the origin of $\mathbf{R}^3$ and an electrically charged particle moving in the field of the monopole in $\mathbf{R}^3 \setminus 0$. Let m, e denote the mass and the charge of the particle. The magnetic field of the monopole is $B = \frac{g}{4\pi}\frac{x}{r^3}$, where g is the monopole strength. Naively one could write the Lagrangian as

$$\int \frac{1}{2} m (\frac{dx(t)}{dt})^2 + e \int A(x(t)) \cdot \frac{x(t)}{dt}$$

where the integral is taken along the path of the particle. However, there is no everywhere smooth vector potential A which satisfies $B = \nabla \times A$. We know that the magnetic field is in fact a curvature form in a principal circle bundle over $\mathbf{R}^3 \setminus 0$. Let Γ be the connection form in the total space of the bundle. We extend the notion of the point particle such that the position of the particle is a point p in the *total space* of the bundle. Then a well-defined Lagrangian is

$$L = \int \frac{1}{2} m (\frac{dx(t)}{dt})^2 + e \int \Gamma (\frac{dp(t)}{dt}), \tag{4.4.1}$$

where x is the projection of the position p to the base space. Using a local section ψ of the bundle and setting A equal to the pull-back $\psi^*\Gamma$ the Lagrangian L can be rewritten as

$$\int \frac{1}{2} m \left[\frac{dx(t)}{dt}\right]^2 + e \int \left[A \cdot \frac{dx(t)}{dt} + \lambda^{-1}\frac{d\lambda}{dt}\right],$$

where $\lambda(t)$ parametrizes the position of p in the vertical S^1 direction [with respect to the given local trivialization; $p(t) = \psi(x(t))\lambda(t)$].

In order to describe the *Wess-Zumino-Witten* model in $3+1$ space-time dimensions we shall simply replace the circle bundle over the 3-space by the circle bundle SE over S^3G and the magnetic field by the curvature form (4.3.16). We consider here only the case $G = SU(N)$, $N > 2$. We can construct a local vector potential for the field F_E as follows. If $U \subset S_0^3 G$ is an open contractible subset we can choose for each $f \in U$ an extension $f : D_4 \to G$ such that the choice of the extension depends smoothly on the element $f \in U$. We define a 1-form A on U by

$$A(X) = \pi i q_6 \int_{D_4} \mathrm{tr}(df f^{-1})^3 dX. \tag{4.4.2}$$

By a direct computation, the exterior derivative of A is the 2-form F_E. The vector potential on other components of S^3G is defined similarly; instead of considering extensions of f to D_4 one has to choose a smooth interpolation to a fixed element f_1 in the respective component.

The interaction of a "particle" moving in the the total space of SE is locally described by the term

$$\int_0^1 A = \pi q_6 \int_{D_4\times[0,1]} \operatorname{tr}(df f^{-1})^3 d(\partial_t f f^{-1})$$
$$= 2\pi i q_6 \left\{ \int_{D_4\times[0,1]} \operatorname{tr}(d' f f^{-1})^5 - 5 \int_{S^3\times[0,1]} (df f^{-1})^3 (\partial_t f f^{-1}) \right\},$$

(4.4.3)

where d' means exterior differentiation with respect to all of the variables in $D_4 \times [0,1]$; the path of the particle has been parametrized by $t \in [0,1]$. The total Lagrangian of the Wess-Zumino-Witten model is

$$L = L_{kin} + L_{int} = \Lambda \int_{S^3\times[0,1]} \operatorname{tr}(\partial^\mu f)(\partial_\mu f^{-1}) + \int_0^1 A, \tag{4.4.4}$$

where Λ is a constant and the kinetic energy term can be defined with respect to either the Euclidean or the Lorentzian metric on $S^3 \times [0,1]$ by choosing the appropriate signs (the standard Riemannian metric is used on S^3). For a closed path the right-hand-side of (4.4.4) does not depend (modulo $2\pi\times$ integer) on the choice of the extensions to D_4. However, for an open path one has to remember that the formula is only locally valid and the correct interaction Lagrangian is defined as the integral of the globally defined connection form in the total space of the circle bundle, along the path of the particle.

Our Lagrangian is not precisely of the form studied in Witten [1983]; it differs by the last term in (4.4.3) If one wants, this term can be cancelled by modifying the curvature form by the exact two-form dA_0, where A_0 is the one-form

$$A_0(X) = \pi i q_6 \int_{S^3} \operatorname{tr}(df f^{-1})^3 X. \tag{4.4.5}$$

Rotating the soliton

Following Witten we shall next look at what happens when we rotate a particle by the angle 2π. We shall first consider a rather special example but we shall later show how to deduce the characteristics of a general case from this example. Let $f : S^3 \to G$ be the soliton defined by the identification $SU(2) = S^3$ and by the embedding of $SU(2)$ in $SU(N)$ as the group of matrices T with $T_{ij} - \delta_{ij} = 0$ for $i > 2$ or $j > 2$. Thinking of S^3 as the one-point compactification of $\mathbf{R}^3$ the map f can be written as

$$f(x) = \exp\left[\pi i h(r)\sigma \cdot \frac{x}{r}\right], \tag{4.4.6}$$

where $\sigma_1, \sigma_2, \sigma_3$ are the Pauli matrices and h is a smooth monotonically increasing real function such that $h(0) = 0$ and $h(+\infty) = 1$. Thus at ∞ the function f approaches the value -1. Consider now the path in S^3G obtained by rotating f around the third coordinate axis. This can be written as $f(t,x) = g(t)f(x)g(t)^{-1}$ where we have chosen

$$g(t) = \begin{pmatrix} 1 & 0 & 0 & \dots & 0 \\ 0 & e^{-2\pi i t} & 0 & \dots & 0 \\ 0 & 0 & e^{2\pi i t} & \dots & 0 \\ \vdots & \vdots & \vdots & \ddots & \vdots \\ 0 & 0 & 0 & \dots & 1 \end{pmatrix}.$$

The conjugation of f by $g(t)$ is the same as the rotation of the argument x by the angle $2\pi t$ in the (x_1, x_2) plane. Let us compute the holonomy corresponding to the closed path $t \mapsto f(t,x)$, $t \in [0,1]$. In order to apply formula (4.4.3) we have to find for each $t \in [0,1]$ an interpolation between the map $x \mapsto f(t,x)$ and $x \mapsto f(x)$. Such an interpolation is given by the formula $f(t,x,s) = g(t,s)f(x)g(t,s)^{-1}$, where

$$g(t,s) = \begin{pmatrix} 1 & 0 & 0 & \dots & 0 \\ 0 & se^{-2\pi i t} & \sqrt{1-s^2} & \dots & 0 \\ 0 & -\sqrt{1-s^2} & se^{2\pi i t} & \dots & 0 \\ \vdots & \vdots & \vdots & \ddots & \vdots \\ 0 & 0 & 0 & \dots & 1 \end{pmatrix}.$$

Inserting $f(t,x,s)$ into (4.4.3) the second term is equal to zero and the first term is $i\pi$. In general, the holonomy around a closed loop γ is equal to $\exp(\int_\gamma A)$. Thus in our case we get $e^{i\pi} = -1$. We have shown that the rotation of the soliton by the angle 2π corresponds to the multiplication by the factor -1 in the fiber above f in the circle bundle SE. This means that the soliton in the Wess-Zumino-Witten model behaves like a fermion.

Next let $f_1 : S^3 \to G$ be an arbitrary smooth map of winding number $=1$. Because f and f_1 are in the same connected component of S^3G we can find a function $G(x,s)$ such that $G(x,o) = f(x)$ and $G(x,1) = f_1(x)$. Let $x \mapsto R_t(x)$ denote the rotation by the angle $2\pi t$. We have to compare the parallel transports along the two loops $t \mapsto G(R_t(x), 0)$ and $t \mapsto G(R_t(x), 1)$.

Exercise 4.4.7. Show by a change of variables in the integral that

$$\int_{S^3 \times [0,1]} \mathrm{tr}(dff^{-1})^3 (\partial_t f f^{-1}) = 0$$

for the function $(t, x) \mapsto f(R_t(x))$.

Since the second term on the right-hand-side of (4.4.3) vanishes, we have to show that

$$\int_{S^3 \times [0,1]} \mathrm{tr}(d' ff^{-1})^5 = \int_{S^3 \times [0,1]} \mathrm{tr}(d' f_1 f_1^{-1})^5. \tag{4.4.8}$$

The difference of the left- and right-hand sides in (4.4.8) is equal to the integral of the exterior derivative of the integrand over a cylinder in S^3G bounded by the loops $t \mapsto f(R_t(x))$ and $t \mapsto f_1(R_t(x))$. Such a cylinder is defined by the map $(t, s) \mapsto G(R_t(x), s)$. But the form $\mathrm{tr}(dff^{-1})^5$ is closed and therefore the difference vanishes. Thus we have shown that the rotation of any element in S^3G with winding number produces a factor -1 in the holonomy group.

If f has winding number 1 then f^n has winding number n; this follows from the formula $\mathrm{tr}[d(fg)(fg)^{-1}]^3 = \mathrm{tr}(dff^{-1})^3 + \mathrm{tr}(dgg^{-1})^3 +$ an exact form. As in the case $n = 1$ one can easily show that the holonomy from a rotation by the angle 2π depends only on the winding number and not on the particular representative of the homotopy class of n. With a little more effort one can show that the holonomy is equal to $(-1)^n$. Thus we have the following proposition.

PROPOSITION 4.4.9. *Let $f \in S^3G$ with winding number n. Then f behaves like a fermion under rotations if n is odd and like a boson if n is even.*

Interchange of two solitons

Next we shall consider what happens when we interchange two elements f_1 and f_2, both with winding number 1, such that the *supports of f_i are spatially separated* . The support of a function $f : S^3 \to G$ is by definition the closure of the set of points x with $f(x) \neq 1$. Spatially separated means that there are open balls B_i (with respect to the natural metric on S^3) such that $B_1 \cap B_2 = \emptyset$ and $\mathit{supp} f_i \subset B_i$.

We assume that the space of states of a quantum field theory carries a representation T of the extension $\widehat{S^3G}$ of the group S^3G of gauge transformations in a Hilbert space $\mathcal{F}$; we assume that the center of $\widehat{S^3G}$ is represented by the operators $\lambda \cdot 1$ with $|\lambda| = 1$. Fix $f \in S^3G$ to be the standard soliton defined above. The classical fields f_i define a ray in the space of linear operators in $\mathcal{F}$: For $i = 1, 2$ we choose a function $g_i(t, x)$ such that $g_i(0, x) = f(x)$ and $g_i(1, x) = f_i(x)$. Let T_i be the operator representing the class of $(g_i, 1)$ in $\widehat{S^3G}$. Choosing another interpolation between f and f_i changes the operator T_i by a multiplicative constant of modulus =1. The physical interpretation of the operators T is the following. The space $\mathcal{F}$ is assumed to contain a vacuum vector ψ_0 (which is the unique vector, up to a multiplicative constant, which has the minimum energy). When acting by T_i on the vacuum ψ_0 one obtains a one-particle state which in a classical approximation is described by the field f_i. We want to show that the two-particle states $T_1T_2\psi_0$ and $T_2T_1\psi_0$ differ by the factor -1.

We consider first the case when f_1, f_2 take values in the subgroup $SU(2)$. Then the functions g_i can also be chosen in such a way that the values are in $SU(2)$. From the fact that any 4-form vanishes on the three-dimensional manifold $SU(2)$ and that the functions f_i have non-overlapping supports it follows that

$$\gamma_3(A; g_1, g_2) \equiv 0. \tag{4.4.10}$$

Thus the multiplication rule in $D_4G \times Map(\mathcal{A}_3, S^1)$ gives

$$\begin{aligned}(g_1, 1)(g_2, 1) &= (g_1g_2, 1) = (g_2g_1g^{-1}, 1)\\ &= (g_2g_1, e^{2\pi iC_5(g)}) = (g_2, 1)(g_1, 1)(1, e^{2\pi iC_5(g)})\end{aligned} \tag{4.4.11}$$

and therefore $T_1T_2 = e^{2\pi iC_5(g)}T_2T_1$, where $g = g_2^{-1}g_1^{-1}g_2g_1$. At the end points $t = 0, 1$ we have $g(t, x) = 1$ and therefore g is really a map from $S^1 \times S^3$ to $SU(2)$. If $g' : S^1 \times S^3 \to SU(2)$ is another map which is homotopic to g, i.e., there is a map $h : S^1 \times S^3 \times [0, 1] \to SU(2)$ such that $h(t, x, 0) = g(t, x)$ and $h(t, x, 1) = g'(t, x)$, then

$$C_5(g) - C_5(g') \equiv q_6 \frac{1}{10} \int \mathrm{tr}(dhh^{-1})^5 \; mod\mathbf{Z}$$

and because any 5-form vanishes on $SU(2)$ we get $C_5(g) = C_5(g')$ mod$\mathbf{Z}$. It follows that $(1, e^{2\pi iC_5(g)})$ depends only on the homotopy class of the map $g : S^1 \times S^3 \to SU(2)$.

It was shown in Witten [1983] that $\exp[2\pi iC_5(g)] = +1$ if the map g represents the identity in the homotopy group $\pi_4 SU(2) = \mathbf{Z}_2$ and is equal to -1 if g represents the nontrivial homotopy class. Thus we have the following theorem.

THEOREM 4.4.12. *The soliton creation operators T_1 and T_2 anticommute for spatially separated solitons.*

This is true provided that we can show that the map g is non-contractible. We leave it as an exercise for the reader to show that g can in fact be continuously deformed to the map $f(t, x)$ studied in the proof of Proposition 4.4.9. This completes the proof of the theorem above in the case when f_1 and f_2 take values in the subgroup $SU(2) \subset SU(N)$.

The general case is reduced to the above by the following trick. If f_1, f_2 are arbitrary spatially separated solitons in $SU(N)$ then we choose paths $t \mapsto g_i(t, x)$ connecting f_i to f such that $g_i(t, x) \in SU(2)$ for $0 \le t \le \frac{1}{2}$ and f_1, f_2 are spatially separated for $\frac{1}{2} \le t \le 1$. (This is possible by a dimensional argument.) The contribution to the commutator of T_1 and T_2 coming from the cocycle $c^{2,4}$ in (4.3.4) when integrated over the domain $\frac{1}{2} \le t \le 1$ vanishes since the supports do not overlap there. Thus the only contribution comes from the path inside $SU(2)$ which we have already computed.

4.5. Chern classes

We shall consider polynomials $P(A)$ of a complex $N \times N$ matrix variable A which are invariant in the sense that $P(gAg^{-1}) = P(A)$ for all $g \in GL(N, \mathbf{C})$. For example, if we expand

$$\det\left(1 + \frac{\lambda}{2\pi i} A\right) = \sum_{n=0}^{N} \lambda^n P_n(A) \tag{4.5.1}$$

then the coefficients $P_n(A)$ are homogeneous invariant polynomials of degree n in A. These polynomials will play a special role in the following discussion.

To each homogeneous polynomial $P(A)$ one can associate a unique *symmetric* multilinear form $P(A_1, \dots A_n)$ such that $P(A, \dots, A) = P(A)$. The general formula for the n linear form in terms of $P(A)$ is

$$\begin{aligned} P(A_1, \dots, A_n) =& \frac{1}{n!}\{P(A_1 + \dots + A_n) \\ &- \sum_j P(A_1 + \dots + \hat{A}_j + \dots + A_n) \\ &- \sum_{j,j'} P(A_1 + \dots A_j + \dots A_{j'} \dots + A_n) - \dots\}. \end{aligned}$$

When $P(A)$ is invariant we clearly have $P(gA_1g^{-1},\dots,gA_ng^{-1})$. Writing $g = g(t) = \exp(tX)$ we get the useful formula

$$\begin{aligned} 0 &= \frac{d}{dt}P(g(t)A_1g(t)^{-1},\dots,g(t)A_ng(t)^{-1})|_{t=0} \\ &= \sum_j P(A_1,\dots,[X,A_j],\dots,A_n). \end{aligned} \tag{4.5.2}$$

If F_i is a $N \times N$ matrix valued differential form of degree k_i on a manifold M, $1 \le i \le n$, and P a symmetric n linear form then we can define a complex valued differential form $P(F_1,\dots,F_n)$ of degree $k_1 + \cdots + k_n = p$ by

$$P(F_1,\dots,F_n)(t_1,\dots,t_p) =$$
$$\left(\prod \frac{1}{k_i!}\right) \sum_\sigma \epsilon(\sigma) P(F_1(t_{\sigma(1)},\dots,t_{\sigma(k_1)}),\dots,F_n(t_{\sigma(p-k_n+1)},\dots,t_{\sigma(p)}))$$

where the sum is taken over all permutations of the indices $1,2,\dots,p$.

Let F be the curvature form of a vector bundle E over M with fiber $\mathbf{C}^N$. The curvature transforms in a change of a local trivialization as $F \mapsto gFg^{-1}$ and therefore $P(F,\dots,F)$ is well-defined, independent of the local trivialization, for any invariant symmetric polynomial P.

PROPOSITION 4.5.3. *The symmetric homogeneous polynomial* $P(F, \dots, F)$ *of degree* n *in the curvature* F *is a closed form of degree* $2n$.

PROOF: Locally we can write $F_{\mu\nu} = \partial_\mu A_\nu - \partial_\nu A_\mu + [A_\mu, A_\nu]$. Using the property $d(\alpha\wedge\beta) = d\alpha\wedge\beta + (-1)^{deg\alpha}\alpha\wedge\beta$ of differential forms we have

$$\begin{aligned} dP(F,\dots,F) &= \sum_j P(F,\dots,dF,\dots,F) \\ &= \sum_j \{P(F,\dots,DF,\dots,F) - P(F,\dots,[A,F],\dots,F)\}. \end{aligned} \tag{4.5.4}$$

The covariant derivative $DF = 0$ by the Bianchi identity and the sum of the terms involving $[A,F]$ is zero by (4.5.2).

In particular, the class in $H^{2n}(M,\mathbf{R})$ defined by the closed $2n$ form $\mathrm{Re}P_n(F)$ is called the nth *Chern class* of the bundle E and is denoted by $c_n(E)$.

THEOREM 4.5.5. *The Chern classes are topological invariants: They do not depend on the choice of connection in the vector bundle E.*

PROOF: Let A_0 and A_1 be two connections in E and F_0, F_1 the corresponding curvatures. Define a one-parameter family $A_t = A_0 + t\eta$ of connections with $\eta = A_1 - A_0$; note that the difference η transforms homogeneously in a change of local trivialization, $\eta \mapsto g\eta g^{-1}$. Let us introduce the notation $Q(A, B) = kP(A, B, \dots, B)$ when B is repeated $k-1$ times. Using

$$F_t = dA_t + \frac{1}{2}[A_t, A_t] = F_0 + tD\eta + \frac{1}{2}t^2[\eta, \eta],$$

where D is the covariant derivative determined by A_0, we get

$$\frac{d}{dt}P(F_t) = Q\left(\frac{d}{dt}F_t, F_t\right) = Q(D\eta, F_t) + tQ([\eta, \eta], F_t). \tag{4.5.6}$$

On the other hand,

$$\begin{aligned} dQ(\eta, F_t) =& Q(d\eta, F_t) - n(n-1)P(\eta, dF_t, F_t, \dots, F_t) \\ =& Q(d\eta, F_t) - n(n-1)P(\eta, dF_t, F_t, \dots, F_t) \\ &+ nP([A_0, \eta], F_t, \dots, F_t) - n(n-1)P(\eta, [A_0, F_t], \dots, F_t) \\ =& Q(D\eta, F_t) - n(n-1)P(\eta, DF_t, F_t, \dots, F_t) \\ =& Q(D\eta, F_t) + tn(n-1)P(\eta, [\eta, F_t], F_t, \dots, F_t) \end{aligned}$$

(4.5.7)

where we have used $DF_t = DF_0 + tD^2\eta + t^2[D\eta, \eta] = t[F_0, \eta] + t^2[D\eta, \eta] = t[F_t, \eta]$. By (4.5.2) we have

$$P([\eta, \eta], F_t, \dots, F_t) - (n-1)P(\eta, [\eta, F_t], F_t, \dots, F_t) = 0$$

or in other words,

$$Q([\eta, \eta], F_t) - n(n-1)P(\eta, [\eta, F_t], F_t, \dots, F_t).$$

Combining this with

$$dQ(\eta, F_t) = Q(D\eta, F_t) + tQ([\eta, \eta], F_t)$$

and with (4.5.6) and (4.5.7) we obtain

$$\frac{d}{dt}P(F_t) = dQ(\eta, F_t). \tag{4.5.8}$$

Integrating this result over the interval $0 \leq t \leq 1$ we get

$$P(F_1) - P(F_0) = d \int_0^1 Q(\eta, F_t) dt$$

which shows explicitly that the difference of the differential forms $P(F_1)$ and $P(F_0)$ is an exact form.

Given a Hermitian inner product in the fibers of the vector bundle E it is always possible to choose a Hermitian connection, that is, a connection such that in an orthonormal basis the vector potential takes values in the Lie algebra of the unitary group $U(N)$. In that case the determinant $\det(1 + \frac{\lambda}{2\pi i} F)$ is real for any real parameter λ and the Chern classes are given by the expansion in powers of λ; the first two positive powers lead to

$$\begin{aligned} c_1(F) &= \frac{1}{2\pi i N} \mathrm{tr} F \\ c_2(F) &= \frac{1}{(2\pi i)^2} [\mathrm{tr} F^2 - (\mathrm{tr} F)^2]. \end{aligned}$$

The coefficients in the expansion can be best computed by diagonalizing the matrix F. Writing $F = diag(\alpha_1, \ldots, \alpha_N)$ one obtains

$$\det\left(1 + \frac{\lambda}{2\pi i} F\right) = \prod_k \left(1 + \frac{\lambda \alpha_k}{2\pi i}\right) = \sum_n \left(\frac{\lambda}{2\pi i}\right)^n S_n(\alpha)$$

with

$$\begin{aligned} S_0 = 1,\ S_1 = \mathrm{tr}\alpha,\ S_2 &= \frac{1}{2}(\mathrm{tr}\alpha)^2 - \frac{1}{2}\mathrm{tr}\alpha^2 \\ S_3 = \frac{1}{6}(\mathrm{tr}\alpha)^3 &- \frac{1}{2}\mathrm{tr}\alpha^2\, \mathrm{tr}\alpha + \frac{1}{3}\mathrm{tr}\alpha^3 \end{aligned}$$

etc. Note that c_n vanishes identically if $n > \frac{1}{2}\dim M$ or $n > N$. If $n = \frac{1}{2}\dim M$ then we can integrate the form $c_n(E)$ over M and the value of the integral is called *the Chern number* associated to the vector bundle E.

Example 4.5.9. Consider a vector bundle E over $M = S^4$ such that the transition functions take values in the group $SU(N)$, $N \geq 2$. Dividing S^4 to the upper and lower hemispheres $S^4_\pm$ the bundle is given by the transition function ϕ along the equator S^3. The vector potentials $A_\pm$ are then related by $A_- = \phi A_+ \phi^{-1} + d\phi\phi^{-1}$ on the equator. Using the formula $\mathrm{tr}F^2 = d\mathrm{tr}(F\wedge A - \frac{1}{3}A^3)$ we can compute the Chern number corresponding to the second Chern class,

$$\begin{aligned}\frac{1}{8\pi^2}\int_{S^4_+}\mathrm{tr}F_+^2 &+ \frac{1}{8\pi^2}\int_{S^4_-}\mathrm{tr}F_-^2 \\ &= \frac{1}{8\pi^2}\int_{S^3}\left[\mathrm{tr}(F_+\wedge A_+ - \tfrac{1}{3}A_+^3) - \mathrm{tr}(F_-\wedge A_- - \tfrac{1}{3}A_-^3)\right] \\ &= \frac{1}{8\pi^2}\int_{S^3}\left[\mathrm{tr}\tfrac{1}{3}(d\phi\phi^{-1})^3 - d\mathrm{tr}(A_+\wedge d\phi\phi^{-1})\right] \\ &= \frac{1}{24\pi^2}\int_{S^3}\mathrm{tr}(d\phi\phi^{-1})^3.\end{aligned}$$

Remark 4.5.10. The value of the integral above is an integer which depends only on the homotopy class of the map $\phi : S^3 \to SU(N)$.

Since the equivalence class of the bundle E depends only on the homotopy class of the transition function ϕ, the Chern number $\int c_2(E)$ gives a complete topological characterization of E.

The *Chern character* $ch(E)$ of a vector bundle is defined as follows. It is a formal sum of differential forms of different degrees,

$$ch(E) = \mathrm{tr}\ \exp\left(\frac{1}{2\pi i}F\right),$$

where again F is the curvature form of E. When the exponential is evaluated as a power series we obtain

$$ch(E) = \sum_{k=0}^{\infty}\frac{1}{(2\pi i)^k k!}\mathrm{tr}F^k.$$

Clearly all the terms can be expressed using the Chern classes; the three first terms are

$$ch(E) = N + c_1(E) + \frac{1}{2}(c_1(E)\wedge c_1(E) - 2c_2(E)) + \dots.$$

The Chern character is a convenient tool because one has

$$ch(E\oplus E') = ch(E) + ch(E')\ \ ch(E\otimes E') = ch(E)\cdot ch(E').$$

This follows immediately from the definition and the elementary properties of the exponential function.

Further reading on characteristic classes: The proof above of the topological invariance of the Chern classes follows Chern [1979] which we recommend for further study. Also: Milnor and Stasheff [1974], Chern and Simons [1974].

CHAPTER 5 THE CHIRAL ANOMALY

5.0. Introduction

The Dirac operator on a manifold M is a first order partial differential operator acting on sections of a spin bundle over M. The Dirac operator is elliptic when the metric of M is positive definite. The main task in this chapter is to study properties of the determinant of the Dirac operator.

The space of sections of the spin bundle is infinite-dimensional. The determinant of a linear operator in a Hilbert space is a priori well-defined only if it is of type 1 + a trace-class operator. However, the Dirac operator is never of this type. In order to define the determinant one must "regularize" the Dirac operator. There is a great freedom in choosing the regularization; the requirement is that the regularized determinant should display the essential information about the spectrum of the original operator (especially the zeros of the operator) and should be continuous in the possible parameters.

We shall study the case when the Dirac operator $D = D(A)$ is parametrized by a vector potential A. The Dirac equation transforms equivariantly with respect to gauge transformations and therefore one would expect that the determinant $\det D(A)$ is gauge invariant. This is indeed the case when D operates on standard Dirac fermions consisting of components of both *chiralities* $\pm$. However, when the Dirac field is massless the components belonging to the opposite chiralities decouple and it is natural to study D_+ and D_- separately. (D_+ is the *Weyl-Dirac* operator which maps positive chirality spinors to negative chirality spinors and D_- goes to the opposite direction.) In this case it turns out that one cannot regularize the determinant in a gauge invariant way but there is a *chiral anomaly*, which measures how the determinant is changed under a gauge transformation.

The anomaly of the Dirac determinant manifests itself as a source for the *axial vector current* . In the same way as the electromagnetic current is associated to gauge tranformations of the electromagnetic vector potential, the gauge transformations acting on Dirac field through a phase transformation, there is a chiral current j^5_μ such that the time component j^5_0 generates the chiral rotations (phase transformations, opposite phases for chiralities $\pm$). Classically, in the case of a massless Dirac field, the chiral current is conserved, $\partial^\mu j^5_\mu = 0$. However, in the case of chiral fermions (only half of the fermion components are coupled to the vector potential) there is an anomaly: The divergence of the chiral current does not vanish. This was the way anomalies were originally found (in perturbation theory computations of the chiral current) [Adler, 1969; Bardeen, 1969; Bell and Jackiw, 1969; Brown, Shi, and Young, 1969].

When the gauge group is non-Abelian there is a similar phenomenon; there is an additional Lie algebra index labelling the different components of the current and classically the divergence equation is replaced by the covariant divergence $\nabla^\mu j^5_\mu = 0$, where $\nabla^\mu = \partial^\mu + [A^\mu, \cdot]$ is the covariant derivative defined by the gauge potential A_μ [Gross and Jackiw, 1972].

It was realized much later that there is geometrical and topological reason for the occurrence of anomalies in the "effective action" (=logarithm of the determinant of the Dirac operator). The anomalies can be derived using Atiyah-Singer index theory. The index of a Dirac operator D is the difference $n_+(D) - n_-(D)$ of the multiplicity n_+ of the zero eigenvalue of D in the positive chirality sector and the number n_- of negative chirality zero modes. According to the Atiyah-Singer index theory that number can be expressed as an integral of certain characteristic class (involving the Chern classes) over the space-time manifold M. The density under the integral is the divergence of the (Abelian) axial vector current [Nielsen, Römer, and Schroer, 1977, 1978; Nielsen and Schroer, 1978; Jackiw and Rebbi, 1977].

In the case of non-Abelian chiral transformations one has to use *families index theory*. Again, characteristic classes are involved. The anomaly can be neatly expressed through the curvature of the space $\mathcal{A}/\mathcal{G}$ of vector potentials modulo gauge transformations. We shall explain this point of view in Section 5.4, following closely the presentation in Atiyah and Singer [1984]; see also Alvarez-Gaume and Ginsparg [1984].

The anomalies manifest themselves also in the Hamiltonian approach in a variety of ways. There is a relation between anomalies and pair production of particles [Alvarez-Gaume and Ginsparg, 1984]. For the main theme for this book the most important consequence of anomalies is the fact that the current algebra will be modified: There are Schwinger terms in the commutators. These commutators are precisely those which we have already studied in the previous chapter!

One can view the chiral anomalies also as the noninvariance of the fermionic path integral measure under gauge transformations; we shall not discuss that point of view here; see Fujikawa [1979].

Our treatment of anomalies will use cohomological methods. However, there are many aspects of this approach which we cannot cover in this book; for more specialized discussions see, e.g., Bonora and Cotta-Ramusino [1983]; Bonora, Cotta-Ramusino, Rinaldi, and Stasheff [1987, 1988], and references therein.

5.1. The Clifford algebra

Let H be a real vector space equipped with an inner product $(\cdot,\cdot)$. The *Clifford algebra* $C(H)$ based on this data is the associative algebra containing the identity 1 and generated by the vectors $x \in H$ subject to the defining relations

$$xy + yx = 2(x, y). \tag{5.1.1}$$

Assume in the following that H is finite-dimensional with an orthonormal basis $\{e_1, \dots e_n\}$. A basis for the algebra $C(H)$ is given by 1 and the products

$$e_{i_1} e_{i_2} \dots e_{i_p},\ 1 \leq i_1 < i_2 < \dots < i_p \leq n \tag{5.1.2}$$

since $e_i^2 = 1$ and $e_i e_j = -e_j e_i$ for $i \neq j$. Thus the dimension of $C(H)$ is $\sum_{p=0}^{n} \binom{n}{p} = 2^n$.

A (reducible) representation of $C(H)$ can be constructed in the vector space $\Lambda(H)$, in the exterior algebra of H, as follows. Denote $dx = x\wedge$, the exterior multiplication by the vector x, and $\imath_x$ the contraction operator defined by linearity and

$$\imath_x(e_{i_1} \wedge \dots \wedge e_{i_p}) = \sum_{k=1}^{p} (-1)^{k-1} (x, e_{i_k}) e_{i_1} \wedge \dots \hat{e}_{i_k} \dots \wedge e_{i_p}$$

where the caret means that e_{i_k} has been deleted. The basic commutation relations are

$$dx\, dy = -dy\, dx,\ \imath_x \imath_y = -\imath_y \imath_x,\ dx\, \imath_y + \imath_y\, dx = (x, y). \tag{5.1.3}$$

A representation of $C(H)$ is obtained as $x \mapsto \gamma(x) = dx + \imath_x$. The dimension of the representation is equal to $\dim\Lambda(H) = \dim C(H) = 2^n$.

Next we shall consider the case when n is even, $n = 2m$. In order to define an irreducible representation of $C(H)$ we shall shall complexify the Clifford algebra, $C(H)_{\mathbf{C}} = C(H) \otimes_{\mathbf{R}} \mathbf{C}$. Let us define

$$a_k = \frac{1}{2}(e_k + ie_{k+m}),\ a_k^* = \frac{1}{2}(e_k - ie_{k+m}), k = 1, 2, \dots, m.$$

They satisfy the canonical anticommutation relations

$$[a_i^*, a_j]_+ = \delta_{ij} \tag{5.1.4}$$

and all other anticommutators =0. We have

(5.1.5.) $$e_k = a_k + a_k^*,\ e_{k+m} = i(a_k^* - a_k) \qquad 1 \le k \le m.$$

The fermionic Fock space $\mathcal{F}$ is by definition the complex Clifford algebra modulo the left ideal generated by the *annihilation operators* a_k. A basis of $\mathcal{F}$ consists of the vectors $a_{i_1}^* \ldots a_{i_p}^* \cdot 1$, where $1 \le i_1 < i_2 < \cdots < i_p \le m$ and the vector $1 \in C(H)_{\mathbf{C}}$ is *the vacuum* in $\mathcal{F}$. The dimension of the Fock space is 2^m. By (5.1.5) the Fock space carries a representation of the Clifford algebra; we denote by γ_i the operator representing e_i in $\mathcal{F}$. It is a simple exercise to show that this representation is irreducible.

Define $\gamma_{2m+1} = -i^m \gamma_1 \ldots \gamma_n$, called the *chirality operator*. From the anticommutation relations $\gamma_i\gamma_j + \gamma_j\gamma_i = 2\delta_{ij}$ follows that γ_{2m+1} anticommutes with each γ_i, $1 \le i \le n$, and $\gamma_{2m+1}^2 = 1$. It follows that we can construct in the same Fock space $\mathcal{F}$ a representation of the Clifford algebra based on the *odd dimensional* vector space $H \oplus \mathbf{R}e_{2m+1}$ by representing e_{2m+1} by the operator γ_{2m+1}.

Orthogonal transformations of H extend to automorphisms of the Clifford algebra, as can be seen from the defining relations (5.1.1). In a given complex representation γ of $C(H)$ in a vector space V we would like to represent the automorphisms $R \in SO(n)$ by linear operators $T(R)$. Let us first consider the even dimensional case, $n = 2m$. A complete basis of the Lie algebra $\mathbf{so}(n)$ is given by the matrices $s_{ij} = e_{ij} - e_{ji}$, $1 \le i < j \le n$, where the e_{ij}'s are elements of the Weyl basis of the general linear algebra $\mathbf{gl}(n)$ with commutation relations $[e_{ij}, e_{kl}] = \delta_{jk}e_{il} - \delta_{il}e_{kj}$. Set

$$T(s_{ij}) = \frac{1}{2}\gamma(e_i)\gamma(e_j) = -T(s_{ji}).$$

Then, by a straightforward computation,

(5.1.6) $$[T(s_{ij}), T(s_{kl})] = \delta_{jk}T(s_{il}) + \delta_{il}T(s_{jk}) - \delta_{ik}T(s_{jl}) - \delta_{jl}T(s_{ik})$$

which are precisely the commutation relations of the matrices s_{ij}. Thus we have a representation of the Lie algebra $\mathbf{so}(n)$ in V. Furthermore,

(5.1.7) $$[T(s_{ij}), \gamma(e_k)] = \delta_{jk}\gamma(e_i) - \delta_{ik}\gamma(e_j)$$

which shows that the generators $\gamma(x)$ transform like vectors under the adjoint action of the representation T of the Lie algebra of $SO(n)$. The question is now: can we exponentiate the infinitesimal generators $T(s_{ij})$ to obtain a representation of the group $SO(n)$ in V? Since we are dealing with a finite-dimensional representation we know that we can do the exponentiation to obtain a representation of the covering group $Spin(n)$. In fact, in the present case we have a spin representation [double valued representation of $SO(n)$]. To see this we compute

$$\begin{aligned} e^{\frac{\alpha}{2}\gamma(e_1)\gamma(e_2)} &= \sum \frac{1}{n!}(\alpha/2)^n(\gamma(e_1)\gamma(e_2))^n \\ &= \cos(\alpha/2) + \gamma(e_1)\gamma(e_2)\sin(\alpha/2) \end{aligned}$$

which shows that $T(e^{2\pi\cdot\frac{1}{2}\gamma(e_1)\gamma(e_2)}) = -1$ whereas $e^{2\pi s_{12}} = +1$.

The representation T of $Spin(n)$ in V is reducible. The operator γ_{2m+1} commutes with all the generators $T(s_{ij})$ and therefore the eigenspaces of the chirality operator are invariant under $Spin(n)$. The square of γ_{2m+1} is one and so the eigenvalues are ± 1. It is easy to construct the corresponding eigenspaces $V_\pm$. Since γ_{2m+1} anticommutes with each γ_k, $1 \le k \le n$, it anticommutes also with the creation operators a_k^*. The vacuum is an eigenvector of γ_{2m+1} corresponding to the eigenvalue $+1$ and consequently the eigenspace V_+ consists of vectors obtained by acting by a polynomial of even degree in the creation operators to the vacuum whereas V_- is generated by polynomials of odd degree. Both subspaces are of dimension 2^{m-1}. For example, if $n = 4$ then the represention of $Spin(4) = SU(2) \times SU(2)$ in V splits into a pair of two-dimensional representations; these representations are just the defining representations of the two $SU(2)$ subgroups.

In the odd dimensional case we can extend the representation of $Spin(2m)$ in V to a representation of $Spin(2m+1)$ in the same vector space by using the chirality operator. The missing elements of the Lie algebra of $Spin(2m+1)$ are

$$T(s_{i,2m+1}) = \frac{1}{2}\gamma_i\gamma_{2m+1}, \quad 1 \le i \le 2m.$$

Exercise 5.1.8. Show that the representation of $Spin(2m+1)$ above is irreducible.

5.2. The Dirac operator

Let (M, g) be an oriented Riemannian manifold of dimension n. Let FM be the bundle of oriented orthonormal frames in the tangent bundle TM. We shall assume that it has *a spin structure*; that means there is a principal $Spin(n)$ bundle P over M and a covering map

$$\phi : P \to FM, \quad \phi(pg) = \phi(p)\pi(g)$$

where $\pi : Spin(n) \to SO(n)$ is the double covering homomorphism. If the frame bundle is trivial we can always choose the trivial spin structure $M \times Spin(n)$ with the obvious covering map. In general a manifold does not need to have a spin structure. A simple example of such a manifold is the complex projective space $\mathbf{C}P^2$ of complex dimension 2; we shall return to this in Section 6.4. A manifold can have several inequivalent spin structures; see Gilkey [1984] for more details.

Let V be the complex vector space carrying the irreducible representation of the Clifford algebra in dimension n. As we saw, there is a spin representation ρ of $Spin(n)$ in V which gives an operator realization for the automorphisms of the Clifford algebra. Let $E = P \times_\rho V$ be the associated vector bundle over M; it is called *a spin bundle* of M.

Let Γ be a metric compatible connection of FM, that is, in local coordinates we have

$$\nabla_\mu g_{\nu\lambda} = \partial_\mu g_{\nu\lambda} + \Gamma^{\eta}_{\mu\nu} g_{\eta\lambda} + \Gamma^{\eta}_{\mu\lambda} g_{\mu\eta} = 0. \tag{5.2.1}$$

Let $\{h^{(1)}, \ldots, h^{(n)}\}$ be a local orthonormal frame in FM. The metric compatibility is equivalent to the condition that the connection expressed in the basis h takes values in the Lie algebra of the orthogonal group, i.e.,

$$\nabla_\mu h^i = \Gamma^j_{\mu i} h^j \text{ with } \Gamma^j_{\mu i} = -\Gamma^i_{\mu j}. \tag{5.2.2}$$

We can now define a connection in the bundle P (and thus canonically also in the associated bundle E) as follows. Let $\hat{h}$ be a local section of P which covers the section h. Since $Spin(n)$ is the double covering of $SO(n)$ their Lie algebras are isomorphic. With respect to the local section $\hat{h}$ the connection is represented by the Lie algebra valued 1-form $\Gamma^j_{\mu i}$. There is a second local section $\hat{h}z$ which covers h, where z belongs to the center of $Spin(n)$. But a gauge transformation of the Lie algebra valued connection form by an element of the center is an identity transformation and we conclude that the connection is well-defined.

Next we define a first order partial differential operator D, *the Dirac operator,* acting on sections of E. Locally, a section of E is a pair $\psi = (\hat{h}, \xi)$, where $\hat{h}$ is a local section of P and ξ is a locally defined function on M with values in V. We set

$$D\psi = (\hat{h}, \gamma_i h^{(i)\mu} \nabla_\mu \xi), \tag{5.2.3}$$

where $\gamma_i = \gamma(e_i)$ and the $h^{(i)\mu}$'s are the coordinates of the frame h in FM (corresponding to $\hat{h}$ via the canonical projection). To simplify the notation we shall write

$$D\psi = \gamma^\mu \nabla_\mu \psi \tag{5.2.4}$$

with $\gamma^\mu = \gamma_i h^{(i)\mu}$, understanding that a choice of $\hat{h}$ has been made.

Exercise 5.2.5. Show that D indeed is a well-defined operator: If $\hat{h}' = \hat{h}g$ and $\xi' = g^{-1}\xi$ then $D(\hat{h}', \xi')$ as defined by (5.2.3) is the same section as $D(\hat{h}, \xi)$.

Example 5.2.6. Let $M = S^1 \times S^1$ with the flat metric. The frame bundle $FM = M \times SO(2)$ is trivial and we can define a spin structure as $\phi : M \times Spin(2) \to M \times SO(2)$, where $Spin(2) = SO(2)$ and $\phi(x, g) = (x, g^2)$. The spinor representation of the Clifford algebra $C(\mathbf{R}^2)$ is two-dimensional; we can define

$$\gamma_1 = \begin{pmatrix} 0 & 1 \\ 1 & 0 \end{pmatrix}, \quad \gamma_2 = \begin{pmatrix} 0 & i \\ -i & 0 \end{pmatrix}$$

and a complete basis of the Clifford algebra is $\{1, \gamma_1, \gamma_2, \gamma_1\gamma_2\}$. The Dirac operator is now

$$D = \gamma_1 \partial_1 + \gamma_2 \partial_2 .$$

A complete set of solutions of the *Dirac equation* $(D + im)\psi = 0$ consists of the $\mathbf{C}^2$ valued functions

$$\psi_p = e^{ip_1\theta_1 + ip_2\theta_2} \begin{pmatrix} 1 \\ \frac{-m}{p_1 + ip_2} \end{pmatrix}, \quad \text{with}\, p_1^2 + p_2^2 = m^2$$

where $p_1, p_2 \in \mathbf{Z}$. Thus the eigenvalues of the Dirac operator (the mass of the Dirac particle) are $\pm i\sqrt{p_1^2 + p_2^2}$ and each eigenvalue has a finite multiplicity; the latter is a general property of a Dirac operator on any compact manifold.

Example 5.2.7. Let $M = (S^1)^4$ again with the flat metric. A representation of the Clifford algebra $C(\mathbf{R}^4)$ in $\mathbf{C}^4$ is generated by the matrices

$$\gamma_k = \begin{pmatrix} 0 & \sigma_k \\ \sigma_k & 0 \end{pmatrix} \text{ with } \sigma_4 = \begin{pmatrix} 1 & 0 \\ 0 & 1 \end{pmatrix} \text{ and}$$
$$\sigma_1 = \begin{pmatrix} 0 & 1 \\ 1 & 0 \end{pmatrix}, \quad \sigma_2 = \begin{pmatrix} 0 & i \\ -i & 0 \end{pmatrix}, \quad \sigma_3 = \begin{pmatrix} 1 & 0 \\ 0 & -1 \end{pmatrix}.$$

The solutions of the Dirac equation are of the form

$$\psi_p(+) = e^{ip\cdot\theta} \begin{pmatrix} 1 \\ 0 \\ \frac{p_4 - p_3}{m} \\ \frac{ip_2 - p_1}{m} \end{pmatrix}, \quad \psi_p(-) = e^{ip\cdot\theta} \begin{pmatrix} 0 \\ 1 \\ \frac{p_1 + ip_2}{-m} \\ \frac{p_3 - p_4}{m} \end{pmatrix}$$

where $m^2 = p \cdot p = p_1^2 + \ldots p_4^2$ and $p_i \in \mathbf{Z}$.

Example 5.2.8. Let $M = S^3$. We shall not use the Levi-Cività connection induced by the standard Riemannian metric on the unit sphere but the *flat connection* which can be constructed by observing that S^3 is as a manifold the group $SU(2)$. Now the matrices $\frac{i}{2}\sigma_k$ form an orthonormal basis of the Lie algebra $\mathbf{su}(2)$; a global section of the frame bundle FM is obtained by left translating the tangent vectors $\frac{i}{2}\sigma_k$ at the identity to a general position on the group manifold; we denote the left invariant vector fields by J_k. This defines a metric on S^3. A metric compatible connection is defined by declaring that the left invariant vector fields are parallel, i.e., $\nabla_i \sigma_j = 0$. The Dirac operator can be written as

$$D = \sigma_i J_i.$$

Let $D^j_{m_1 m_2}(g)$ be the matrix elements of $SU(2)$ in an irreducible representation characterized by the spin $j \in \frac{1}{2}\mathbf{N}$ (see Section 3.5). We have then

$$\begin{aligned} iJ_3 D^j_{m_1 m_2} &= m_1 D^j_{m_1 m_2}, \\ iJ_\pm D^j_{m_1 m_2} &= \sqrt{(j \mp m_1)(j \pm m_1 + 1)} D^j_{m_1 \pm 1, m_2}, \end{aligned}$$

where $J_\pm = J_1 \pm iJ_2$. Denoting $\sigma_\pm = \sigma_1 \pm i\sigma_2$ we can write $2D = 2\sigma_3 J_3 + \sigma_+ J_- + \sigma_- J_+$ and we can check that the function

$$\psi_\pm^{j m_1 m_2} = \begin{pmatrix} D^j_{m_1 m_2} \\ \alpha_\pm D^j_{m_1 - 1, m_2} \end{pmatrix}, \; j > 0$$

is an eigenvector of D with eigenvalue

$$\lambda_+ = ij, \; \lambda_- = -i(j+1)$$

when $\alpha_\pm = [(\frac{1}{2} - m_1) \pm (j + \frac{1}{2})] \cdot [(j + m_1)(j - m_1 + 1)]^{-\frac{1}{2}}$. Note that the set of eigenvalues consists of all points $\frac{i}{2}\mathbf{Z}$ *except the point* $-\frac{i}{2}$. The multiplicity of the eigenvalue $\lambda_\pm(j)$ is $(2j+1)^2$ for $j > 0$ but the multiplicity of $\lambda = 0$ is 2 (the eigenvectors being the constant spinors on S^3).

We shall need later the following generalization of the Dirac operator. Let E be a spin bundle over M and let F be some other vector bundle over M with structure group G and connection ω. The tensor product bundle $E \otimes F$ has a connection which is given in terms of a local trivialization through the covariant differentiation $\psi \mapsto (\nabla_\mu + A_\mu)\psi$, where ∇_μ corresponds to the spin connection and A is the $\mathbf{g}$ valued local one-form on M defining the connection ω. A_μ acts only on the second factor of the section $\psi = \psi_E \otimes \psi_F$ of $E \otimes F$. The Dirac operator in the present setting is defined as

$$D = \gamma_i h^{(i)\mu}(\nabla_\mu + A_\mu). \tag{5.2.9}$$

Exercise 5.2.10. Check that (5.2.9) indeed gives a well-defined operator in the space of sections of $E \otimes F$, i.e., $\psi \mapsto D\psi$ does not depend on the gauge choice.

5.3. The determinant of a Dirac operator

The massive Dirac operator

In this subsection we want to clarify what is meant by the determinant of the infinite-dimensional operator $iD + m_0$ in a space $\Gamma(E \otimes F)$ of sections of an extended spin bundle. Here m_0 is a real constant, the mass of the Dirac particle. The reason why the determinant is important in physical applications is that it can be interpreted as *the effective action* for a coupled Dirac-Yang-Mills system. Let us first consider a simple finite-dimensional example. Let T be a positive linear operator in the Euclidean space $\mathbf{R}^n$. Let us compute the integral

$$I(T) = \int e^{-(x,Tx)} d^n x.$$

After the diagonalization $T = S\text{diag}(\lambda_1, \ldots, \lambda_n)S^{-1}$ and the change of variables $y = S^{-1}x$ the integral becomes a product of the Gaussian integrals $\int \exp(-\lambda_i y_i^2) dy_i$ and thus we get

$$I(T) = \prod \sqrt{\frac{\pi}{\lambda_i}} = (\pi)^{n/2} \cdot (\det T)^{-1/2}. \tag{5.3.1}$$

This computation applies only when T is positive. However, we want to apply an infinite-dimensional generalization of (5.3.1) to the Dirac operator which has both positive and negative eigenvalues. We would like to make sense of the integral

$$I = \int e^{-\int \psi^*(iD+m_0)\psi d^n x} d\psi. \tag{5.3.2}$$

When the Dirac operator D is parametrized by a vector potential A this is the effective action $I(A)$ describing a Dirac particle in an external gauge field. In quantum theory the Dirac field ψ is supposed to describe fermions. This means that the components of the spinor ψ should *anti-commute* among themselves instead of commuting like the components of ordinary vectors. Thus the correct finite-dimensional analog of (5.3.2) is not the integral (5.3.1) but something like

$$\int_{\psi^*,\psi\in\mathbf{C}^n} e^{-\psi^* T\psi} \tag{5.3.3}$$

where ψ_i^*, ψ_i are elements of a Grassmann algebra. The top element (element of highest degree) in the Grassmann algebra is $\psi_1^* \dots \psi_n^* \psi_1 \dots \psi_n$. We shall *define* the integral (5.3.3) as the coefficient of the top term in the expansion of the exponential; this is

$$\begin{aligned}\frac{(-1)^n}{n!}\sum(\psi_{i_1}^* T_{i_1 j_1}\psi_{j_1})\dots(\psi_{i_n}^* T_{i_n j_n}\psi_{j_n})& \\ = -\sum \psi_1^* \dots \psi_n^* T_{1j_1}\dots T_{nj_n}\psi_{j_1}\dots\psi_{j_n}& \\ = -\det T\,\psi_1^*\dots\psi_n^*\psi_1\dots\psi_n.&\end{aligned}$$

Motivated by the finite-dimensional example we define the effective action $I(A)$ to be the determinant of the Dirac operator. However, we still have to define what we mean by $\det(iD + m_0)$. The determinant of a linear operator T in a Hilbert space is well-defined if $T = 1 + S$ where S is a trace-class operator, i.e., the sum of the eigenvalues λ_i of S form an absolutely convergent series. In that case the determinant of T is $\prod(1 + \lambda_i)$ and this converges. In the case of a Dirac operator the eigenvalues do not even form a bounded sequence.

We must introduce a *regularized determinant.* There is a variety of ways to regularize the infinite product of the eigenvalues. We would like to have some sort of continuity of the determinant: The determinant should be expressible as a function $f(\lambda)$ of the set of eigenvalues $\lambda = \{\lambda_k\}$ of $iD + m_0$ such that f is continuous in each argument λ_i (of course f must be such that its value does not depend on the order of the arguments). In addition, we require that f is zero if and only if one of the eigenvalues λ_i vanishes. One simple choice for f is the "cutoff regularization" of the determinant,

$$\det{}^{(M)}(iD + m_0) = \prod_{|\lambda_i|<M} \lambda_i,$$

where $M > 0$ is a "large" cutoff parameter. The obvious disadvantage of this determinant is that it gives no information about the large part of the spectrum. One can introduce the cutoff in a smoother way by choosing a function h on the reals such that $h(x) = x$ for $|x| < M$ and $h(x) \to 1$ very fast as $|x| \to \infty$. A regularized determinant can then be defined as

$$\det{}^h(iD + m_0) = \prod h(\lambda_i).$$

For an elliptic differential operator on a compact manifold such that the real parts of the eigenvalues λ_i are bounded from below (e.g., the Laplace operator) there is a slightly more sophisticated (and more symmetric) way to define the determinant, by the so-called ζ function regularization. Consider the function

$$\zeta(s, \lambda) = \sum_{Re\lambda_i > 0} \lambda^{-s}$$

One can show that this function is holomorphic in the half-plane $\mathrm{Re}\, s > s_0$ for large enough s_0. Furthermore, it extends holomorphically to a regular function at the point $s = 0$. Let $\lambda_1, \ldots, \lambda_p$ be the set of eigenvalues with $\mathrm{Re}\, \lambda_i \leq 0$. The ζ-regularized determinant is then defined as

$$\lambda_1 \lambda_2 \ldots \lambda_p \exp\left[-\frac{d}{ds}\zeta(s, \lambda)\right]\Big|_{s=0} \tag{5.3.4}$$

It is a simple exercise to show that in the finite-dimensional case (5.3.4) gives the usual determinant.

The chiral case

Let the dimension n of M be even, $n = 2m$. The massless Dirac operator ($m_0 = 0$) D anticommutes with the chirality operator γ_{2m+1}. Let S_+ (respectively, S_-) be the subspace of positive (respectively, negative) chirality spinor fields; *a chiral field*, or a *Weyl spinor field*, is a Dirac spinor field which takes values only in one of the two different eigenspaces of the chirality operator. By the remark above, the Dirac operator maps S_+ to S_- and S_- to S_+. We can thus write $D = D_+ + D_-$, with $D_\pm : S_\pm \to S_\mp$. $D_\pm$ are the *Weyl operators* on M. In the massless case we can study the coupling of a vector potential A independently to either of the chiral fields and therefore it would be natural to define the determinant (effective action) for the Weyl operators. However, besides the regularization, we run immediately into a problem: The Weyl operators are operators between *different spaces* $S_\pm$ and therefore it does not make sense to speak about the eigenvalue problem for $D_\pm$.

There is a way out of this dilemma. Instead of trying to define the determinant of D_+ directly we define it relative to a fixed operator T :

$S_- \to S_+$. For example, if the vector bundle F is trivial (we keep the notation of Section 5.2) we could take $T = D_-^0$ as the free Weyl operator determined by the flat connection $A = 0$ in F. Since T does not depend on the vector potential A the determinant $\det TD_+(A)$, considered as a function of A, is a good replacement for the ill-defined determinant $\det D_+(A)$. Of course, in order that the determinant does not vanish identically, we must choose T such that it has no zero modes.

Example 5.3.5. Let $M = S^1 \times S^1$ with the flat metric and let $F = M \times \mathbf{C}^N$. As the gauge group we take $G = U(N)$ with the obvious action in F. The fibers of $S_\pm$ can now be identified as $\mathbf{C} \otimes \mathbf{C}^N = \mathbf{C}^N$. The free Weyl operator is now

$$D_-^0 = \frac{\partial}{\partial \theta_1} + i \frac{\partial}{\partial \theta_2}$$

and its only zero modes on the torus are the constant functions. Adding to D_-^0 a small "mass" $0 < \epsilon$, $T = D_-^0 + \epsilon$, we obtain an operator $T : S_- \to S_+$ without zero modes. We have

$$TD_+(A) = \partial_1^2 + \partial_2^2 + \text{lower order terms}.$$

The spectrum of this elliptic operator behaves essentially like the spectrum of the Laplace operator (for large eigenvalues). We can apply for example the ζ function regularization to define the determinant of $TD_+(A)$. The actual computation of the determinant is a complicated matter and no simple closed form of it is known.

Remark. If it happens that the Weyl operator D_-^0 does not have zero modes then one can use instead of $D_-^0 D_+$ the operator $(D_-^0)^{-1} D_+ = 1 + (D_-^0)^{-1} \gamma^\mu A_\mu$. The advantage with this operator is that it has a generalized determinant $\det_p$ for high enough $p > 1$; we shall discuss these determinants in detail in the next chapter.

For a fixed metric on M but for a variable connection ω in F the regularized determinant $\det_{reg}$ of the Dirac operator is a complex valued function in the space $\mathcal{A}$ of $\mathbf{g}$ valued connections in F. Next we want to discuss the *gauge dependence* of this function. In order to avoid some unessential technical complications we shall assume that the bundle F is trivial. Now a connection in F is a globally defined $\mathbf{g}$ valued 1-form A on M. A gauge transformation is a mapping of A onto itself, $A \mapsto A^g = gAg^{-1} + dgg^{-1}$, where $g : M \to G$ is a smooth function. Under a gauge transformation g a Dirac spinor field ψ is mapped to the field $[T(g)\psi](x) = g(x)\psi(x)$. Since the Dirac operator is constructed in terms of covariant derivatives we have

$$T(g)D(A)T(g^{-1}) = D(A^g)$$

and therefore, if ψ is an eigenvector of $D(A)$ corresponding to the eigenvalue λ, then $T(g)\psi$ is an eigenvector of $D(A^g)$ corresponding to the same eigenvalue. The spectra of $D(A)$ and $D(A^g)$ are identical and consequently the regularized determinants of $D(A)$ and $D(A^g)$ are the same. The Dirac determinant is thus really a function in the quotient $\mathcal{A}/\mathcal{G}$, the space of vector potentials modulo gauge transformations.

The situation is different with the Weyl operator $D_+(A)$ [or $D_-(A)$]. The Weyl operator $D_+(A)$ between S_+ and S_- transforms equivariantly with respect to the gauge transformations but we have to consider the operator $TD_+(A)$ and this does not transform equivariantly in general. Thus there is no reason why the "determinant" of the Weyl operator should be gauge invariant. In physics literature the variation of the determinant under the group of gauge transformations is called the *chiral anomaly.* The manner in which the determinant transforms under the gauge group depends on the regularization chosen. There is an important invariant which does not depend on the regularization, namely, *the cohomology class* of the chiral anomaly: Let $f(A)$ be the Weyl determinant, defined by some regularization. Then

$$c(A;g) = \frac{f(A^g)}{f(A)}$$

is a function on $\mathcal{A}\times\mathcal{G}$ taking values in the multiplicative group $\mathbf{C}^\times$. The quotient is well-defined since $D_+(A^g)$ has precisely the same zero-modes as the operator $D_+(A)$. By definition, c is a one-cocycle of $\mathcal{G}$ with values in the group of $\mathbf{C}^\times$ valued functions on $\mathcal{A}$,

$$c(A^g;g')c(A;g) = c(A;gg').$$

It is the class $[c] \in H^1(\mathcal{G})$ which is regularization independent. We shall sketch the proof below.

The Dirac determinant bundle

A cocycle c defines a line bundle DET over $\mathcal{A}/\mathcal{G}$, the *determinant bundle*; the sections of the dual determinant bundle DET^* are complex valued functions ψ on $\mathcal{A}$ such that

$$\psi(A^g) = \psi(A)\,c(A;g). \tag{5.3.6}$$

The equivalence classes of line bundles correspond to cohomology classes of cocycles c. In order to classify these line bundles we shall study the topology of the base space $\mathcal{A}/\mathcal{G}$. The topology depends on the topology of the manifold M but we shall consider here only the simple case when $M = S^n$. Choose a point $N \in S^n$ and a disk $D' \subset S^n$ with N as the center. For any $A \in \mathcal{A}$ there is a smooth $f_A : S^n \to G$ such that $f(N) = 1$ and such that the radial component of the gauge transformed vector potential A^f vanishes in the disk D'. The function f_A is uniquely defined *inside of the disk* D'. The complement of D' in S^n is also a disk, to be denoted by D. Denote by h_A the restriction of f_A to the disk D. Now h_A is not uniquely defined; it depends on the choice of f_A. However, the class $[h_A]$, h_A modulo the group $\mathcal{G}_D$ of gauge transformations $g : D \to G$, with $g = 1$ on the boundary of D, is uniquely defined. We have thus a mapping $\mathcal{A} \to \mathcal{A}_D/\mathcal{G}_D$, $A \mapsto q(A) = A^{h_A}|_D \bmod \mathcal{G}_D$.

Let $g \in \mathcal{G}$ and $A \in \mathcal{A}$. Now $f_{A^g} = f_A g^{-1}$ but $q(A^g) = q(A)$ and therefore the map $q : \mathcal{A} \to \mathcal{A}_D/\mathcal{G}_D$ gives also a well-defined map $q : \mathcal{A}/\mathcal{G} \to \mathcal{A}_D/\mathcal{G}_D$. This map is a *homotopy equivalence.* This means that there is a map p in the opposite direction such that qp and pq can be continuously deformed to the identity maps. The map p is constructed as follows. Let B be a vector potential in the disk D. Extend this to a vector potential A in S^n by defining a potential B' in D' as $B'(r,\theta) = rB(1,\theta)$, where θ denotes collectively the angular coordinates in the disk D' and the boundary values of B' are defined to be equal to the boundary values of B on the common boundary. Now A is defined by patching together B and B'. Suppose that $C = B^h$ for some $h \in \mathcal{G}_D$. The boundary values of C and B are equal and therefore the vector potential $p(B) = A$ is obtained from $p(C)$ by a gauge transformation g which is equal to 1 in D' and coincides with h in D'. It follows that p gives really a map from $\mathcal{A}_D/\mathcal{G}_D$ to $\mathcal{A}/\mathcal{G}$. We shall show that pq is homotopic to the identity map on $\mathcal{A}/\mathcal{G}$. Consider the following one-parameter family of mappings on $\mathcal{A}/\mathcal{G}$:

$$\rho_t(A) = \begin{cases} A^{f_A}|_D \text{ on } D \\ tA^{f_A}|_{D'} + (1-t)rA^{f_A}|_{\partial D} \text{ on } D' \end{cases}.$$

We have $\rho_1 = id$ and $\rho_0 = pq$. This is the required homotopy.

Exercise 5.3.7. Show that qp is homotopic to the identity mapping on $\mathcal{A}_D/\mathcal{G}_D$.

Next we shall show that $\mathcal{A}_D/\mathcal{G}_D$ is homotopic to $S^{n-1}G$. (In general $S^m G$ is disconnected; we shall reserve here this notation for the connected component of identity.) For any vector potential A in the disk D we choose a mapping $u_A : D \to G$ such that A^{u_A} is in the radial gauge (the radial component of the transformed potential vanishes) and that $u_A(0) = 1$, where the center of D is denoted by 0. The function u_A is uniquely defined. A gauge transformation of A by an element g

of $\mathcal{G}_D$ multiplies the map u_A by g^{-1} from the right and it follows that $A \mapsto u_A|_{\partial D}$ defines a map $i : \mathcal{A}_D/\mathcal{G}_D \to S^{n-1}G$. This is a homotopy equivalence. The homotopy inverse j is obtained as

$$j(f) = d\hat{f}\hat{f}^{-1}, \text{ with } \hat{f} \text{ any extension of } f \text{ to } D.$$

The family $\rho_t(A) = tA + (1-t)du_A u_A^{-1}$, $0 \le t \le 1$, of mappings of $\mathcal{A}_D/\mathcal{G}_D$ onto itself is a homotopy connecting ji to the identity mapping.

Exercise 5.3.8. Show that ij is homotopic to the identity on $S^{n-1}G$.

Combining the homotopy equivalences q and i we obtain a homotopy equivalence $\mathcal{A}/\mathcal{G} \simeq S^{n-1}G$. It follows that line bundles over $\mathcal{A}/\mathcal{G}$ can be classified in terms of line bundles over $S^{n-1}G$; the (equivalence classes of) line bundles over the former space are obtained as pull-backs of line bundles over the latter space. The problem reduces then to a study of the second de Rham cohomology of $S^{n-1}G$. The cohomology depends both on n and the group G. If $n=2$ the first homotopy group of $S^{n-1}G$ is isomorphic with the second homotopy group of G which is always zero. By the Hurewicz theorem the second cohomology of $S^{n-1}G$ is now isomorphic with the second homotopy which in turn is equal to the third homotopy of G. For any compact simple Lie group $\pi_3 G = \mathbf{Z}$. The classification of the determinant bundles in the case of $M = S^2$ corresponds to the classification of the central extensions of the loop group S^1G. If $n=4$ then the first homotopy group of $S^{n-1}G$ is equal to the fourth homotopy of G which is zero if G is a simple compact group of rank bigger than one. In that case again the second cohomology of $S^{n-1}G$ is equal to the second homotopy which is equal to the fifth homotopy group of G which is equal to $\mathbf{Z}$ for $G = SU(N), N > 2$.

We have considered here only the case when the bundle F is trivial. However, it is not difficult to show that in the general case $\mathcal{A}/\mathcal{G}$ is homotopic to an appropriate component of the disconnected group of *all* smooth maps $S^3 \to G$.

In fact, we already have an explicit construction of the nontrivial line bundles over S^3G in Section 4.3. Over the identity component S^3_0G of S^3G the sections of the line bundle are by definition complex valued functions ψ on D_4G which satisfy the condition

$$\psi(fg) = \psi(f) \cdot e^{2\pi i \omega(f;g)}$$

where ω is a 1-cocycle for the natural right action of S^4G on D_4G (we have again identified S^4G as the subgroup of D_4G consisting of functions g which are equal to 1 on the boundary). The cocycle is given by

$$\omega(f;g) = \int c^{1,5}(f^{-1}df; g),$$

where the integral is computed over a five-dimensional disk D_5 with boundary S^4. The sections of the pull-back bundle DET^* over $\mathcal{A}/\mathcal{G}$ are functions on $\mathcal{A}$ which satisfy the cocycle condition

$$\psi(A^g) = \psi(A) \cdot \exp\left[2\pi i \int c^{1,5}(A;g)\right], \quad g \in \mathcal{G} = S^4G.$$

The first Chern class c_1 of the line bundle DET^* is the generator of the cohomology group $H^2(\mathcal{A}/\mathcal{G};\mathbf{Z})$. (Elements of the second cohomology group with integral coefficients can be thought of here as equivalence classes of closed two-forms such that the integral of the form over any compact surface without boundary is 2π times an integer.) The line bundle with Chern class equal to n times this basic Chern class is obtained simply by replacing $c^{1,5}$ by $nc^{1,5}$.

To show that the determinant bundle of the Weyl operator really is twisted one must also do some analysis. But the only thing necessary is to show that the phase of the regularized determinant $\det_{reg}(D_0^* D_A)$ winds around zero for a suitable loop of gauge transformations $A \mapsto A^{g(t)}$, $0 \leq t \leq 2\pi$; let us assume that this is the case. (This follows from an explicit evaluation of the variation of the determinant; see Atiyah and Singer [1984].) Now if there is a function $h : \mathcal{A} \to \mathbf{C}^\times$ such that the gauge variation of the regularized determinant can be written as $h(A^g)/h(A)$ then the phase could not wind around zero along any loop in the space of gauge transformations since the space $\mathcal{A}$ is contractible. This shows that the cocycle $\exp[2\pi i\omega(A;g)]$ is nontrivial. To show that the deteminant bundle corresponds to a value n of the Chern class one has to show that the determinant of the Weyl operator winds exactly n times around zero along a noncontractible loop in $\mathcal{G}$ which generates $\pi_1(\mathcal{G})$.

5.4. On the geometry of the Dirac determinant bundle

Curvature and anomalies

There is a close connection between the chiral anomaly and the curvature of the moduli space $\mathcal{A}/\mathcal{G}$. We shall explain the main results without proofs.

Let M be a compact connected oriented spin manifold with a fixed Riemannian metric g. We assume that the dimension of M is even, say $2n$. Let P be a principal G bundle over M with G a compact Lie group. The group $\mathcal{G}$ of gauge transformations consists by definition of all automorphisms of the bundle P which descend to an identity transformation on the base M. If P is trivial, $P = M \times G$, then the elements of $\mathcal{G}$ are just G valued functions f on M; the action of f in P is then given by $(x,g) \mapsto (x, f(x)g)$. In order that the moduli space $\mathcal{A}/\mathcal{G}$ is a smooth manifold we shall fix a point $p_0 \in P$ and consider only those gauge transformations which leave p_0 fixed. The group $\mathcal{G}$ acts in the space $\mathcal{A}$ of connections in P in the natural way, i.e., through pull-back of the one forms $A \in \mathcal{A}$. We have then an action of $\mathcal{G}$ in $P \times \mathcal{A}$ which commutes with the action of the structure group G in P. We have now a principal $\mathcal{G}$ bundle $P \times \mathcal{A}$ over $\mathcal{L} = (P \times \mathcal{A})/\mathcal{G}$. We can further divide by G (since G commutes with $\mathcal{G}$) and we obtain a principal G bundle $\mathcal{L}$ over $M \times \mathcal{A}/\mathcal{G}$.

Let us define a connection ω in the bundle $\mathcal{L}$. Note first that the space $P \times \mathcal{A}$ has a natural metric: If (p, A) is a point in $P \times \mathcal{A}$ and (t_i, B_i) $(i = 1,2)$ is a pair of tangent vectors at (p, A) then the inner product is

$$
\begin{aligned}
&< (t_1,B_1),(t_2,B_2) > \\
&\qquad = g(h(t_1),h(t_2)) + (v(t_1),v(t_2)) + \int_M (B_1^{\mu}, B_{2\,\mu}) d(vol_M),
\end{aligned}
\tag{5.4.1}
$$

where we have used a fixed invariant inner product $(\cdot,\cdot)$ in the Lie algebra of G and h, v are the horizontal and vertical projections defined by the connection A. The Riemannian metric is written as $g(t_1,t_2) = g^{\mu\nu} t_{1\,\mu} t_{2\,\nu} = t_1^{\mu} t_{2\,\mu}$. The integration can be actually carried out in the base manifold M since the variations B_i of a connection transform homogeneously in gauge transformations, $B_i \mapsto f B_i f^{-1}$. The vertical vectors are identified as elements of $\mathbf{g}$ in the usual way. The metric (5.4.1) descends to a G invariant metric on $\mathcal{L}$: With respect to a local trivialization the sum of the first two terms on the right-hand-side of (5.4.1) is

$$
g(u_1,u_2) + (X_1 - u_{1\mu} A_{\mu}, X_2 - u_{2\mu} A_{\mu})
$$

where $u_1, u_2 \in T_{\pi(p)}M$, X_1, X_2 are in $\mathbf{g}$ and (A_{μ}) is a local one-form on M describing the connection A. In a gauge transformation f the point x is left invariant, X transforms as $X \mapsto f(x) X f(x)^{-1} + t_{\mu} A_{\mu}(x)$, and $A_{\mu} \mapsto f A_{\mu} f^{-1} + \partial_{\mu} f\, f^{-1}$. The difference $X - t \cdot A$ transforms according to the adjoint representation and by the invariance of the inner product in $\mathbf{g}$, (5.4.1) is invariant under $\mathcal{G}$. The invariance under G follows immediately from the equivariantness of the connection forms and the invariance of the inner product in $\mathbf{g}$.

The connection ω in $\mathcal{L}$ is given by declaring that the horizontal subspace at a given point is the orthogonal complement of the vertical subspace with respect to the Riemanniann metric $< \cdot,\cdot >$ on $\mathcal{L}$.

Exercise 5.4.2. Show that the distribution of horizontal subspaces given above really defines a connection.

Let us compute the curvature of the connection ω. Let $(x, [A])$ be a point in $M \times \mathcal{A}/\mathcal{G}$ and let (u_i, B_i) be a pair of tangent vectors at $(x, [A])$. The tangent vectors to $\mathcal{A}/\mathcal{G}$ can be represented by vectors in the space of $\mathbf{g}$ valued one-forms on M by fixing the gauge: For a given representative A of the gauge class $[A]$ a tangent vector is a form B_μ such that $\partial_\mu B^\mu - [A_\mu, B^\mu] = 0$. This is precisely the condition that the form B_μ is orthogonal to the gauge orbit through the point A. Namely, a tangent vector along the gauge orbit is a form $C = [X, A_\mu] + \partial_\mu X$ (where X is a $\mathbf{g}$ valued function on M) and by partial integration

$$\begin{aligned} < B, C > &= \int (B^\mu, [X, A_\mu] + \partial_\mu X) d(vol_M) \\ &= \int (X, [A_\mu, B^\mu] - \partial_\mu B^\mu) d(vol_M). \end{aligned}$$

This has to be zero for all X which implies the *background gauge* condition above. We can split the curvature Ω as $\Omega^{2,0} + \Omega^{1,1} + \Omega^{0,2}$ corresponding to the splitting of the tangent spaces of $M \times \mathcal{A}/\mathcal{G}$ to the tangent space of M and $\mathcal{A}/\mathcal{G}$. It is not difficult to see that

$$\Omega^{2,0}(u_1, u_2) = F(u_1, u_2), \qquad \Omega^{1,1}(u, B) = u_\mu B_\mu$$

where B is a tangent vector at A (in the background gauge) to $\mathcal{A}/\mathcal{G}$ and the u's are tangent vectors at $x \in M$; F is the curvature form corresponding to the vector potential A.

In order to evaluate $\Omega^{0,2}$ we construct the connection form in the $\mathcal{G}$ bundle $\mathcal{A} \to \mathcal{A}/\mathcal{G}$. The vertical subspace at A consists of all one-forms of the type $[X, A_\mu] + \partial_\mu X$, where $X : M \to \mathbf{g}$ is an infinitesimal gauge transformation. The horizontal subspace consists of the vector potentials B in the background gauge with respect to the base point A. The horizontal projection of an arbitrary tangent vector B at A is

$$h(B_\mu) = B_\mu - D_\mu^A \Delta_A^{-1} D_\nu^A B^\nu$$

where $D_\mu^A Z \equiv \partial_\mu Z - [A_\mu, Z]$ and $\Delta_A = D_\mu^A D^{A\mu}$ is the covariant Laplacian. It follows that the connection form with respect to a given local trivialization of the bundle $\mathcal{A}$ is given by

$$B \mapsto \Delta_A^{-1} D_\mu^A B^\mu. \tag{5.4.3}$$

Evaluating this at the point $x \in M$ we get the value of the connection form in $\mathcal{L}$ in the direction of the tangent vector $(0, B)$. The curvature of this gives

$$\Omega^{0,2}(B, B') = \Delta_A^{-1}[B, B'], \tag{5.4.4}$$

where B, B' are in the background gauge.

The curvature formula at hand we can compute the Chern classis of the vector bundle $\mathcal{E} = \mathcal{L} \times_\rho \mathbf{C}^N$ over $M \times \mathcal{A}/\mathcal{G}$, where ρ is a representation of G in $\mathbf{C}^N$. Let us assume for simplicity that $M = S^{2n}$. We shall use some results from Atiyah-Singer index theory. Let $D(A)_\pm$ be the Weyl operators on M defined by a vector potential A. Consider the kernels $\ker D(A)_\pm$. For a given A these are finite-dimensional vector spaces. However, the family of vector spaces $\ker D(A)_\pm$ do not form a vector bundle over $\mathcal{A}/\mathcal{G}$ since the dimension of the kernels may jump. Instead, their formal difference $K = \ker D_+ - \ker D_-$ is defined in the sense of K theory; the difference is not a vector bundle in the ordinary sense but it makes sense to speak about characteristic classes of K. It follows from the families index theorem in Atiyah and Singer [1971] that the Chern classes of K can be evaluated by integrating the Chern character of $\mathcal{E}$ over the first factor M. To be more precise, one has the following theorem [Atiyah and Singer, 1984]:

THEOREM 5.4.5. *The Chern character of K is*

$$ch(K) = \int_M ch(\mathcal{E}).$$

If M is not a sphere then there is a correction to the above formula from a characteristic class associated to the spin bundle of M. In order to illustrate the use of this result we give more explicit formulas in two particular cases. Let $M = S^4$ and $G = SU(N)$ (so that $\operatorname{tr} F = 0$.) The 0:th Chern class of K is simply

$$\frac{-1}{8\pi^2}\int_M \operatorname{tr} F \wedge F$$

which is thus the Chern number computed from the second Chern class of the bundle $P \times_G \mathbf{C}^N$. The first Chern class of K is the 2-form

$$\begin{aligned} c_1(B, B') &= \frac{-i}{24\pi^3}\int_M \operatorname{tr} \Omega^{4,2}(B, B') \\ &= \frac{-i}{24\pi^3}\int_M \epsilon^{\alpha\beta\gamma\delta} \operatorname{tr}\left\{F_{\alpha\beta}F_{\gamma\delta}\Delta_A^{-1}[B^\mu, B'_\mu]\right. \\ &\quad \left. + F_{\alpha\beta}\Delta_A^{-1}[B^\mu, B'_\mu]F_{\gamma\delta} + F_{\alpha\beta}(B_\gamma B'_\delta + B'_\gamma B_\delta)\right\}. \end{aligned}$$

To any multiple $nc_1(K)$ of the first Chern class one can associate a complex line bundle over the parameter space $\mathcal{A}/\mathcal{G}$. The curvature of the line bundle is the differential form $nc_1(K)$. Because of $H^2(\mathcal{A}/\mathcal{G};\mathbf{Z}) = \mathbf{Z}$ [in the case $M = S^4$ and $G = SU(N)$, $N > 2$] these line bundles must be the same as the bundles described by the cocycles c in (5.3.6), corresponding to the chiral anomaly of the Dirac operator. Thus we have a direct relation between the curvature of $\mathcal{A}/\mathcal{G}$ and the chiral anomaly.

The commutator anomaly

There is a close relation between the commutator anomalies (Schwinger terms) in the Hamiltonian approach in odd dimensional space and the chiral anomaly in even dimensional space-time. For understanding the essential features of this correspondence it is sufficient to consider the case of a trivial bundle P. Let the physical space be the sphere S^{2n-1} and space-time the tube $M = S^{2n-1} \times I$ where $I = [t_0, t_1] \subset \mathbf{R}$ is a compact interval. We may interpret the value of a vector potential A at $t = t_0$ as the boundary value of a vector potential in the distant past $t \to -\infty$ in the physical space-time and the value of A at $t = t_1$ as the present observed value. Let DG be the group of smooth gauge transformations $f : M \to G$ which leave the boundary values at $t = t_0$ invariant, i.e., $f = 1$ at $t = t_0$. Elements of DG can be naturally thought of as maps from the $n-$dimensional disk to the gauge group G, the sphere at $t = t_0$ being the center of the disk and the sphere S^{n-1} at $t = t_1$ being the boundary of the disk. For any A in the space $\mathcal{A}$ of smooth vector potentials there is a unique $f \in DG$ such that $A^f = fAf^{-1} + dff^{-1}$ is in the *temporal gauge*, that is, the time component A_t vanishes. Denote by f_A this gauge transformation.

The group DG contains as a subgroup the group $\mathcal{G}$ consisting of gauge transformations g which are equal to one at $t = t_0$ and constant at $t = t_1$. The group $\mathcal{G}$ can be naturally identified as the group of based gauge transformations $S^n \to G$ studied earlier. We have now a diffeomorphism between $\mathcal{A}$ and $\mathcal{P}\mathcal{A}^{n-1} \times DG$, where $\mathcal{A}^{n-1}$ is the space of vector potentials in S^{n-1} and $\mathcal{P}\mathcal{A}^{n-1}$ denotes the space of smooth paths in $\mathcal{A}^{n-1}$ parametrized by $t_0 \leq t \leq t_1$. The diffeomorphism ϕ is given by $A \mapsto (A^{f_A}, f_A)$. Because of this the group DG acts in *two ways* in $\mathcal{A}$, corresponding to the left and right multiplications in the second factor DG. A section of the determinant bundle over $\mathcal{A}/\mathcal{G}$ is a complex valued function ψ on $\mathcal{A}$ which transforms according to the chiral anomaly (5.3.6). The gauge transformations $g \in \mathcal{G}$ act from the right on DG and they commute with the left action of DG on DG. It follows that we might try to define a representation of DG in the space $\Gamma(DET^*)$ of sections of DET^* by $(T(f)\psi)(A) = \psi(f^{-1} \cdot A)$ using the left action. However, this first guess is wrong because the function $\psi'(A) = \psi(f^{-1} \cdot A)$ does not satisfy the anomaly condition (5.3.6). We can cure this problem by defining

$$[T(f)\psi](A) = e^{2\pi i\beta(A;f)}\psi(f^{-1} \cdot A) \tag{5.4.6}$$

for a suitable function γ. In order that $T(f)\psi$ is really a section of DET^* we must have

$$\exp[2\pi i\beta(A^g; f)] \cdot c(f^{-1}A; g) = \exp[2\pi i\beta(A; f)] \cdot c(A; g). \tag{5.4.7}$$

An explicit formula for $c(A;g)$ can in principle be found by the transgression method explained in (4.1.8)-(4.1.10). The equation (5.4.7) has a solution β since for any fixed f the cocycles $(A,g) \mapsto c(A;g)$ and $(A,g) \mapsto c(f^{-1};g)$ must represent the same cohomology class. Instead of trying to solve for (the rather complicated function) β we can use the fact that the determinant bundle is the pull-back of a bundle over DG. Using $\mathcal{A} = \mathcal{P}\mathcal{A}^{n-1} \times DG$ we may think (up to a bundle equivalence) of the sections ψ as functions of $(A, f) \in \mathcal{P}\mathcal{A}^{n-1} \times DG$ such that

$$\psi(A, fg^{-1}) = \psi(A, f)c(f^{-1}df; g). \tag{5.4.8}$$

Since c here does not depend on A we may drop it and look for β as function of f and g only. Instead of writing the general formula we consider the cases $n = 2$ and $n = 4$ as illustration. When $n = 2$ the cocycle c is equal to $\gamma(f, g^{-1}) + C(g^{-1})$, where γ is given by the formula (4.2.1) and $C(g)$ by (4.2.4). One checks that

$$\beta(f; g) = \gamma(f^{-1}, g)$$

is a solution of (5.4.7), when $A = f^{-1}df$. This gives *a projectice* representation of the group DG. By a simple computation,

$$T(f)T(f') = T(ff') \cdot e^{2\pi i\gamma(f,f')}. \tag{5.4.9}$$

We have thus a true representation of the *central extension* of DG, discussed in Section 4.2, in the vector space $\Gamma(DET^*)$. The subgroup N consisting of elements $(g, \exp[2\pi iC(g)]) \in DG \times S^1$ with $g \in \mathcal{G}$ acts trivially in $\Gamma(DET^*)$: For $g \in \mathcal{G}$ we have

$$\begin{aligned}[T(g)\psi](h) &= \psi(g^{-1}h) \cdot e^{2\pi i\gamma(h^{-1},g)} \\ &= \psi(h) \cdot e^{2\pi i[\gamma(h^{-1},g)+\gamma(h,h^{-1}gh)+C(h^{-1}g^{-1}h)]} = \psi(h).\end{aligned}$$

The last equation follows by a simple computation from the defining relations. We can divide out the trivially represented normal subgroup N and we obtain a representation of the central extension $\widehat{LG}$ of the loop group. We conclude that the chiral anomaly manifests itself in the Hamiltonian approach as the anomaly in the commutation relations of the infinitesimal gauge transformations.

In the case $n = 4$ we can use the formula (4.3.2) for the chiral anomaly $c(A; g)$. Using again the homotopy equivalence of $\mathcal{A}/\mathcal{G} \equiv S^3G$ we can replace the variable $A \in \mathcal{A}$ by $f \in D_4G$, $c(A; g) \mapsto c(f^{-1}df; g)$, and $\omega(f; b)$ can be written as $\gamma_3(0; f^{-1}, g)$ where γ_3 is given by (4.3.3) and (4.3.4). Using (4.3.8)-(4.3.12) we can make the following observations. (1) The extension $D_4G \times Map(\mathcal{A}^3, S^1)$ acts in $\Gamma(DET^*)$ through the formula

$$[T(f,\lambda)\psi](h) = \lambda(f^{-1}h)\psi(f^{-1}h) \cdot e^{2\pi i\gamma_3(0;f^{-1},h)}.$$

(2) The normal subgroup N consisting of elements $(g, \exp[2\pi i C_5(g)])$ acts trivially and therefore we obtain a representation of the quotient $\widehat{S^3G}$ in $\Gamma(DET^*)$. (3) On the Lie algebra level there is a Schwinger term (4.3.11) in the commutation relations of the infinitesimal gauge transformations (at the time $t = t_1$).

CHAPTER 6 DETERMINANT BUNDLES OVER GRASSMANNIANS

6.0. Introduction

Denoting by H the Hilbert space of square-integrable Dirac spinor fields on a manifold M, transforming according to a unitary representation ρ of a gauge group G, we have a linear representation of the group $\mathcal{G}$ of gauge transformations in the space H. If ρ is faithful we can consider $\mathcal{G}$ as a subgroup of the general linear group $GL(H)$. By constructing representations of $GL(H)$ we automatically obtain representations of $\mathcal{G}$. It turns out that in the case when the dimension d of M is odd, $\mathcal{G}$ is contained in a smaller group $GL_p \subset GL(H)$ which has the property that it perturbs the subspace $H_+ \subset H$ consisting of eigenvectors of a Dirac operator belonging to positive eigenvalues, by an operator A for which the trace $\text{tr}|A|^{2p}$ exists. The *Schatten index* depends on the dimension of M. The statement above is true when $p \geq (d+1)/2$.

The representation of $\mathcal{G}$ in H has the serious drawback that the spectrum of the Dirac Hamiltonian is not bounded below; this is considered as an unacceptable property in quantum field theory. Following the usual path in QFT we try to transport the representation of $\mathcal{G}$ to the fermionic Fock space. In the case $d = 1$ we may choose $p = 1$ and the Fock space can be realized as the space of holomorphic sections of a complex line bundle DET_1^* over an infinite-dimensional Grassmannian manifold Gr_1. The group GL_1 acts on the manifold Gr_1 and there is an action of *a central extension* $\widehat{GL}_1$ in the bundle DET_1^* which projects on the natural action of GL_1 in Gr_1. The action of the extension is holomorphic and it follows that we have a representation of $\widehat{GL}_1$ in the Fock space. However, when $d > 1$ we must take $p > 1$ and the group GL_p cannot be represented in the Fock space even projectively.

We can define the bundles DET_p^* for any $p \geq 1$ using generalized Fredholm determinants. (We have to define determinants for operators of the type $1 + A$ with $\text{tr}|A|^p < \infty$; the usual determinant is defined only in the case $p = 1$.) There is an extension $\widehat{GL}_p$ of GL_p by an infinite-dimensional Abelian ideal which acts in DET_p^*; but this action is not holomorphic and this is one way to see that the group of gauge transformations does not act in the Fock space when the dimension of space is bigger than one. Nevertheless, we have a representation of $\widehat{GL}_p$ in the space $\Gamma(DET_p^*)$ of all smooth sections. This representation is not unitary with respect to any inner product. In fact, the group $\widehat{GL}_p$ does not have any unitary faithful representations for $p > 1$; this has been shown recently in Pickrell [1988]. It is possible that this reflects some incurable disease of the Hamiltonian approach to quantum field theory in space-time dimension bigger than two. We shall return to these questions in Chapter 12, where a possible way out of this dilemma is proposed.

The embedding $\mathcal{G} \subset GL_p$ defines by pull-back an Abelian extension of $\mathcal{G}$. For cohomological reasons this has to be equivalent (in the cases $M = S^1$ and $M = S^3$) to the extensions we have constructed in Chapter 4. In fact, in the one-dimensional case we obtain precisely the central extension of the loop algebras we are familiar with. In the case $G = SU(N)$ this construction gives the basic representation (central charge $k = 1$) of the affine algebra. In general, we get a finitely reducible representation.

The case $p = 1$ has been extensively discussed in Pressley and Segal [1986] which is warmly recommended for further study. The generalization to the case $p > 1$ is taken from Mickelsson and Rajeev [1988]. The theory of Fredholm modules, which is the starting point of our approach, is a basic tool in the noncommutative geometry of A. Connes. Our discussion is at many points parallel to Connes [1986].

We have included also a discussion on the spin structure of the Grassmannian Gr_1 because the Grassmannian is in many ways the simplest infinite-dimensional manifold with interesting topological and geometric structure; also it appears in certain contexts as a "universal moduli space" (see Chapter 12). The construction of a Dirac operator on Gr_1 is also closely related to the construction of quantized supersymmetric sigma models in 1+1 space-time dimensions; this is because of the fact that loop groups can be embedded in Gr_1 and the loop groups are the configuration spaces for the so-called principal sigma models. We shall return to the discussion of sigma models in Chapter 10.

We complete this chapter by a discussion on a generalization of the Plücker embedding to the Grassmannians Gr_p; the case $p = 1$ has been studied by Pressley and Segal. The Plücker embedding for infinite-dimensional Grassmannians was first used by M. Sato and Y. Sato for constructing solutions of the KP hierarchy. We shall return to these matters in Chapter 11. It is an open question whether the machinery of this chapter can be used to solve some categories of "exactly solvable systems" in higher space dimensions in the same way as the theory of group actions on DET_1^* is used to solve the KP equation (in one space dimension).

6.1. Embedding S^dG in the general linear group GL_p modelled by Schatten ideals

For any pair (H_1, H_2) of Hilbert spaces and an integer $p \geq 1$ we denote by $L_p = L_p(H_1, H_2)$ the space of linear operators $A : H_1 \to H_2$ such that

$$\| A \|_p^p = \text{tr}(A^* A)^{p/2} < \infty.$$

In particular, L_1 is the space of trace class operators when $H_1 = H_2$ and L_2 consists of Hilbert-Schmidt operators. Each L_p is a complete linear space with respect to the norm $\| \cdot \|_p$. The basic properties of the *Schatten ideals* which we shall use are

(1) $BAC \in L_p(H_1, H_2)$, for $A \in L_p(H_1, H_2)$, $B \in L(H_2)$, $C \in L(H_1)$
(2) if $A \in L_p(H_1, H_2)$ and $B \in L_q(H_2, H_3)$ then $BA \in L_r(H_1, H_3)$ with $\frac{1}{r} = \frac{1}{q} + \frac{1}{p}$
(3) $L_1 \subset L_2 \subset L_3 \subset \ldots$ and each L_p is dense in L_q with respect to the L_q norm when $p < q$

where $L(H)$ denotes the space of all bounded linear operators in H. For proofs we refer to Simon [1979].

Let $H = H_+ \oplus H_-$ be a complex separable Hilbert space, where $H_\pm$ is a pair of closed infinite-dimensional subspaces. For any $p \geq 1$ let GL_p denote the group consisting of invertible bounded linear operators of the form

$$g = \begin{pmatrix} a & b \\ c & d \end{pmatrix},$$

where $a : H_+ \to H_+$, $d : H_- \to H_-$, $c : H_+ \to H_-$ and $b : H_- \to H_+$ are linear operators such that $b, c \in L_{2p}$. (That GL_p is really closed under multiplication and matrix inversion follows from the fact that the product of an element of L_{2p} and a bounded operator is in L_{2p}.) Let $\begin{pmatrix} u & v \\ x & y \end{pmatrix}$ be the inverse of g. Then we have $au = 1 - bx$. Since bx is compact, the right-hand-side is a Fredholm operator. It follows that au, and thus also a, has a finite-dimensional kernel and cokernel so that a is a Fredholm operator. From the equation $dy = 1 - cv$ follows similarly that d is Fredholm.

We define a metric topology in GL_p by

$$d(g, g') = \| a - a' \| + \| d - d' \| + \| b - b' \|_{2p} + \| c - c' \|_{2p} .$$

The usual operator norm has been used for the diagonal blocks. With this topology GL_p has a Banach-Lie structure, modelled by the Banach space $L(H_+) \oplus L_{2p}(H_+, H_-) \oplus L(H_-) \oplus L_{2p}(H_-, H_+)$. The chain of embeddings of the Schatten ideals leads to

$$GL_0 \subset GL_1 \subset \cdots \subset GL_\infty,$$

where each GL_p is dense in GL_q for $p \leq q$; the group GL_0 consists of operators g such that the blocks b, c are finite-rank operators.

Next we discuss an embedding of the gauge group $Map(M, G)$, studied in Chapter 4, to the general linear group GL_p when $p \geq \frac{1}{2}(d+1)$ and $d = \dim M$ is an *odd integer* and M is a compact spin manifold (This discussion is based on Pressley and Segal [1986] and Connes [1986].) Let V be a complex finite-dimensional inner product space carrying a linear representation ρ of the group G. Let E be the tensor product of the trivial vector bundle $M \times V$ and the spin bundle over M. Let $H = L^2(E)$ denote the completion of the space of smooth sections of E with respect to the L^2 inner product

$$(\psi, \psi') = \int_M (\psi(x), \psi'(x))dx.$$

The group $Map(M, G)$ acts linearly in $L^2(E)$ through

$$[T(f)\psi](x) = \rho(f(x)) \cdot \psi(x),$$

where $f \in Map(M, G)$. If the representation ρ is unitary then also the representation T of $Map(M, G)$ is unitary. We could also consider a more general vector bundle than the trivial bundle $M \times V$ by replacing the group $Map(M, G)$ by the group of automorphisms of the vector bundle, but we shall avoid any unessential complications and stick to the case $M \times V$.

Let D be the *massless free Dirac operator* on M. In local coordinates,

$$D = \sum_{\mu=1}^{d} \sum_{i=1}^{d} \gamma^i h_i^\mu \frac{\partial}{\partial x^\mu},$$

where $x^1, \ldots, x^d$ are the local coordinates, the h_i^μ's are the components of a local orthonormal frame $\{h_1, \ldots, h_d\}$ in the tangent bundle, and $\gamma^1, \ldots, \gamma^d$ are the Dirac matrices acting on the spinor indices,

$$\gamma^i \gamma^j + \gamma^j \gamma^i = 2\delta_{ij}.$$

We decompose H into two complementary subspaces $H_\pm$. The space H_+ (respectively, H_-) consists of linear combinations of the eigenvectors of D corresponding to non-negative (respectively, negative) eigenvalues.

PROPOSITION 6.1.1. *The operators $T(f)$ belong to GL_p for any smooth function $f : M \to G$ when $p \geq \frac{1}{2}(d+1)$.*

PROOF: This is essentially a problem of studying the "ultraviolet" properties of the operators $T(f)$ and the global topology or the metric structure of M is not so important, as we shall see below. Therefore, in order to simplify the discussion and to make the basic idea more transparent we shall carry out the proof only in the case when $M = S^1 \times S^1 \times \cdots \times S^1$ (d times) is the torus. (For the general case, see, e.g., Mickelsson and Rajeev [1988].) We denote $\epsilon = \begin{pmatrix} 1 & 0 \\ 0 & -1 \end{pmatrix}$ on $H_+ \oplus H_-$. We write $T(f) = \begin{pmatrix} a & b \\ c & d \end{pmatrix}$, so that

$$[\epsilon, T(f)] = 2\begin{pmatrix} 0 & b \\ -c & 0 \end{pmatrix}$$

and therefore the problem is to prove that $\| [\epsilon, T(f)] \|_{2p} < \infty$. Denote by ψ_k ($k \in \mathbf{Z}^d$) the Fourier components of the Dirac field $\psi : M \to \mathbf{C}^d \otimes V$. Then

$$(D\psi)_k = \not{k}\psi_k,$$

where $\not{k} = \sum_{i=1}^d \gamma^i k_i$, and

$$\begin{aligned} (\epsilon\psi)_k &= \frac{\not{k}}{|k|}\psi_k, \qquad & k \neq 0 \\ &= \psi_k, & k = 0. \end{aligned}$$

Setting

$$f_k = \int e^{-ik\cdot x}\rho(f(x))dx$$

we have

$$(T(f)\psi)_k = \sum_q f_{k-q}\psi_q.$$

Thus

$$([\epsilon, T(f)]\psi)_k = \sum_q \left(\frac{\not{k}}{|k|} - \frac{\not{q}}{|q|}\right) f_{k-q}\psi_q.$$

Proving $T(f) \in GL_p$ is thus reduced to showing that

$$C = \sum_{k,q} \text{tr}\left[f_{k-q}\left(\frac{\not{k}}{|k|} - \frac{\not{q}}{|q|}\right)^2 f^*_{k-q}\right]^p < \infty,$$

where "tr" means the trace in the finite-dimensional space V. By the algebra of the Dirac matrices we have

$$\left(\frac{\not k}{|k|} - \frac{\not q}{|q|}\right)^2 = 2\left(1 - \frac{k \cdot q}{|k||q|}\right).$$

Redefining $k \to k + q$ in the summation we obtain

$$C = \sum_k \mathrm{tr}(f_k^* f_k)^p \sum_q \left[1 - \frac{(k+q)\cdot q}{|k+q||q|}\right]^p.$$

Let us first look at the second summation, $S_p(k) = \sum_q [1 - \frac{(k+q)\cdot q}{|k+q||q|}]^p$. Since each term in this sum is ≤ 1 and the sum $\sum \mathrm{tr}(f_k^* f_k)^p$ converges, we need to worry only about the "ultraviolet" behavior $q \to \infty$ of the sum $S_p(k)$ because any finite number of terms will produce a convergent sum in k. Now

$$1 - \frac{(k+q)\cdot q}{|k+q||q|} \sim -\frac{(q \cdot k)^2}{|q|^4} + \frac{k^2}{2|q|^2} + O(1/|q|^3)$$

and therefore for large q we get

$$\begin{aligned} S_p(k) &\sim |k|^{2p} \sum_q \left(\frac{1}{|q|^2}\right)^p \sim |k|^{2p} \int \frac{1}{|q|^{2p}} d^d q \\ &\sim |k|^{2p} \int^{\infty} |q|^{-2p} |q|^{d-1} d|q|. \end{aligned}$$

The last integral is convergent at infinity when $2p > d$. In this case

$$C \sim \sum_k \mathrm{tr}(k^2 f_k^* f_k)^p.$$

The sum is convergent for smooth functions f because $|f_k|$ decreases faster than any power of $|k|$ as $|k| \to \infty$.

Exercise 6.1.2. Show that the mapping $T : Map(M, G) \to GL_p$ is a *continuous* homomorphism in the case when M is the torus, $2p > d$.

We shall now study the topological properties of the groups GL_p.

PROPOSITION 6.1.3. *Let $\mathcal{F}(H_+)$ be the space of Fredholm operators in H_+. Then the map $\Lambda : GL_p \to \mathcal{F}(H_+)$, $\Lambda(g) = a$ is a homotopy equivalence. In particular, the group GL_p decomposes to connected* *components GL_p^n labelled by the Fredholm index of a ($= -\text{ind}\, d$), $\text{ind}\, a =$ dim $\ker a -$ dim $\text{coker}\, a$.*

PROOF: The subgroup B of GL_p consisting of all elements of the form

$$\begin{pmatrix} 1 & b \\ 0 & d \end{pmatrix}$$

is contractible, therefore the canonical projection $GL_p \to GL_p/B$ is a homotopy equivalence. The projection can be written as

$$\begin{pmatrix} a & b \\ c & d \end{pmatrix} \mapsto \begin{pmatrix} a \\ c \end{pmatrix},$$

so GL_p/B is an open subset of $\mathcal{F}(H_+) \times L_2(H_+, H_-)$. Since $L_2(H_+, H_-)$ is a vector space, $\left(\begin{smallmatrix} a \\ c \end{smallmatrix}\right) \mapsto a$ is also a homotopy equivalence. For the second part note that $g_t = \left(\begin{smallmatrix} a & tb \\ tc & d \end{smallmatrix}\right)$ is a Fredholm operator for all $0 \leq t \leq 1$ and so $\text{ind}\, g_0 = \text{ind}\, g = 0$. It follows that $0 = \text{ind}\, a + \text{ind}\, d$.

We shall need also another infinite chain of linear groups,

$$GL^0 \subset GL^1 \subset \cdots \subset GL^\infty,$$

where GL^p consists of all invertible linear operators A in a given infinite dimensional Hilbert space such that $A - 1 \in L_p$. The topology in GL^p is defined by the metric $d(g, g') = \| g - g' \|_p$. Obviously GL^p is a Banach-Lie group modelled by L_p. There is a close relation between the groups GL_p and GL^p. If $g \in GL_p^0$ then $\text{ind}\, a = 0$ and therefore $a = q + t$, where q is an invertible operator and t is an operator of finite rank. Consequently, for any $g \in GL_p^0$ there is an invertible operator $q \in GL(H_+)$ such that $aq^{-1} - 1 \in L_p$. We define the group

$$\mathcal{E}_p = \{(g, q) \mid g \in GL_p^0,\ q \in GL(H_+),\ aq^{-1} - 1 \in L_p\} \subset GL_p \times GL(H_+).$$

The group multiplication in $\mathcal{E}_p$ is $(g, q)(g', q') = (gg', qq')$ but the topology is defined by the norm

$$\| (g, q) \| = \| a \| + \| d \| + \| b \|_{2p} + \| c \|_{2p} + \| a - q \|_p .$$

Exercise 6.1.4. Show that $\mathcal{E}_p$ is closed under multiplication.

The group GL^p acts from the right on $\mathcal{E}_p$ by $(g, q) \cdot t = (g, qt)$. The quotient $\mathcal{E}_p / GL^p$ is GL_p^0. Since the right action is free, $\mathcal{E}_p$ can be viewed as a principal GL^p bundle over GL_p^0. From the next proposition follows that

$$\pi_i(GL_p^0) \cong \pi_{i-1}(GL^p),\ i = 1, 2, \ldots$$

On the other hand, according to Palais [1965],

$$\pi_i(GL^p) \cong \pi_i(GL(n, \mathbf{C})),\ i < 2n.$$

Thus $\pi_0(GL_p) = \mathbf{Z}$, $\pi_1(GL_p^0) = 0$, $\pi_2(GL_p) = \mathbf{Z}$. From Hurewicz' theorem follows that $H^2(GL_p^0, \mathbf{Z}) = \mathbf{Z}$, so inequivalent complex line bundles (and principal $\mathbf{C}^\times$ bundles) over GL_p^0 are classified by integers (the value of the first Chern class).

PROPOSITION 6.1.5. *The group $\mathcal{E}_p$ is contractible.*

PROOF: The proof is essentially same as for Proposition 6.6.2 in Pressley and Segal [1986], only L_1 is replaced by L_p.

6.2. The determinant bundle over Gr_p

The Grassmannian Gr_p and the Stiefel manifold St_p

Let $m < n$ be positive integers. The *complex Grassmannian manifold* $Gr(m, n)$ is the set of m-dimensional linear subspaces of $\mathbf{C}^n$. The differentiable and topological structure of $Gr(m, n)$ is defined by writing $Gr(m, n) = GL(n, \mathbf{C})/B$, where B is the stability subgroup at some point in $Gr(m, n)$. For example, taking V to be the plane spanned by the m first basis vectors in the standard basis of $\mathbf{C}^n$, then the stability subgroup B at V consists of block triangular matrices of the type

$$\begin{pmatrix} a & b \\ 0 & d \end{pmatrix},$$

where a is any invertible $m \times m$ matrix, b is any $m \times (n - m)$ matrix and d is any invertible $(n - m) \times (n - m)$ matrix. As a quotient of two complex groups $Gr(m, n)$ is naturally a complex manifold of dimension $m \times (n - m)$.

In this section we shall study certain infinite-dimensional generalizations of $Gr(m, n)$. Let $H = H_+ \oplus H_-$ as in previous section. Let B_p be the subgroup consisting of the block triangular operators in GL_p with $c = 0$. We define the Grassmannian Gr_p by

$$Gr_p = GL_p/B_p.$$

Now B_p is the stability subgroup of the plane H_+. Let $W = g \cdot H_+$ for some $g \in GL_p$ and let

$$pr_\pm : W \to H_\pm$$

be the orthogonal projections. From the fact that the diagonal blocks in g are Fredholm operators imply that the projection pr_+ is a Fredholm operator. Since the off-diagonal blocks are in L_{2p} it follows that pr_- is a L_{2p} operator.

Exercise 6.2.1. Let $W \subset H$ be a closed subspace such that the orthogonal projections $pr_\pm : W \to H_\pm$ have the two properties mentioned above. Show that there exists $g \in GL_p$ such that $W = g \cdot H_+$.

The inclusions $GL_0 \subset GL_1 \subset \ldots$ lead to the natural embeddings

$$Gr_0 \subset Gr_1 \subset Gr_2 \subset \ldots Gr_\infty,$$

where each Gr_p is dense in Gr_q for $p < q$. The triangular subgroup B_p is contractible and therefore the topology of Gr_p is similar to that of the group GL_p. In particular, the homotopy equivalence $GL_p \cong GL_q$ implies the homotopy equivalence $Gr_p \cong Gr_q$ for all p, q. All the Grassmannians Gr_p are simply connected and the second homotopy group is $\pi_2(Gr_p) = \mathbf{Z}$. It follows from the Hurewicz theorem that the second cohomology of Gr_p is also equal to $\mathbf{Z}$. The inequivalent complex line bundles over any connected component of Gr_p are classified by the integers (the first Chern class of the bundle). The connected components of Gr_p correspond to the components of GL_p: The Fredholm index of the projection pr_+ is equal to the index of the a-block in g. Let $Gr_p^n \subset Gr_p$ consist of those planes W for which $\operatorname{ind} pr_+$ is equal to n.

Let $\{e_k\}_{k \in \mathbf{Z}}$ be an orthonormal basis of H such that $e_k \in H_-$ for $k < 0$ and $e_k \in H_+$ for $k \geq 0$. Let $W \in Gr_p$ with $\operatorname{ind} W = n$. A sequence of vectors $w = \{w_k\}_{k \geq -n}$ W is called *an admissible basis* of W if (1) it can be reached from an orthonormal basis of W by a linear isomorphism, (2) the matrix w_+ defined by

$$pr_+ w_k = \sum_{j \geq -n} (w_+)_{jk} e_j$$

is in $1 + L_p$, where pr_+ is now the orthogonal projection of W to the subspace of H spanned by the vectors e_j, $j \geq -n$. If w and w' are two admissible basis of W then by (2) the operator t transforming w to w', $w' = wt$, must be in GL^p.

Let St_p denote the space of all admissible basis for all $W \in Gr_p$, St_p is an infinite-dimensional *Stiefel manifold* ; it is in a natural way a principal GL^p bundle over Gr_p, the GL^p action being given by the basis transformations and the canonical projection $St_p \to Gr_p$ is the mapping associating to the basis w the plane W spanned by the vectors w. We still have to prove that each $W \in Gr_p$ has an admissible basis. We carry out the proof in the case $n = 0$ and leave the general case as an exercise. When $n = 0$ we can write $W = g \cdot H_+$ for some $g = \begin{pmatrix} a & b \\ c & d \end{pmatrix} \in GL_p$ with $\operatorname{ind} a = 0$. Therefore there exists an invertible operator q and a finite-rank operator t such that $a = q + t$. Now $w_i = \sum_{j \geq 0} e_j (aq^{-1})_{ji}$, $i = 0, 1, 2, \dots$ is an admissible basis of W (the matrix w_+ is equal to aq^{-1}.)

Exercise 6.2.2. Prove the existence of an admissible basis for arbitrary index n. (Imitate the proof for the case $p = 1$ in Pressley and Segal [1986].)

The differentiable structure in St_p is defined essentially in the same way as for finite-dimensional Stiefel manifolds. The space St_p decomposes to connected components labeled by the index n. The connected component St_p^0 is easiest described as the homogeneous space $St_p^0 = \mathcal{E}_p / B_p$, where B_p denotes the subgroup of $\mathcal{E}_p$ consisting of the elements (g, q), where $g \in GL_p$ is an upper block-triangular operator and $q = a$. The group B_p is the stability subgroup at the point $w = \{e_0, e_1, \dots\}$. The topology of St_p can be also given by the metric

$$d(w, w') = \| w_+ - w'_+ \|_p + \| w_- - w'_- \|_{2p},$$

where the matrix w_- is defined by $pr_- w_i = \sum_{j<n} (w_-)_{ji} e_j$, pr_- being the projection to the subspace spanned by the vectors e_i, $i < n$. As a quotient of two complex groups St_p is naturally a complex manifold.

Generalized Fredholm determinants

We need a generalization of the determinant $\det S$ which is normally defined only for operators of the form $1 + A$, $A \in L_1$. Let $S = 1 + A$ be a bounded linear operator in the Hilbert space H. The ordinary determinant can be computed from the power series expansion (assuming that $A \in L_1$ is near zero)

$$\log \det S = \operatorname{tr} \log S = \operatorname{tr} \left(A - \frac{A^2}{2} + \frac{A^3}{3} - \dots \right).$$

However, when $A \in L_p$ we can use a "renormalized" expression to define a generalized determinant $\det_p S$. We set

$$R_p(A) = -1 + (1 + A) \exp \left[\sum_{j=1}^{p-1} (-1)^j \frac{A^j}{j} \right]$$

and we define

$$\det_p(1+A) = \det(1+R_p(A)).$$

The operator $R_p(A)$ is in L_1 by a repeated use of the property (3) of the Schatten ideals. It follows that $\det_p S$ really converges. Note that $\det_1$ is just the ordinary determinant det.

PROPOSITION 6.2.3.

(1) *The mapping $A \mapsto det_p(1+A)$ is continuous in the topology of L_p.*

(2) *If $A \in L_p$ then $1+A$ is invertible if and only if $det_p(1+A) \neq 0$.*

(3) *If $A \in L_{p-1}$ then*

$$det_p(1+A) = det_{p-1}(1+A) \cdot \exp\left[(-1)^{p-1} tr \frac{A^{(p-1)}}{p-1}\right].$$

(4) *There is a symmetric function $\gamma_p(A,B)$ such that*

$$det_p AB = det_p A \, det_p B \cdot \exp \gamma_p(A,B)$$

for $A, B \in L_p$. Futhermore, $\gamma_p(A,B)$ is a trace of a polynomial (with real coefficients) in the operators A and B.

PROOF: (1) Expanding $R_p(A)$ as a power series of A one sees that the first nonzero term is a constant times A^p; using the analyticity of the exponential mapping we conclude that $R_p : L_p \to L_1$ is continuous. Combining with the continuity of the ordinary determinant we arrive at (1).

(3) By definition, $1 + R_p(A) = [1 + R_{p-1}(A)]\exp\left[(-1)^{p-1}\frac{A^{p-1}}{p-1}\right]$, so $\det(1+R_p(A)) = \det(1+R_{p-1}(A))\exp\left[(-1)^{p-1}\mathrm{tr}\frac{A^{p-1}}{p-1}\right]$.

(4) If $p=1$ there is nothing to prove, $\gamma_1 = 0$. We use induction on p. Suppose that $A, B \in L_{p-1}$. Then $\det_p(AB)$ is equal to

$$\begin{aligned}
&\det_{p-1} AB \cdot \exp\left[(-1)^{p-1}\mathrm{tr}\frac{(AB-1)^{p-1}}{p-1}\right] \\
&= \det_{p-1} A \cdot \det_{p-1} B \cdot \exp\left[\gamma_{p-1}(A,B) + (-1)^{p-1}\mathrm{tr}\frac{(AB-1)^{p-1}}{p-1}\right] \\
&= \det_p A \cdot \det_p B \cdot \exp\Bigg[\gamma_{p-1}(A,B) + (-1)^{p-1}\mathrm{tr}\frac{(AB-1)^{p-1}}{p-1} \\
&\quad - (-1)^{p-1}\mathrm{tr}\frac{(A-1)^{p-1}}{p-1} - (-1)^{p-1}\mathrm{tr}\frac{(B-1)^{p-1}}{p-1}\Bigg],
\end{aligned}$$

by the induction hypothesis. Denoting the expression in the square brackets by $\gamma_p(A,B)$ we have proven the claim for the index p in the case $A,B \in 1+L_{p-1}$. Using the continuity of $\det_p$ the same must hold for all $A,B \in 1+L_p$, since $L_{p-1} \subset L_p$ is dense.

(2) Suppose first that $A \in L_1$. Now

$$\det_p(1+A) = \det(1+A)\det\exp[\dots].$$

Since $\exp T$ is invertible for any T, we see that $\det_p(1+A) \neq 0$ iff $1+A$ is invertible. Now $L_1 \subset L_p$ is dense in the topology of L_p and by (1) the determinant $\det_p$ is continuous; on the other hand the set of invertible linear operators is an open subset of $L(H)$ and thus $\det_p(1+A) \neq 0$ implies that $1+A$ is invertible. Conversely, assume that $1+A$ is invertible. By (4) we have

$$1 = \det_p 1 = \det_p(1+A)\det_p(1+A)^{-1} \cdot e^{\gamma_p(1+A,1+B)}$$

which implies $\det_p(1+A) \neq 0$.

In the special case $p=2$ we have

$$\det_2(1+A) = \det(1+A)\cdot\exp(-\mathrm{tr}A),\ A \in L_1$$
$$\det_2(1+A(1+B)) = \det_2(1+A)\cdot\det_2(1+B)\cdot\exp(-\mathrm{tr}AB).$$

PROPOSITION 6.2.4. *Define* $\omega_p(A,B) = \det_p B \cdot e^{\gamma_p(A,B)}$ *for* $A,B \in 1+L_p$. *Then* $\omega_p(A,BC) = \omega_p(AB,C)\cdot\omega_p(A,B)$ *for all* $A,B,C \in 1+L_p$.

PROOF: If A is invertible we may write

$$\omega_p(A,B) = \frac{\det_p AB}{\det_p A}$$

from which the asserted *cocycle property* follows immediately. Both sides of the equation to be proven are continuous functions of the variables A,B,C. Since the set of invertible linear operators is dense in $L(H)$, the equation holds for all $A,B,C \in 1+L_p$.

The determinant bundle DET_p

We now define *the determinant line bundle* DET_p over Gr_p. There is a right action of GL^p on $St_p \times \mathbf{C}$ given by

$$(w,\lambda)\cdot t = (wt, \lambda\omega_p(w_+,t)^{-1}).$$

The associativity of the action follows directly from the cocycle property of ω_p, Proposition 6.2.4. Since the action is free, we can define a smooth manifold by

$$DET_p = (St_p \times \mathbf{C})/GL^p.$$

We can define a projection $DET_p \to Gr_p$ by $(w, \lambda) \mapsto$ the plane W spanned by the vectors w_i. Thus DET_p is a complex line bundle over Gr_p, the action of $\mathbf{C}$ in DET_p being given by $(w, \lambda) \cdot \mu = (w, \lambda\mu)$. Since DET_p is obtained by dividing the complex manifold $St_p \times \mathbf{C}$ by the holomorphic action of the complex group GL^p, it is a complex manifold, too.

Next we shall define a fiber metric in DET_p. Define first a function $\ell : St_p \to \mathbf{R}_+$ by $\ell(w) = \exp -\frac{1}{2}\gamma_p(w_+, w_+^*)$. Using Proposition 6.2.3 we get

$$\begin{aligned} \frac{|\det_p w_+|}{|\det_p w_+ w_+^*|^{1/2}} &= \frac{|\det_p w_+|}{|\det_p w_+|^{1/2}|\det_p w_+^*|^{1/2}} e^{-\frac{1}{2}\gamma_p(w_+, w_+^*)} \\ &= e^{-\frac{1}{2}\gamma_p(w_+, w_+^*)} = \ell(w), \end{aligned}$$

where we have also used the fact that $\det_p A^* = \overline{\det_p A}$; this follows from Proposition 6.2.3 (3). If $t \in GL^p$ is *unitary* we get immediately from the equation above,

$$\ell(wt)\left|\frac{\det_p w_+}{\det_p w_+ t}\right| = \ell(w).$$

Thus for unitary t we obtain $\ell(w)|\lambda| = \ell(w')|\lambda'|$ when $w' = wt$ and $\lambda' = \lambda\omega_p(w, t)^{-1}$. This means that the norm

$$|(w, \lambda)| := |\lambda|\ell(w)$$

is well-defined in DET_p provided that we use only *orthonormal frames* w when evaluating the norm. Namely, if w and w' are orthonormal then automatically the basis transformation t is unitary.

Holomorphic sections

We shall next study the space Γ_p of *holomorphic* sections of the dual bundle DET_p^*. The reason why we choose the dual DET_p^* instead of DET_p is that the latter does not have any nonzero holomorphic sections (this is true already in the finite-dimensional case). A holomorphic section of DET_p would be a holomorphic function $\psi : St_p \to \mathbf{C}$ such that $\psi(wt) = \psi(w)\omega_p(w, t)^{-1}$, which does not have everywhere nonsingular solutions except $\psi = 0$ due to the -1 in the exponent. On the other hand, a holomorphic section of DET_p^* is a holomorphic function

$$\psi : St_p \to \mathbf{C}, \ \psi(wt) = \psi(w)\omega_p(w,t), \ t \in GL^p.$$

This equation has solutions; for example,

$$\psi_0(w) = \det{}_p w$$

is an element of Γ_p, *the vacuum vector* in Γ_p. Other sections can be constructed as follows. Let $S = \{i_0, i_1, i_2, \dots\} \subset \mathbf{Z}$ such that $i_\nu = \nu$ for large enough $\nu \in \mathbf{Z}$. This means that S contains all the positive integers except a finite set and only a finite set of negative integers. Any $w \in St_p$ can be thought of as a matrix $w = \begin{pmatrix} w_+ \\ w_- \end{pmatrix}$ where the column index labels the different vectors w_i of the basis w and the row index labels the coordinates in the standard basis $\{e_i\}$ of H. Let w_S be the submatrix obtained by selecting the rows labeled by the integers S. Put

$$\psi_S(w) = \det{}_p w_S \, e^{\beta_p(w_+) - \beta_p(w_S)}$$

where β_p is the polynomial defined by the formula $\det_p A = \det A \cdot \exp \beta_p(A)$, $A \in 1 + L_1$. Note that the difference $w_+ - w_S$ is a finite-rank matrix and therefore the traces $\mathrm{tr}(w_+^k - w_S^k) = \mathrm{tr}(w_+ - w_S)(w_+^{k-1} + w_+^{k-2} w_S + \cdots + w_S^{k-1})$ occurring in the formula for $\beta_p(w_+) - \beta_p(w_S)$ are all traces of finite-rank operators and thereby converge. For $S = \{0, 1, 2, \dots\}$ we get $\psi_S = \psi_0$.

Exercise 6.2.5. Show that ψ_S is really a section of DET_p^*. *Hint:* Consider first the case $t, w_+ \in 1 + L_1$ and use then a density argument.

We shall consider the case $p = 1$ in more detail. Now all the determinants are ordinary determinants and $\beta_1 = 0$. We define an inner product in the space of sections spanned by the vectors ψ_S by setting $(\psi_S, \psi'_S) = 0$ if $S \neq S'$ and $(\psi_S, \psi_S) = 1$. One can show that the Hilbert space with the orthonormal basis $\{\psi_S\}$ is exactly the space of square integrable holomorphic sections of DET_p^* [Pickrell, 1987]. We shall not dwell in measure theoretic questions in this paper. Instead, we shall discuss here the sections of DET_p^* in terms of *semi-infinite cohomology*. This is important in the physical applications since the semi-infinite forms are the state vectors in the Fock representation for a quantized Dirac field in 1+1 space-time dimensions. We shall write the vacuum vector formally as

$$\psi_0 = f_0 \wedge f_{-1} \wedge f_{-2} \wedge \dots,$$

where f_k is the one-form on H defined by $f_k(e_j) = \delta_{-k,j}$. In a finite-dimensional case $0 \leq j \leq N$ we could write $\det w_+ = (f_0 \wedge \cdots \wedge f_{-N})(w_0, w_1, \ldots, w_N)$. In the infinite-dimensional case the infinite wedge product is *defined* to mean the determinant function in $1 + L_1$. More generally we put

$$\psi_S = f_{i_0} \wedge f_{i_1} \wedge f_{i_2} \cdots$$

where $i_0 > i_1 > i_2 > \ldots$ is the complement of the set S in $\mathbf{Z}$. For each $i \in \mathbf{Z}$ define the *creation operator* a_i^* in Γ_1 by

$$a_i^* \cdot f_{i_0} \wedge f_{i_1} \wedge \cdots = f_i \wedge f_{i_0} \wedge f_{i_1}.$$

The usual rules of computation with the wedge product are taken over to the semi-infinite case. The *annihilation operator* a_i is defined by

$$a_i \cdot f_{i_0} \wedge f_{i_1} \wedge \cdots = (-1)^k f_{i_0} \wedge \ldots f_{i_{k-1}} \wedge f_{i_{k+1}} \wedge \ldots$$

if $i = i_k$ for some $k \geq 0$ and $a_i \psi_S = 0$ otherwise. The vacuum vector ψ_0 is characterized by the properties

$$a_k \psi_0 = 0, \; k > 0, \qquad a_k^* \psi_0 = 0, \; k \leq 0.$$

All the basis vectors ψ_S are obtained by acting on the vacuum by a product of the creation and annihilation operators. With respect to the inner product given above, the operator a_i^* is the Hermitian conjugate of a_i. We shall return to these matters in connection with the group actions in the determinant line bundle in the next section.

The complexification $\mathbf{C}Gr_p$

The Grassmannian Gr_p is a complex manifold as a quotient of the complex group GL_p by a complex subgroup. However, there is a further complexification $\mathbf{C}Gr_p$ of Gr_p which will be of interest. The points of the real submanifold $Gr_p \subset \mathbf{C}Gr_p$ are parametrized by Hermitian operators $F : H \to H$ such that

(1) $F^2 = 1$
(2) $F - \epsilon$ is in L_{2p}.

By definition, the space $\mathbf{C}Gr_p$ consists of *all* operators F which satisfy the two conditions above. Geometrically, a point on $\mathbf{C}Gr_p$ is a pair (W, W') of planes in H such that $W \in Gr_p$ and W' is a complementary plane (i.e. $W \cap W' = 0$ and $W + W' = H$) such that the orthogonal projection of W' onto H_+ is a L_{2p} operator. The correspondence between the operators F and the pairs (W, W') is as follows. We can write

$$F = g\epsilon g^{-1} \text{ with } g \in GL_p.$$

The group element is not uniquely defined but if $F = g'\epsilon g'^{-1}$ then $g' = gh$ for some bounded operator h with vanishing off-diagonal blocks. The pair (W, W') is defined by

$$W = g \cdot H_+, \;\; W' = g \cdot H_-.$$

There is no ambiguity here since $H_+ = h \cdot H_+$ and $H_- = h \cdot H_-$.

The determinant bundle can be naturally extended over the complexification $\mathbf{C}Gr_p$. Points in $\mathbf{C}DET_p$ are equivalence classes of triplets (w, W', λ), where w is an admissible basis of $W \in Gr_p$ such that $(W, W') \in \mathbf{C}Gr_p$ and $\lambda \in \mathbf{C}$. The equivalence relation is defined as in the case of DET_p, $(w, W', \lambda) \sim (wt, W', \lambda\omega_p(w_+, t)^{-1})$ with $t \in GL^p$.

6.3. Lifting the action of GL_p in Gr_p to an action of the extension $\widehat{GL_p}$ in DET_p

The extension of GL_p^0

The group GL_p acts in a natural way on the manifold Gr_p since Gr_p is a homogeneous space for the group. We could try to lift this action to the determinant bundle DET_p by putting $g \cdot (w, \lambda) = (g \cdot w, \lambda)$, where w is an admissible basis and $\lambda \in \mathbf{C}$. However, this does not work because in general the basis gw is not admissible. Instead, we shall construct a group extension $\widehat{GL_p}$ which acts in DET_p. We shall first study the action of the connected component GL_p^0 of identity. In the following we shall write $Gr_p, St_p, \ldots$ instead of $Gr_p^0, St_p^0, \ldots$.

LEMMA 6.3.1. *There exists a smooth function $\alpha : \mathcal{E}_p \times St_p \to \mathbf{C}^\times$ such that*

$$\frac{\alpha(g, q; wt)}{\alpha(g, q; w)} = \frac{\omega_p(w_+, t)}{\omega_p((gwq^{-1})_+, qtq^{-1})}$$

for all $(g, q) \in \mathcal{E}_p$, $w \in St_p$ and $t \in GL^p$. Let W be the plane spanned by the basis w and let $F = \begin{pmatrix} F_{11} & F_{12} \\ F_{21} & F_{22} \end{pmatrix}$ be the linear operator in $H = H_+ \oplus H_-$ such that $F|_W = +1$ and $F|_{W^\perp} = -1$. The general solution can be written as

$$\begin{aligned} \alpha(g, q; w) =& f(g, q; W) \frac{det_p w_+}{det_p(gwq_+^{-1})} \\ & \times \frac{det_{p\frac{1}{2}}(q^{-1}a(F_{11} + 1 + q^{-1}bF_{21}))}{det_{p\frac{1}{2}}(F_{11} + 1)}, \end{aligned}$$

where $f : \mathcal{E}_p \times Gr_p \to \mathbf{C}^\times$ is an arbitrary smooth function.

PROOF: An arbitrary solution of the problem is clearly obtained by multiplying a special solution by a suitable function f. Formally $\det_p w_+ \times \det_p(gwq_+^{-1})^{-1}$ is a solution. However, this function has singularities (due the zeros of the denominator) and zeros. We can regularize it by multiplying by a certain function on $\mathcal{E}_p \times Gr_p$. Let

$$h = \begin{pmatrix} w_+ & \alpha \\ w_- & \beta \end{pmatrix}$$

be an invertible operator in H such that $W = h \cdot H_+$ and $W^\perp = h \cdot H_-$. Denote

$$h^{-1} = \begin{pmatrix} x & y \\ u & v \end{pmatrix}.$$

Now $F = h\epsilon h^{-1}$ and in particular $F_{11} = w_+x - \alpha u = 2w_+x - 1$, $F_{21} = w_-x - \beta u = 2w_-x$. Thus we obtain

$$\begin{aligned}
&\exp\left\{\gamma_p(xw_+, (gwq^{-1})_+) - \gamma_p(w_+, \tfrac{1}{2}q^{-1}[a(F_{11}+1) + bF_{21}])\right\} \\
&= \exp\left[\gamma_p(xw_+, (gwq^{-1})_+) - \gamma_p(w_+, q^{-1}aw_+x + q^{-1}bw_-x)\right] \\
&= \frac{\det_p(q^{-1}aw_+xw_+ + q^{-1}bw_-xw_+)}{\det_p(q^{-1}aw_+ + q^{-1}bw_-)\det_p(xw_+)} \\
&\quad \times \frac{\det_p w_+ \cdot \det_p(q^{-1}aw_+x + q^{-1}bw_-x)}{\det_p(q^{-1}aw_+xw_+ + q^{-1}bw_-xw_+)} \\
&= \frac{\det_p w_+}{\det_p(gwq_+^{-1})} \cdot \frac{\det_p(q^{-1}a \cdot \frac{1}{2}(F_{11}+1) + q^{-1}b \cdot \frac{1}{2}F_{21})}{\det_p \frac{1}{2}(F_{11}+1)},
\end{aligned}$$

where we have used repeatedly Proposition 6.2.3. We have now shown that the ratio of the determinants in 6.3.1 is a regular function.

We define an action of $\mathcal{E}_p \times Map(Gr_p, \mathbf{C}^\times)$ in $St_p \times \mathbf{C}$. We set

$$\text{(A0)} \qquad (g, q, \mu) \cdot (w, \lambda) = (gwq^{-1}, \lambda\alpha(g, q; w)\mu(\pi(w))),$$

where α corresponds to any fixed choice of f in Lemma 6.3.1 and $\pi : St_p \to Gr_p$ is the canonical projection. The group multiplication law in $\mathcal{E}_p \times Map(Gr_p, \mathbf{C}^\times)$ is not that of a direct product but

$$\text{(A1)} \qquad (g, q, \mu) \cdot (g', q', \mu') = (gg', qq', \mu'')$$

where

$$\begin{aligned}
&\mu''(W) = \mu(g' \cdot W)\mu'(W)\Omega_p(g, q, g', q'; W), \\
\text{(A2)} \qquad &\Omega_p = \alpha(g, q; g'wq'^{-1})\alpha(g', q'; w)\alpha(gg', qq'; w)^{-1}.
\end{aligned}$$

Next we want to show that the group action on $St_p \times \mathbf{C}$ can be pushed to an action on $(St_p \times \mathbf{C})/GL^p = DET_p$. Let $(w', \lambda') = (w, \lambda) \cdot t$, $t \in GL^p$. We have to show that $(g, q, \mu) \cdot (w, \lambda)$ and $(g, q, \mu) \cdot (w', \lambda')$ are equal modulo the right action of GL^p. Using the property of α stated in 6.3.1 we obtain

$$\begin{aligned}
(g,q,\mu)(w',\lambda') &= (g,q,\mu)(wt, \lambda\omega_p(w_+,t)^{-1}) \\
&= (gwtq^{-1}, \mu(\pi(w))\lambda\omega_p(w_+,t)^{-1}\alpha(g,q;w)) \\
&= (gwq^{-1}, \mu(\pi(w))\lambda\omega_p(w_+,t)^{-1}\omega_p((gwq^{-1})_+, qtq^{-1}) \\
&\quad \times \alpha(g,q;wt)) \cdot qtq^{-1} \\
&= (gwq^{-1}, \mu(\pi(w))\lambda\alpha(g,q;w)) \cdot qtq^{-1} \\
&= [(g,q,\mu)(w,\lambda)] \cdot qtq^{-1}.
\end{aligned}$$

The action of $\mathcal{E}_p \times Map(Gr_p, \mathbf{C}^\times)$ in DET_p is not faithful. We compute the kernel of the action. Suppose $(g, q, \mu)(w, \lambda) = (w, \lambda)$ mod GL^p. Then there is $t \in GL^p$ such that $(gwq^{-1}, \lambda\mu(\pi(w))\alpha(g, q; w)) = (wt, \lambda\omega_p(w_+, t)^{-1})$. This means that $g = 1$, $q = t^{-1}$ and $\mu(W) = \mu_q(W) = \alpha(1, q; w)^{-1} \cdot \omega_p(w_+, q^{-1})^{-1}$. Thus we have proven the following theorem:

THEOREM 6.3.2. *The formulas (A) define a faithful action of the group*

$$\widehat{GL_p^0} = [\mathcal{E}_p \times Map(Gr_p, \mathbf{C}^\times)]/N$$

in DET_p, where N is the normal subgroup consisting of the elements $(1, q, \mu_q)$.

It follows immediately from the definition that $\widehat{GL_p^0}$ is a principal bundle over GL_p^0 with fiber $Map(Gr_p, \mathbf{C}^\times)$. It contains a normal Abelian subgroup consisting of the elements $(1, 1, \mu)\text{mod}N$, $\mu \in Map(Gr_p, \mathbf{C}^\times)$. Dividing $\widehat{GL_p^0}$ by $Map(Gr_p, \mathbf{C}^\times)$ we get GL_p^0. We say that the group $\widehat{GL_p^0}$ is *an extension* of GL_p^0 by $Map(Gr_p, \mathbf{C}^\times)$. The bundle projection $\widehat{GL_p^0} \to GL_p^0$ is given by $(g, q, \mu) \mapsto g$.

Let $U \subset GL_p$ be a neighborhood of unity consisting of elements $g = \begin{pmatrix} a & b \\ c & d \end{pmatrix}$ such that a is invertible. We define a local section $\phi : U \to \widehat{GL_p^0}$ by $\phi(g) = (g, a, 1)$. We compute the *local two-cocycle* $\xi(g_1, g_2)$ defined by the equation

$$\phi(g_1)\phi(g_2) = \phi(g_1 g_2) \cdot (1, 1, \xi(g_1, g_2)),$$

where $\xi(g_1, g_2) \in Map(Gr_p, \mathbf{C}^\times)$. The right-hand-side is equal to $(g_1 g_2, a_{12}, \xi(g_1, g_2)\alpha(1, 1; w))$, where a_{12} is the a-block in $g_1 g_2$. The left-hand-side is equal to

$$(g_1 g_2, a_1 a_2, \alpha(g_1, a_1; g_2 w a_2^{-1})\alpha(g_2, a_2; w)\alpha(g_1 g_2, a_1 a_2; w)^{-1})$$

which is equal, modulo the right action by the element $(1, q, \mu_q) \in N$, $q = a_2^{-1} a_1^{-1} a_{12}$, to the group element $(g_1 g_2, a_{12}, \alpha(g_1, a_1; g_2 w a_2^{-1}) \times \alpha(g_2, a_2; w)\alpha(g_1 g_2, a_1 a_2; w)^{-1} \mu_q(W) \alpha(g_1 g_2, a_1 a_2; w q^{-1})\alpha(1, q; w) \times \alpha(g_1 g_2, a_{12}; w)^{-1})$. Thus $\xi(g_1, g_2)$ is equal to the third component of this element divided by $\alpha(1, 1; w)$.

The case $p = 1$. Now we can choose $\alpha \equiv 1$, since $\omega_1(w_+, t) = \det t$. In particular, $\mu_q(W) = \det q$ and we obtain for the cocycle the simple formula

$$\xi(g_1, g_2) = \det a_2^{-1} a_1^{-1} a_{12}.$$

The case $p = 2$. One can show that the function

$$\alpha(g, q; w) = \exp -\mathrm{tr}\left[(1 - q^{-1}a)(w_+ - 1) + q^{-1}b(\frac{1}{2}F_{21} - w_-)\right]$$

satisfies the requirement of 6.3.1. Now $\alpha(1, q; w) = \exp[-\mathrm{tr}(1 - q^{-1})(w_+ - 1)] = \det_2 w_+ \cdot \det_2 q^{-1} / \det_2 w_+ q^{-1} = \omega_2(w_+, q^{-1}) \cdot \det_2 q^{-1}$. Inserting to the general formula we get

$$\begin{aligned}\xi(g_1, g_2) =& \det_2(a_2^{-1} a_1^{-1} a_{12} \cdot \alpha(g_1 g_2, a_{12}; w a_2^{-1} a_1^{-1} a_{12})^{-1} \\ & \times \alpha(1, a_{12}^{-1} a_1 a_2; w)\alpha(g_1, a_1; g_2 w a_2^{-1})\alpha(g_1 g_2, a_1 a_2; w).\end{aligned}$$

Exercise 6.3.3. Show that α given above is a correct choice. Show that for this α one gets $\xi(g_1, g_2) = 1$, when both g_1 and g_2 are lower triangular, $b_1 = b_2 = 0$.

The extension of the Lie algebra $\mathbf{gl}_p$

We shall study next the Lie algebra of the group $\widehat{GL_p^0}$. Since the group $\widehat{GL_p^0}$ is an extension of GL_p^0 by the normal subgroup $Map(Gr_p, \mathbf{C}^\times)$, the Lie algebra $\widehat{\mathbf{gl}_p}$ must be an extension of the Lie algebra $\mathbf{gl}_p$ by the Abelian ideal $Map(Gr_p, \mathbf{C})$. Corresponding to the group cocycle ξ there is a Lie algebra two-cocycle $\eta : \mathbf{gl}_p \times \mathbf{gl}_p \to Map(Gr_p, \mathbf{C})$ such that the commutator in $\widehat{\mathbf{gl}_p} = \mathbf{gl}_p \oplus Map(Gr_p, \mathbf{C})$ is given by

$$[(X, \mu), (Y, \nu)] = ([X, Y], X \cdot \nu - Y \cdot \mu + \eta(X, Y)).$$

Here $X \cdot \nu$ stands for the Lie derivative of the function ν in the direction of the vector field X on Gr_p. The vector field is generated by the natural action of the one-parameter subgroup $\exp(sX)$ of GL_p on Gr_p. The condition $\exp(sX) \in GL_p, \forall s \in \mathbf{R}$ is equivalent to the requirement that the off-diagonal blocks of the bounded linear operator X are in L_{2p}. Since the Lie product is antisymmetric we must have $\eta(X,Y) = -\eta(Y,X)$. The Jacobi identity implies that

$$\eta([X,Y],Z) + \eta([Y,Z],X) + \eta([Z,X],Y) \\ - Z \cdot \eta(X,Y) - X \cdot \eta(Y,Z) - Y \cdot \eta(Z,X) = 0.$$

Let $\mathbf{n}$ be the Lie algebra of N. Then we find the commutator in the standard way from the group multiplication rule by writing

$$\frac{d^2}{dtds}\phi(e^{tX})\phi(e^{sY})\phi(e^{-tX})\phi(e^{-sY})|_{t=s=0} = ([X,Y],0,\eta(X,Y)) \bmod \mathbf{n}$$

where $\mathbf{n}$ is the Lie algebra of N. For general p the formula for η is rather complicated. We shall do the computation only in the cases $p = 1, 2$.

The case $p = 1$. Now

$$\phi(e^{tX})\phi(e^{sY})\phi(e^{-tX})\phi(e^{-sY}) \\ = (e^{tX}e^{sY}e^{-tX}e^{-sY}, a(e^{tX}e^{sY}e^{-tX}e^{-sY}), \\ \det(a(e^{-sY}a(e^{-tX})a(e^{sY})a(e^{tX})a(e^{tX}e^{sY}e^{-tX}e^{-sY}))) \bmod N.$$

Using the formula $\det e^A = 1 + \mathrm{tr} A + \ldots$ and applying the derivative $\frac{d^2}{dtds}$ to both sides we get after a few lines of algebra the result

$$\eta(X,Y) = \mathrm{tr}(b(X)c(Y) - b(Y)c(X)).$$

Using the embedding $Map(S^1, G) = S^1G \subset GL_1$ described in Section 6.1. we can pull back the extension $\widehat{GL_p^0}$ of GL_p^0 to get an extension $\widehat{S^1G}$ of S^1G. Let $f, g : S^1 \to \mathbf{g}$ be a pair of elements of the Lie algebra $S^1\mathbf{g}$. Denoting by f_k, g_k, $k \in \mathbf{Z}$ the Fourier components we get

$$\eta(f,g) = \sum_{k>q\geq 0} \mathrm{tr}(f_k g_{-k} - g_k f_{-k}) \\ = \sum_{k>0} k\,\mathrm{tr}(f_k g_{-k} - g_k f_{-k}) = \frac{1}{2\pi}\int_0^{2\pi} \mathrm{tr} f(\phi)\frac{d}{d\phi}g(\phi)d\phi.$$

Here "tr" means the finite-dimensional matrix trace in the representation space V of G; see Section 6.1. For a simple Lie algebra $\mathbf{g}$ we therefore get as the central extension of the loop algebra an affine Kac-Moody algebra.

The cocycle η has also a geometric meaning. Namely, it reflects the curvature of the canonical line budle DET_1 over Gr_1. For a given point on the Grassmannian, parametrized by the operator F, the tangent space $T_F Gr_1$ consists of the Hermitian Hilbert-Schmidt operators X such that $FX + XF = 0$; this relation is obtained from the constraint $F^2 = 1$ by taking a directional derivative at the point F. A closed two-form is defined by

$$\chi_F(X, Y) = \mathrm{tr} FXY. \tag{6.3.4}$$

This form is the first Chern class of the bundle DET_1. At the origin $F = \epsilon$ we have $\chi(X, Y) = \eta(X, Y)$.

The case $p = 2$. Let X, Y be elements of the Lie algebra $\mathbf{u}_2$ of the *unitary* subgroup of GL_2 (therefore X, Y are anti-Hermitian). Using the choice of α introduced earlier for $p = 2$ we get after a lengthy but straightforward computation

$$\eta(X, Y) = \frac{1}{8} \mathrm{tr}\, [[\epsilon, X], [\epsilon, Y]](F - \epsilon).$$

In the unitary case the action of GL_p on Gr_p can be written as $g \cdot F = gFg^{-1}$. However, for nonunitary g the action is a more complicated nonlinear function of F. The formula $F \mapsto gFg^{-1}$ cannot be valid for arbitrary g, since the matrix F parametrizing the Grassmannian planes is always Hermitian. The formula above does *not* give the Lie algebra cocycle for the complex group GL_2, derived from our choice for α. Nonetheless, we can *define* a Lie algebra cocycle on $\mathbf{gl}_2$ by complex linearity from our formula for the real subalgebra $\mathbf{u}_2 \subset \mathbf{gl}_2$. Of course, the two-cocycle $\eta_2 = \eta$ of $\mathbf{gl}_2$ is also a two-cocycle of the subalgebra $\mathbf{gl}_1$. However, it is cohomologous to the cocycle $\eta_1(X, Y) = \mathrm{tr}(b(X)c(Y) - b(Y)c(X))$. Namely, let

$$\beta(X)(F) = -\frac{1}{16} \mathrm{tr}[X, \epsilon][F, \epsilon].$$

Then we have $X \cdot \beta(Y) - Y \cdot \beta(X) - \beta([X, Y]) = \frac{1}{16} \mathrm{tr}([Y, \epsilon][[X, F], \epsilon] - [X, \epsilon][[Y, F], \epsilon] + [[X, Y], \epsilon][F, \epsilon]) = -\frac{1}{8} \mathrm{tr}[[\epsilon, X], [\epsilon, Y]]F$ so that $\eta_1 = \eta_2 + \delta\beta$. This does *not* mean that η_1 and η_2 are cohomologous in $\mathbf{gl}_2$; η_1 is not well-defined in the larger Lie algebra, since the traces converge only for operators X, Y with trace class off-diagonal blocks. The same holds for the form β. Thus in the physics language we have introduced an "infinite renormalization" β to regularize the otherwise infinite two-cocycle η_1 in $\mathbf{gl}_2$.

Exercise 6.3.5. Prove that the Lie algebra $\mathbf{gl}_p$ does not have any nontrivial *central* extension when $p > 1$.

In contrast to the rather complicated nonlinear action of GL_p on the operator $F = F(W)$ parametrizing points of Gr_p there is a simple natural action of the same group in the complexification $\mathbf{C}Gr_p$. The linear transformation g sends $F \in \mathbf{C}Gr_p$ to gFg^{-1}. In the same way as in the case of DET_p there is an extension of GL_p by the Abelian ideal $Map(\mathbf{C}Gr_p, \mathbf{C}^\times)$ which lifts the action of GL_p to the complexification $\mathbf{C}DET_p$.

As in the case $p = 1$, the cocycle η can be interpreted as a closed 2-form on the manifold Gr_2. It is the Chern class of the determinant bundle. For a complete treatise on the Chern classes on the Grassmannians Gr_p we refer to Quillen [1988].

Exercise 6.3.6. Work out in detail the action of the group extension in $\mathbf{C}DET_p$. Show that the extension can be defined such that the action in $\mathbf{C}DET_p$ is *holomorphic* .

The extension of GL_p

To complete this section we shall construct the extension of the full group GL_p. Let $\sigma \in GL_p$ be an element such that the Fredholm index of the block $a(\sigma)$ is equal to one. We shall choose σ to be unitary, for example $\sigma(e_i) = e_{i+1}\, \forall i \in \mathbf{Z}$. There is a map $GL(H_+) \to GL(H_+)$, $q \mapsto q_\sigma$ such that the map $\mathcal{E}_p \to \mathcal{E}_p$, $(g,q) \mapsto (\sigma g \sigma^{-1}, q_\sigma)$ covers the map $g \mapsto \sigma g \sigma^{-1}$ on the base space GL_p^0. For example, we can define

$$q_\sigma = \begin{cases} \sigma q \sigma^{-1} & \text{on } \sigma(H_+) \\ 1 & \text{on } H_+ \ominus \sigma(H_+). \end{cases}$$

Denote by $\mathbf{Z}$ the subgroup of GL_p generated by σ. Now

$$GL_p \cong \mathbf{Z} \ltimes GL_p^0$$

is a semidirect product, where the action of σ on GL_p^0 is defined by $g \mapsto \sigma g \sigma^{-1}$; the composition rule is therefore $(n,g)(n',g') = (n + n', g\sigma^n g' \sigma^{-n})$. We lift the action of σ to an endomorphism of $\mathcal{E}_p \times Map(Gr_p, \mathbf{C}^\times)$ by

$$\sigma(g,q,\lambda) = (\sigma g \sigma^{-1}, q_\sigma, \lambda_\sigma h(g,q;g^{-1}W)),$$

where h is a smooth function on $\mathcal{E}_p \times Gr_p$ to be determined below, and for any $\lambda \in Map(Gr_p, \mathbf{C}^\times)$ we set $\lambda_\sigma(W) = \lambda(\sigma^{-1}W)$. From the requirement that the normal subgroup N of $\mathcal{E}_p \times Map(Gr_p, \mathbf{C}^\times)$ is mapped on to itself we get the condition

$$h(1, q; W) = \mu_{q_\sigma}/(\mu_q)_\sigma$$

and from the requirement that we get an automorphism on the quotient $(\mathcal{E}_p \times Map(Gr_p, \mathbf{C}^\times))/N$ we get

$$\frac{\Omega_p(g_1, q_1, g_2, q_2; W)}{\Omega_p(\sigma g_1 \sigma^{-1}, q_{1\sigma}, \sigma g_2 \sigma^{-1}, q_{2\sigma}; \sigma W)} \\ = h(g_1, q_1; g_2 W) h(g_2, q_2; W) h(g_1 g_2, q_1 q_2; W)^{-1}.$$

This equation says that the quotient on the left should be a coboundary of a 1-chain h of the group GL_p^0 (with respect to the natural action on Gr_p). Thus, there is a solution h if and only iff Ω_p and the two-cocycle $\Omega_p^{(\sigma)}$ in the denominator represent the same cohomology class. The cohomology classes of the different group extensions are determined by the de Rham cohomology classes obtained by evaluating the corresponding Lie algebra cocycle. We give the proof of the invariance of the cohomology classes in the case $p = 2$ (in the case $p = 1$, $\Omega_p = 1$ and we can take $h = 1$); the case $p > 2$ requires more computation but is essentially straight-forward. The Lie algebra cocycle $\eta^{(\sigma)}$ obtained from $\eta = \eta_2$ through the automorphism σ of GL_2 is

$$\begin{aligned} \eta^{(\sigma)}(X, Y) &= \frac{1}{8} \mathrm{tr} \left[\left[\epsilon, \sigma X \sigma^{-1} \right], \left[\epsilon, \sigma Y \sigma^{-1} \right] \right] \left(\epsilon - \sigma F \sigma^{-1} \right) \\ &= \frac{1}{8} \mathrm{tr} \left[\left[\sigma^{-1} \epsilon \sigma, X \right], \left[\sigma^{-1} \epsilon \sigma, Y \right] \right] \left(\sigma^{-1} \epsilon - F \right). \end{aligned}$$

Thus $\eta^{(\sigma)}$ is obtained from η by substituting $\epsilon \mapsto \epsilon_\sigma = \sigma^{-1}\epsilon\sigma$. The difference $\epsilon - \epsilon_\sigma$ is of finite rank. Therefore,

$$\begin{aligned} \beta(X; F) = &\frac{1}{16} \mathrm{tr} \left([X, \epsilon] \left[F, \epsilon - \epsilon_\sigma \right] + \left[X, \epsilon - \epsilon_\sigma \right] \left[F, \epsilon_\sigma \right] \right) \\ &+ \frac{1}{2} \mathrm{tr} X \left(\epsilon - \epsilon_\sigma \right) \end{aligned}$$

is well-defined; by a simple computation,

$$\eta - \eta^{(\sigma)} = \delta\beta.$$

The extension of $\widehat{GL_p}$ for the whole group GL_p can now be defined as

$$\widehat{GL_p} = \mathbf{Z} \ltimes \widehat{GL_p}^0$$

where the action of $\mathbf{Z}$ on $\widehat{GL_p}^0$ is defined by the action of its generator σ.

6.4. The Dirac field on Gr_1

A finite-dimensional example: $\mathbf{C}P^2$

Our first task is to construct a generalized spin structure on the infinite-dimensional Grassmannian Gr_1. In fact, Gr_1 does not have a spin structure in the usual sense but one has to consider the so-called $Spin^{\mathbf{C}}$ structure. This means that the bundle of orthonormal frames does not have a double covering but there is a bundle $P = Spin^{\mathbf{C}} Gr_1$ with a structure group $Spin^{\mathbf{C}}$ and a bundle map $\phi : P \to OF(Gr_1)$ onto the bundle of oriented orthonormal frames of Gr_1 such that $\phi(pg) = \phi(p)\rho(g)$, where ρ is a homomorphism from the group $Spin^{\mathbf{C}}$ onto a *restricted orthogonal group* , to be defined later, such that the kernel of ϕ is equal to $\mathbf{C}^\times$.

The necessity to deal with the extended spin structures is not something connected to only infinite dimensions; there are many interesting finite-dimensional manifolds which have only a $Spin^{\mathbf{C}}$ structure. Before going to the infinite-dimensional case of Gr_1 let us look at the finite-dimensional projective space $M = \mathbf{C}P^2$ consisting of complex lines in $\mathbf{C}^3$. The unitary group (we use the standard Hermitian inner product in $\mathbf{C}^3$) acts transitively on M and we can write $M = U(3)/(U(2) \times U(1))$. The structure group $SO(4)$ of the bundle OFM of orthonormal frames reduces to a subgroup isomorphic with $U(2)$. To see this fix a complex line $\ell \subset \mathbf{C}^3$. Choose a unit vector $z_1 \in \ell$ and an orthonormal basis z_2, z_3 in the complement $\ell^\perp$. Think of the line ℓ as an operator F in $\mathbf{C}^3$ such that its restriction to ℓ is -1 and the restriction to $\ell^\perp$ is +1. We have $F^* = F$, $F^2 = 1$ and the tangent space $T_\ell M$ can be identified as the space of Hermitian operators A such that $AF + FA = 0$. In the basis $\{z_i\}$ we can write then

$$A = \begin{pmatrix} 0 & 0 & a_1 \\ 0 & 0 & a_2 \\ \bar{a}_1 & \bar{a}_2 & 0 \end{pmatrix}. \tag{6.4.1}$$

An orthonormal basis f for $T_\ell M$ is defined by the four Hermitian matrices $e_{13}+e_{31}, e_{23}+e_{32}, ie_{13}-ie_{31}, ie_{23}-ie_{32}$. If we change the basis z_1

of ℓ and the basis z_2, z_3 of $\ell^\perp$ by a unitary transformation $g = (g_1, g_2) \in U(1) \times U(2)$ then it is easy to see that the basis f is transformed by a real orthogonal transformation $\sigma(g)$. In fact, the vector a is transformed by

$$a = \begin{pmatrix} a_1 \\ a_2 \end{pmatrix} \mapsto g_2 \begin{pmatrix} a_1 \\ a_2 \end{pmatrix} g_1^{-1}$$

and thus, writing $SO(4) = SU(2) \times_{\mathbf{Z}_2} SU(2)$, we obtain

(6.4.2) $$\sigma(g) = \sigma(g_1, \lambda h) = (h, \lambda g_1^{-1}) \bmod \mathbf{Z}_2$$

where we have chosen a decomposition $g_2 = \lambda h$ with $\lambda \in U(1)$, $h \in SU(2)$. The kernel of σ is the diagonal $U(1)$ in $U(1) \times U(2)$. We conclude that for each $\ell \in M$ we can define a preferred set of orthonormal frames f such that the frames in this set are transformed among themselves by elements of the group $[U(1) \times U(2)]/U(1)_{diag} \simeq U(2)$. In other words, we have reduced the structure group $SO(4)$ to a subgroup $U(2)$.

The spin group in four dimensions is $SU(2) \times SU(2)$. It contains the subgroup $SU(2) \times U(1)$ which is a double covering of $U(2)$. If we could build a principal $SU(2) \times U(1)$ bundle which is a double covering of the principal $U(2)$ bundle which we just constructed then we would have a spin structure such that the structure group of the spin bundle is $SU(2) \times U(1)$. However, it is known that this is not possible. Instead, we can easily construct a covering with fiber S^1. In four dimensions,

$$Spin^{\mathbf{C}} = (SU(2) \times SU(2)) \times_{\mathbf{Z}_2} U(1)$$

and we can define a homomorphism $\tilde{\sigma} : U(2) \to Spin^{\mathbf{C}}$ by

(6.4.3) $$\tilde{\sigma}(g) = (h, \lambda, \lambda), \; g = \lambda h, \; \lambda \in U(1), \; h \in SU(2).$$

The covering homomorphism $\rho : Spin^{\mathbf{C}} \to SO(4)$ is $\rho(g, g', g'') = (g, g') \bmod \mathbf{Z}_2$ and therefore $\sigma = \rho \circ \tilde{\sigma}$ which implies that the frame bundle OFM, with the reduced structure group $U(2)$, can be lifted to a $Spin^{\mathbf{C}}$ bundle with the reduced structure group also equal to $U(2)$. Let $F^\ell M$ denote the set of unitary basis of $\ell^\perp$. The fiber of the extended spin bundle over M at ℓ is

(6.4.4) $$F^\ell M \times_{U(2)} Spin^{\mathbf{C}}$$

where the group $U(2)$ acts naturally in $F^\ell M$ from the right and it acts in $Spin^{\mathbf{C}}$ from the left through the homomorphism $\tilde{\sigma}$.

The central extension of the orthogonal group O_1

Next we shall define the infinite-dimensional orthogonal groups O_p and discuss their extensions. Let H be a *real* Hilbert space with complexification $H^{\mathbf{C}} = H \otimes_{\mathbf{R}} \mathbf{C}$ and let $< \cdot, \cdot >$ be a Hermitian inner product in $H^{\mathbf{C}}$. We also assume that a complex structure J has been given in H such that $H^{\mathbf{C}}$ splits to a direct sum $W \oplus \overline{W}$, where W is the eigenspace of the antisymmetric operator J corresponding to the eigenvalue $+i$ and $\overline{W}$ is the complex conjugate space (corresponding to the eigenvalue $-i$). Define the complex bilinear (non-Hermitian) form

$$B(x,y) = < x, \overline{y} > \tag{6.4.5}$$

in $H^{\mathbf{C}}$. By definition, the orthogonal group $O_p(H^{\mathbf{C}})$ consists of invertible bounded linear operators A in $H^{\mathbf{C}}$ such that

$$[A, J] \in L_{2p} \text{ and } B(Ax, Ay) = B(x,y) \forall x, y. \tag{6.4.6}$$

The operator J plays here the role of ϵ earlier; using the decomposition $W \oplus \overline{W}$ we can write the operators $A \in O_p(H^{\mathbf{C}})$ in the block matrix form

$$A = \begin{pmatrix} a & b \\ c & d \end{pmatrix}$$

where the off-diagonal blocks are of type L_{2p}, a is a Fredholm operator in W, and d is a Fredholm operator in $\overline{W}$. The *real orthogonal* group $O_p(H)$ consists of all real matrices which satisfy (6.4.6). Let $\mathcal{S}_p(W)$ be the space of L_{2p} operators $S : W \to \overline{W}$ such that

$$B(Sx, y) = -B(Sy, x) \ \forall x, y \in W.$$

If we fix an orthonormal basis $\{e_1, e_2, \dots\}$ in W we can think of S as an antisymmetric matrix labelled by the positive integers, $S_{ij} = B(Se_i, e_j)$.

Exercise 6.4.7. Show that for each $g \in O_p(H^{\mathbf{C}})$ there is an element $S \in \mathcal{S}_p$ such that the operator $d + cS$ is invertible.

We shall now concentrate on the case $p = 1$. The spin extensions in the cases $p > 1$ are not central and one has to deal with infinite-dimensional Abelian extensions in a similar way a s for the linear groups GL_p. Since we do not have much to say about the representation theory we shall not consider these cases.

We need some information about the *Pfaffians* of antisymmetric operators. If A is an antisymmetric $2n \times 2n$ matrix we define

$$\mathrm{Pf}(A) = \frac{1}{2^n n!} \sum_{\sigma \in S_{2n}} (-1)^{\epsilon(\sigma)} A_{\sigma(1)\sigma(2)} \cdots A_{\sigma(2n-1)\sigma(2n)},$$

where $\epsilon(\sigma)$ is the parity of the permutation σ. It is known that $\mathrm{Pf}(A)^2 = \det(A)$; see Pressley and Segal [1986], or Jaffe, Lesniewski, and Weitsman [1989] for a proof involving fermionic functional integration. We shall use also the *relative Pfaffians* defined as follows. First let A be an invertible antisymmetric matrix and B an arbitrary antisymmetric matrix. Set

$$\mathrm{Pf}(A, B) = \frac{\mathrm{Pf}(A^{-1} - B)}{\mathrm{Pf}(A^{-1})}.$$

Now the square of $\mathrm{Pf}(A, B)$ is equal to $\det(1 - AB)$ which is well-defined even when A is singular. It follows that the square root $\mathrm{Pf}(A, B)$ is also a regular polynomial of the matrix elements for all A. In fact, one can show that

$$\mathrm{Pf}(A, B) = \sum_S \mathrm{Pf}(A_S)\mathrm{Pf}(B_S),$$

where the sum is taken over all minors of the square matrices, labelled by the index sets $S \subset \{1, 2, \ldots, 2n\}$. These definitions carry over to the infinite-dimensional case when one restricts to the operators A, B which differ from the identity operator by a trace class operator. Furthermore, in the case of L_p operators one can define generalized Pfaffians Pf_p by

$$\mathrm{Pf}_p(A, B) = \mathrm{Pf}(A, B) \exp\left(\frac{1}{2} \sum_{j=1}^{p-1} \frac{1}{j} \mathrm{tr}(AB)^j\right).$$

The separate factors on the right are well-defined only when $AB \in L_1$ but the product converges in the limit $AB \to L_p$; see Jaffe, Lesniewski, and Weitsman [1989] for proofs. The square of $\mathrm{Pf}_p(A, B)$ is easily seen to be $\det_p(1 - AB)$.

Consider the following fiber bundle over O_1. The fiber over a point $g \in O_1$ consists of all functions $f : \mathcal{S}_1 \to \mathbf{C}$ of the form

$$f(S) = \mathrm{Pf}((a + bS_0)^{-1}(a + bS))$$

where $S_0 \in \mathcal{S}_1$ is any element such that $a + bS_0$ is invertible. We have to show that the argument of the Pfaffian is of the form $1 - AB$, where

A and B are antisymmetric operators in L_2. Choosing an orthonormal basis $\{e_1, e_2, \ldots, \bar{e}_1, \bar{e}_2, \ldots\}$ in $W \oplus \overline{W}$ we can write the orthogonality conditions for g in this basis in the matrix form $APA^t = P$, where P is the block matrix

$$\begin{pmatrix} 0 & 1 \\ 1 & 0 \end{pmatrix}.$$

In particular, $ba^t + ab^t = 0$ for $g \in O_1$. Using this relation we observe that $A = (a + bS_0)^{-1}b \in L_2$ is an antisymmetric operator. Setting $B = S_0 - S$ we have $(a + bS_0)^{-1}(a + bS) = 1 - AB$. The fiber is equal to $\mathbf{C}^\times$; this follows from the fact that $(\mathrm{Pf}T)^2 = \det T$ and from the multiplicative property of the usual determinant. We want to define a multiplication in $\hat{O}_1$ such that it becomes a group with a covering homomorphism onto O_1 given by the projection onto the base manifold. Let us consider the product

$$(g, f)(g', f') = (g'', f'') \tag{6.4.8}$$

where $g'' = gg'$ and f'' is the function defined by

$$f''(S) = f(g'.S)f'(S), \; g'.S = (c' + d'S)(a' + b'S)^{-1}. \tag{6.4.9}$$

Exercise 6.4.10. Show that the operator $g.S$ is antisymmetric (whenever it is well-defined) for all $g \in O_1, S \in \mathcal{S}_1$.

Note that $a' + b'S$ may be a singular operator; however, the singularity $(a' + b'S)^{-1}$ is cancelled by the factor $a' + b'S$ in the expression for $f'(S)$ since

$$d'' + c''S = [a + b(g.S)](a' + b'S)$$

and therefore

$$\begin{aligned} f(g'.S)f'(S) = &\mathrm{Pf}((a + bS_0)^{-1}(a'' + b''S)(a' + b'S)^{-1}) \\ &\times \cdot \mathrm{Pf}((a' + b'S_0')^{-1}(a' + b'S)) \end{aligned}$$

and so

$$[f(g'.S)f'(S)]^2 = \det[(a + bS_0)^{-1}(a'' + b''S)(a' + b'S_0')^{-1}]$$

which is a regular function. Except for the twisting of the bundle $\hat{O}_1$ the product in (6.4.8) and (6.4.9) is the semidirect product of O_1 with the commutative fiber and therefore the multiplication is associative.

Since O_1 is a subgroup of GL_1 the central extension $\widehat{GL_1}$ can be pulled back to give a central extension of O_1. We want to show that the extension $\hat{O}_1$ is a double covering of the pull-back extension $\tilde{O}_1$. Define a map $\xi : \hat{O}_1 \to \widehat{GL}_1$ by

$$\xi(g,f) = (g,q,1)$$

where $f(S) = \mathrm{Pf}(a+bS_0)^{-1}(a+bS)$ and $q = a+bS_0$. The right-hand side indeed represents an element of $\widehat{GL}_1$ since $a-q \in L_1$. If $f' = \mathrm{Pf}(a+S_0')^{-1}(a+bS)$ then we can write $(g,f') = (g,f)\cdot\lambda$, where $\lambda = \mathrm{Pf}(a+bS_0')^{-1}(a+bS_0)$. Now $\xi(g,f') = (g,q(q^{-1}q'),1) \equiv (g,q,\det(q^{-1}q'))$ in $\widehat{GL}_1$. Taking account $\det A = (\mathrm{Pf}\, A)^2$ we get

$$\xi((g,f)\lambda) = \xi(g,f)\cdot\lambda^2 \tag{6.4.11}$$

which shows that ξ preserves the fibers and that it is $2-1$. By a simple computation one can show that ξ is a homomorphism and so the claim above has been proved. The Lie algebra cocycle determining the central extension $\hat{O}_1$ is $\frac{1}{2}\times$ the cocycle for $\mathbf{gl}_1$ when restricted to the Lie algebra of the orthogonal group.

The spin representation of $\hat{O}_1$

Next we want to determine the spin representation of $\hat{O}_1$ in a Fock space $\mathcal{F}(W)$ and and relate it to the action of $\hat{O}_1$ in a line bundle PF which is in a certain sense the square root of the determinant bundle DET. Let $\mathcal{J}_1(H)$ denote the space of all complex structures in H, defined by complex isotropic subspaces $V \subset H^{\mathbf{C}}$ such that $V \in Gr_1(W \oplus \overline{W})$; isotropic means here that $B(x,y) = 0$ for all $x,y \in V$. Both of the groups $O_1(H)$ and $O_1(H^{\mathbf{C}})$ act transitively on $\mathcal{J}_1$. We can write

$$\mathcal{J}_1 = O_1(H)/U(W) = O_1(H^{\mathbf{C}})/GL(W)$$

where the general linear group $GL(W)$ is embedded in the complex orthogonal group as the group of matrices

$$GL(W) \ni a \mapsto \begin{pmatrix} a & 0 \\ 0 & (a^t)^{-1} \end{pmatrix}.$$

The unitary group of W is embedded in the same way. Since $\mathcal{J}_1$ is a quotient of complex groups it is naturally a complex manifold. By the construction, $\mathcal{J}_1$ is a submanifold of Gr_1. The antisymmetric operators $S \in \mathcal{S}_1$ parametrize a dense open submanifold of the connected component of $\mathcal{J}_1$ containing the plane W. The plane corresponding to S is the graph of the operator $S : W \to \overline{W}$ and it is spanned by the vectors $w_i = e_i + Se_i$.

We want to show that there is a complex line bundle PF over $\mathcal{J}_1$ with a bundle map $\phi : PF \to DET$ such that ϕ projects to the embedding $\mathcal{J}_1 \to Gr_1$ on the base and that $\phi(z\lambda) = \phi(z)\lambda^2$ for $z \in PF$, $\lambda \in \mathbb{C}$. Let $\{e_1, e_2, \dots\}$ be a fixed orthonormal basis of W. We define a Fock space $\mathcal{F} = \mathcal{F}(W)$ such that an orthonormal basis for $\mathcal{F}$ is given by the exterior products

$$\psi(i) = e_{i_1} \wedge e_{i_2} \wedge \cdots \wedge e_{i_s}$$

where (i) is any increasing finite sequence of positive integers i_k. The creation and annihilation operators are

$$a_i^* = e_i \wedge \qquad\qquad a_i = \imath(e_i)$$

where $\imath(x)$, $x \in W$, is the contraction operator. The element $1 \in \mathcal{F}$ is the vacuum characterized by $a_i \cdot 1 = 0$ for all i. The only nonzero anticommutation relations are

$$a_i a_j^* + a_j^* a_i = \delta_{ij}. \tag{6.4.12}$$

The Lie algebra of O_1 consists of operators of the form

$$x = \begin{pmatrix} a & b \\ c & d \end{pmatrix}$$

such that

$$c^t = -c \qquad b^t = -b \qquad d = -a^t \qquad b, c \in L_2. \tag{6.4.13}$$

Each element x of the Lie algebra can be represented by a linear operator $T(x)$ in the Fock space $\mathcal{F}$,

$$T(x) = d_{ij} a_i^* a_j + \frac{1}{2} b_{ij} a_i a_j + \frac{1}{2} c_{ij} a_i^* a_j^*. \tag{6.4.14}$$

This is not a true representation of the Lie algebra; there is a 2-cocycle in the commutation relations,

$$[T(x), T(x')] = T([x, x']) + \frac{1}{2}\mathrm{tr}(cb' - c'b). \tag{6.4.15}$$

Except for the factor 1/2 this is same as the cocycle η of $\mathbf{gl}_1$ we met in the previous section. The representation T can be exponentiated to give a projective representation of the connected component of identity

$SO_1 \subset O_1$. It is reducible; the irreducible components correspond to the subspaces consisting of polynomials of even and odd degree in the creation operators a_i^* [Pressley and Segal, 1986].

Let $V \in \mathcal{J}_1$ be in the connected component which contains the element W and choose $g \in SO_1$ such that $V = g \cdot W$. We associate to V the line PF_V in $\mathcal{F}$ spanned by the vector $T(g) \cdot 1$. If $V = g' \cdot W$ then $g' = gk$, where k is the orthogonal transformation $\begin{pmatrix} a & 0 \\ 0 & (a^t)^{-1} \end{pmatrix}$. The vacuum $1 \in \mathcal{F}$ is invariant under the action of k [this can be seen from the formula (6.4.14)] and so $T(g') \cdot 1$ is the same as $T(g) \cdot 1$ up to a phase. We have now defined a complex holomorphic line bundle PF over the component of $\mathcal{J}_1$ containing W.

The vectors in $\mathcal{F}(W)$ can be thought of as sections of the dual bundle PF^*. Namely, for a given $\psi \in \mathcal{F}(W)$ there is a linear map $PF_V \to \mathbf{C}$ given by $PF_V \ni x \mapsto (x, \psi)$; here $(\cdot,\cdot)$ is the inner product in the Fock space determined by the condition $(\psi_0, \psi_0) = 1$ and that each a_i is the adjoint of a_i^*. Only the even elements of the Fock space give nonzero elements in $\Gamma(PF^*)$ since all the vectors in PF are of even degree.

The line $PF_{g \cdot W}$ in $\mathcal{F}(W)$ is characterized by the property that it is annihilated by the transformed annihilation operators

$$a_i' = a_{ij} a_j + b_{ij} a_j^*, \text{for } g = \begin{pmatrix} a & b \\ c & d \end{pmatrix}.$$

If $V = g \cdot W \in \mathcal{J}_1$ corresponds to an operator $S \in \mathcal{S}_1$ then the expression for the vector generating the line PF_V can be simplified as follows. Let

$$\psi_V = e^{\frac{1}{2} S_{ij} a_i^* a_j^*} \cdot 1.$$

Now the orthogonal matrix g can be chosen such that $a = d = 1$, $b = 0$ and $c = S$. Thus $g = \exp\begin{pmatrix} 0 & 0 \\ S & 0 \end{pmatrix}$ and so PF_V is spanned by ψ_V.

Exercise 6.4.16. Show that

$$\psi_V = \sum_{(i)} \mathrm{Pf}(S(i)) e_{i_1} \wedge e_{i_2} \wedge \ldots e_{i_{2N}}$$

where the sum is over all sequences $(i) = \{i_1 < i_2 < \cdots < i_{2N}\}$ of positive integers with even number of elements and $S(i)$ is the submatrix of S obtained by choosing the rows and columns labelled by (i).

We want to show that PF indeed can be understood as the square root of DET restricted to the submanifold $\mathcal{J}_1 \subset Gr_1$. The group $\hat{O}_1$ acts transitively in $PF^\times$ (the zero section deleted) and the stability group at the fiber over $V = W$ is $GL(W)$. Thus $PF^\times = \hat{O}_1 / GL(W)$ and we can define the covering map $sq : PF \to DET$ as follows. An arbitrary point $p \in PF^\times$ can be represented as $p = g \cdot 1$ where $g \in \hat{O}_1$. The bundle $DET^\times$ is the homogeneous space $\widehat{GL_1}/K$ where K consists of the operators with $c = 0$. We set $sq(p) = \xi(g) \bmod K$. This is well-defined since $GL(W) \subset K$. The map sq can be described more explicitly over points in $\mathcal{J}_1$ parametrized by the operators S. For that purpose we shall first describe the *Plücker embedding* of DET into the Fock space $\mathcal{F}(H)$.

Let $\mathcal{F}(H)$ be the space defined using the creation and annihilation operators a_n^*, a_n labelled by both negative and positive integers. The vacuum vector ψ_0 in $\mathcal{F}(H)$ is characterized by the properties

$$a_n\psi_0 = a_{-n}^*\psi_0 = 0, \ \forall n > 0. \tag{6.4.17}$$

The line bundle DET_1 is embedded in $\mathcal{F}(H)$ by the Plücker coordinates; If an element of DET_1 is represented by the pair $(w, \lambda) \in St_1 \times \mathbf{C}$ then the image in $\mathcal{F}(H)$ is the vector (following the notation in 6.2)

$$\lambda \sum_{(i)\in\mathcal{N}} \det w_{(i)} f_{j_1} \wedge f_{j_2} \wedge \dots \tag{6.4.18}$$

where $\mathcal{N}$ consists of all increasing sequences $(i) = \{i_1, i_2, \dots\}$ of integers such that $(i) \cap \mathbf{N}$ and $\mathbf{N} \setminus (i)$ are finite, $(j) = \mathbf{Z} \setminus (i)$, $j_1 > j_2 > \dots$ and $w_{(i)}$ is the matrix formed from the rows of $w = \begin{pmatrix} w_+ \\ w_- \end{pmatrix}$ labelled by the integers (i). Each sequence (j) contains only a finite number of positive integers and almost all the negative integers. If w' differs from w by a basis transformation $t \in GL^1$ then the image of $(w', 1)$ in $\mathcal{F}(H)$ differs from (6.4.18) only by the determinant of t. It follows that we have a well-defined map from DET to $\mathcal{F}(H)$ which is linear in the fibers. This is the Plücker embedding.

Let again $V \in \mathcal{J}_1$ correspond to an operator $S \in \mathcal{S}_1$. The restriction of $sq : PF \to DET$ to the fiber over V can now be described as follows. The generator ψ_V of PF_V is mapped onto the vector ψ'_V represented by the pair $(w, 1) \in St_1 \times \mathbf{C}$, where $w_i = e_i + Se_i$ is the basis of V. The Plücker coordinates of ψ' are then obtained from the subdeterminants of the matrix $\begin{pmatrix} 1 \\ S \end{pmatrix}$.

The spin bundle, Clifford algebra, and the Dirac operator

There is an infinite-dimensional Clifford algebra Cl which acts in the Fock space $\mathcal{F}(W)$. To each basis vector $e_i \in W$ and $\bar{e}_i \in \overline{W}$ there corresponds a "gamma matrix",

$$\gamma(e_i) = a_i, \ \gamma(\bar{e}_i) = a_i^* \quad (i = 1, 2, \dots).$$

They satisfy the anticommutation relations

$$(6.4.19) \qquad \gamma(x)\gamma(y) + \gamma(y)\gamma(x) = B(x, y).$$

The operators $\gamma(x)$ are defined by linearity for each $x \in H$. As in the finite-dimensional case, by (6.4.14) the Lie algebra of $\hat{O}_1$ is the linear span of the operators $[\gamma(x), \gamma(y)]$. The Clifford algebra Cl is simply the algebra generated by the creation and annihilation operators.

We shall now describe the (extended) spin structure on the Grassmannian Gr_1. Let W be a point on the Grassmannian; Gr_1 is defined by a splitting $H = H_+ \oplus H_-$ of a complex Hilbert space H. We shall consider here only the connected component of the Grassmannian containing the element $W = H_+$ and we shall denote by Gr_1 that component. Using the notation of the previous section the plane W is a $+1$ eigenspace of a Hermitian operator F. The tangent space $T_W Gr_1$ consists of all Hermitian operators $G \in L_2$ such that $FG + GF = 0$. A Riemannian metric on Gr_1 is defined by

$$< G, G' >= \operatorname{tr} GG'.$$

Let $\{w_1, w_2, \dots\}$ be any orthonormal basis of W and $\{w_n\}_{n<0}$ an orthonormal basis in $W^\perp$. The complexified tangent space $T_W^{\mathbf{C}} Gr_1$ splits into subspaces T_+ and $T_- = \overline{T}_+$ such that T_+ (respectively, T_-) consists of the complex tangent vectors G such that $FG = +G$ (respectively, $FG = -G$.) Let G_{ij} be the operator in H defined by $< w_k, G_{ij} w_l >= \delta_{ki}\delta_{jl}$. The operators $\{G_{n,-m}\}_{n,m>0}$ form an orthonormal basis of T_+ with respect to the Hermitian inner product

$$< G, G' >= \operatorname{tr} GG'^*$$

in $T_W^{\mathbf{C}}$. Similarly, the operators $\{G_{-n,m}\}_{n,m>0}$ define an orthonormal basis of T_-. A basis of $T_W^{\mathbf{C}}$ is by definition admissible if it is related to the basis $\{G_{n,-m}, G_{-n,m}\}_{n,m>0}$ by an orthogonal transformation $A \in O_1$; the orthogonality is defined with respect to the non-Hermitian inner product $B(G, G') =< G, G'^* >= \operatorname{tr} GG'$ in $T_W^{\mathbf{C}}$; note that in the definition of the (complex or real) orthogonal group O_1 the space T_+ plays here the role of the space which was denoted by W in the construction of the spin groups and T_- corresponds to $\overline{W}$ there. It is easy to see that a change of the basis in W or $W^\perp$ corresponds to a rotation of the G_{ij} basis in $T_W^{\mathbf{C}}$ by an orthogonal matrix of the form

$$\begin{pmatrix} a & 0 \\ 0 & (a^t)^{-1} \end{pmatrix}$$

where a is a unitary operator. Thus the notion of an admissible basis in $T_W^{\mathbf{C}}$ does not depend on the choice of orthonormal basis in $W, W^\perp$. The restricted frame bundle OF_1 of Gr_1 consists of all admissible frames over points $W \in Gr_1$.

As in the finite-dimensional example discussed in the beginning of this section we use the fact that Gr_1 is the homogeneous space $U_1/[U(H_+) \times U(H_-)]$ and thus the structure group O_1 of OF_1 reduces to the subgroup $K = U(H_+) \times U(H_-)$. The group K is embedded in O_1 by the homomorphism σ which sends an element $(g_+, g_-) \in K$ to the element $\begin{pmatrix} a & 0 \\ 0 & (a^t)^{-1} \end{pmatrix}$ with

$$a_{(n,-m),(k,-l)} = (g_+)_{kn}(g_-)_{-m,-l}$$

where the labelling of the matrix elements of a by the set $\mathbf{N} \times (-\mathbf{N})$ corresponds to the labelling of the basis $G_{n,-m}$ in $T_W^{\mathbf{C}}$. The frame bundle OF_1 can now be thought of as an associated bundle to the principal bundle $U_1 \to U_1/K = Gr_1$ through the homomorphism $\sigma : K \to O_1$.

The homomorphism σ can be lifted to a homomorphism $\tilde{\sigma} : K \to \hat{O}_1$ by $\tilde{\sigma}(g) = (\sigma(g), 1)$, in the notation around (6.4.8). We can now define the extended spin structure $Spin_1$ on Gr_1 as the associated bundle

$$Spin_1(Gr_1) = U_1(H) \times_{\tilde{\sigma}} \hat{O}_1(\mathcal{H}).$$

The orthogonal group $O_1(\mathcal{H})$ acts in the Hilbert space $\mathcal{H} = \mathcal{W} \oplus \overline{\mathcal{W}}$, where $\mathcal{W} = \ell_2(\mathbf{N} \times (-\mathbf{N}))$ and $\overline{\mathcal{W}} = \ell_2((-\mathbf{N}) \times \mathbf{N})$ are the Hilbert spaces of square summable sequences over the given indexing sets. The covering map $Spin_1 \to OF_1$ is defined by $U_1 \times \hat{O}_1 \ni (u, g) \mapsto (u, \rho(g))$ where $\rho : \hat{O}_1 \to O_1$ is the canonical projection. The fiber of the covering is equal to S^1 in the real case and $\mathbf{C}^\times$ in case of the complex orthogonal group.

With the spin bundle at hand, it would be natural to try to define the Dirac operator on the Grassmannian Gr_1. In the infinite-dimensional case it is not a priori clear what is the correct domain of definition of the Dirac operator D. Let us first define the operator D in a formal way. The γ matrices are the creation and annihilation operators a_{ij}^*, a_{ij} in $\mathcal{F}(\mathcal{W})$ $(i, j > 0)$ corresponding to the basis $e_{i,-j}$ of $\mathcal{W}$ and $e_{-i,j}$ of $\overline{\mathcal{W}}$. We identify the space $\mathcal{H}$ as the tangent space of Gr_1 at the point $W = H_+$; an orthonormal basis in the complexified tangent space is generated by the action of the elements $e_{i,-j}, e_{-i,j}$ of the complexified Lie algebra of GL_1. Define the linear operators

$$L_{ij} = e_{ij} - a_{jk}^* a_{ik}, \; L_{-i,-j} = e_{-i,-j} + a_{ki}^* a_{kj}$$

for $i, j > 0$. The operators L_{ij} generate the Lie algebra of $GL(H_+)$ and the $L_{-i,-j}$'s the Lie algebra of $GL(H_-)$.

A section of of the bundle E of spinors over Gr_1 is a function $\psi : U_1 \to \mathcal{F}(\mathcal{W})$ such that $\psi(gh) = T(h^{-1})\psi(g)$ for all $g \in U_1$ and $h \in K$, where T is the spin representation of $\hat{O}_1$ in $\mathcal{F}(\mathcal{W})$ and elements of $K = U(H_+) \times U(H_-)$ are considered as elements of $\hat{O}_1$ via the homomorphism $\tilde{\sigma}$. Equivalently, sections of E are functions $\psi : U_1 \to \mathcal{F}(\mathcal{W})$ such that

$$L_{ij}\psi = L_{-i,-j}\psi = 0, \text{ for all } i, j > 0. \tag{6.4.20}$$

It is agreed that the generators x of U_1 act on functions on the group U_1 through the right action, $(x \cdot \psi)(g) = \frac{d}{dt}\psi(ge^{tx})|_{t=0}$. The Dirac operator can be written as

$$D = e_{n,-m}a^*_{nm} + e_{-n,m}a_{mn}. \tag{6.4.21}$$

By a simple computation,

$$[L_{ij}, D] = [L_{-i,-j}, D] = 0$$

and therefore the first order differential operator D indeed maps the space of sections of E into itself. However, we have to take account that D is an infinite sum and the action on ψ diverges in general. Further "renormalizations" are needed in order to make sense of the Dirac operator. It is an open and very interesting problem to find out what is precisely the mathematically correct definition of the Dirac operator in this infinite-dimensional context. The case of a Dirac operator on a loop space, coupled to a polynomial potential, has been recently attacked in Jaffe, Lesniewski, and Weitsman [1987]; it is conceivable that similar methods could work here because of the close relation between Grassmannians and loop spaces.

6.5. The Plücker embedding and a spherical function

The case $p = 1$

In Section 6.4 we met already the Plücker embedding of the determinant bundle DET_1 into the Fock space $\mathcal{F}(H)$. In this section we shall continue the discussion on Plücker embedding from a slightly different standpoint which makes possible a generalization to the case of DET_p for $p > 1$. We shall first start from the case $p = 1$ and then we shall examine to what extent the results can be generalized to $p > 1$.

Let us define a map $\phi : St_1 \times St_1 \to \mathbf{C}$ by

$$\phi(z, w) = \det z^* w = \det(z_+^* w_+ + z_-^* w_-).$$

For each $z \in St_1$ the function $w \mapsto \phi(z, w)$ is a section of DET_1^* and therefore we have a map $St_1 \to \Gamma(DET_1^*)$. We can now define a map $\xi : DET_1 \to \Gamma(DET_1^*)$ by

$$\xi(z, \lambda) = \overline{\lambda} \cdot \phi(z, \cdot), \tag{6.5.1}$$

where $\lambda \in \mathbf{C}$ and $z \in St_1$. Deleting the zero section from DET_1 we get a map $\xi : DET_1^\times \to \Gamma(DET_1^*)$; this latter map is clearly an injection.

Let W be a point in the Grassmannian Gr_1. We can associate a complex line ℓ in $\Gamma(DET_1^*)$ to the point W: Choose any admissible basis w for W and let ℓ be the line spanned by $\xi(w, 1)$. This is the Plücker embedding. The set of *Plücker coordinates* of W are defined as follows. Let $\mathcal{S}$ be the set of increasing sequences $S = (i_1, i_2, \dots)$ of integers such that $S \cap -\mathbf{N}$ and $\mathbf{N} \setminus S$ are finite. For each $S \in \mathcal{S}$ let $e_S \in St_1$ be the basis consisting of vectors $\{e_i\}_{i \in S}$. The Plücker coordinates of w is then the collection of complex numbers $\phi(e_S, w)$, $S \in \mathcal{S}$. In fact, the Plücker coordinate defined by S is the same as $\psi_S(w)$ in the notation of 6.2.

If A is any $n \times m$ matrix and B a $m \times n$ matrix with $n \le m$, then we have the following identity:

$$\det AB = \sum_{(i)} \det A(i) \det B(i), \tag{6.5.2}$$

where the sum is over all sets of integers $1 \le i_1 < i_2 < \cdots < i_n \le m$, $A(i)$ is the matrix obtained from A by selecting the columns labelled by the integers (i) and $B(i)$ is the matrix obtained by selecting from B the rows labelled by the set (i). Using (6.5.2) we obtain

$$\begin{aligned} \phi(z, w) &= \sum_{S \in \mathcal{S}} \overline{\psi_S(z)} \psi_S(w) \\ &= \sum_{S \in \mathcal{S}} \overline{\xi_z(e_S)} \xi_w(e_S), \end{aligned} \tag{6.5.3}$$

where ξ_w denotes the section $\xi(w, 1)$ of DET_1^*, i.e., $\xi_w(w') = \phi(w, w')$. In particular,

$$\phi(e_S, e_{S'}) = \delta_{SS'}, \qquad S, S' \in \mathcal{S}. \tag{6.5.4}$$

Let V be the vector space consisting of all finite linear combinations of the sections ξ_z. We define a Hermitian form in V by setting

$$(6.5.5) \qquad < \xi_z, \xi_w >= \phi(z,w)$$

and extending this by linearity to all vectors in V. The Hermitian form is positive semidefinite: if $\psi = \sum \alpha_n \xi_{z_n}$ then

$$< \psi, \psi > = \sum_{n,m} \overline{\alpha_n} \alpha_m \det z_n^* z_m$$

$$= \sum_{S \in \mathcal{S}} \left(\sum_n \overline{\alpha_n} \det z_n(i)^* \right) \left(\sum_m \alpha_m \det z_m(i) \right)$$

which is a sum of squared absolute values and thus non-negative. Let V_0 be the subspace consisting of vectors with zero norm. Now V/V_0 has a positive definite inner product defined by (6.5.5) and so its completion $\mathcal{F}$ is a Hilbert space.

The proof of positivity of the kernel $\phi(z,w)$ shows that ϕ can be viewed as the "Green's function" for the holomorphicity equations for sections in DET_1^*. Namely, for each fixed z $\phi(z,\cdot)$ is a holomorphic section and ϕ is *complete* in the sense that all holomorphic sections are linear combinations of the $\phi(z,\cdot)$'s with different values of z. In particular, $\psi_S = \phi(e_S,\cdot)$. With a little effort one can show that the finite linear combinations of the sections ψ_S are dense in the space of holomorphic sections with respect to the topology of uniform convergence on compact subsets of Gr_1 [Pressley and Segal, 1986, Chapter 10].

PROPOSITION 6.5.6. *The mapping* $\xi_w \mapsto \{\xi_w(e_S)\}_{S\in\mathcal{S}}$ *extends by linearity to an isometric isomorphism* $\kappa : \mathcal{F} \to \ell_2$, *where the coordinates of vectors in* ℓ_2 *are labelled by the sets* S.

PROOF: By (6.5.4) the vectors ξ_{e_S} form an orthonormal set in $\mathcal{F}$. This set is mapped by κ onto the standard orthonormal basis of ℓ_2. It remains to show that the set $\{\xi_{e_S}\}_{S\in\mathcal{S}}$ is complete. Suppose that ξ_w is orthogonal to each ξ_{e_S}. Then

$$||\xi_w||^2 = \det w^* w = \sum_S \det w(i)^* \det w(i)$$

$$= \sum_S < w, e_S >< e_S, w >= 0$$

and thus $\xi_w \in V_0$. It follows that each ξ_w is an (infinite) linear combination of the vectors ξ_{e_S} and, since V is generated by the vectors ξ_w, each vector in V is a linear combination of the vectors ξ_{e_S}.

In Section 6.2 we defined a fiber metric in the bundle DET_p in terms of orthonormal admissible basis. We shall need a formula for the metric written in such a way that it is valid for an arbitrary admissible basis. In the case $p = 1$ we can set

$$|(w, \lambda)|^2 = |\lambda|^2 \det w^* w = |\lambda|^2 \phi(w, w). \tag{6.5.7}$$

Directly from the definitions we have the following proposition:

PROPOSITION 6.5.8. *The mapping $\xi : DET_1 \to \mathcal{F}$ is an isometry with respect to the fiber metric in DET_1.*

The group $\widehat{GL_1}$ acts in $\mathcal{F}$ by

$$T(g, q, \lambda)\xi_w = \lambda^{-1}\xi_{w'}, \; w' = (g^*)^{-1} w q^*.$$

This action is just the action of $\widehat{GL_1}$ on sections of DET_1^*. Namely,

$$(T(g, q, \lambda)\xi_w)(z) = \lambda^{-1}\xi_w(g^{-1} z q).$$

The restriction to unitary g and q and $|\lambda| = 1$ gives a unitary representation of $\widehat{U}_1$ in $\mathcal{F}$.

The case $p = 2$

Next we shall discuss the Plücker embedding for Gr_p when $p = 2$. The case $p > 2$ does not lead to essentially new phenomena and we want to keep $p = 2$ for simplicity. First we have to modify the definition of ϕ in an appropriate way since the determinant does not converge. We set

$$\phi(z, w) = \det{}_2(z^* w) \exp[\mathrm{tr}(z^* w - z_+^* - w_+ + 1 - \tfrac{1}{4} G_{21}^* F_{21})],$$

where F is the operator which is $+1$ on the plane spanned by w and -1 on the complement of the w-plane, and G corresponds similarly to the plane spanned by the basis z. It is not completely obvious that the trace really converges and we shall prove it before proceeding. Write $w_+ = 1 + x$ and $z_+ = 1 + y$ where x and y are Hilbert-Schmidt operators. Then

$$z^* w = z_+^* w_+ + z_-^* w_- \equiv 1 + x + y^* + z_-^* w_-$$

modulo a trace-class operator. Thus the operator in the exponent is

$$A = z_-^* w_- - \frac{1}{4} G_{21}^* F_{21}$$

modulo trace class operators. But we know from Section 6.3 that $\frac{1}{2}F_{21} - w_-$ is in $L_{4/3}$ and therefore $z_-^*(w_- - \frac{1}{2}F_{21})$ and $(z_- - \frac{1}{2}G_{21})^*\frac{1}{2}F_{21}$ are trace class operators (since $z_-, F_{21} \in L_4$); it follows that their sum A is also a trace class operator.

A simple computation shows that

$$\phi(z, wt) = \phi(z, w)\omega_2(w_+, t) \text{ for } t \in GL^2$$

and therefore $z \mapsto \phi(z, \cdot)$ maps St_2 into $\Gamma(DET_2^*)$. Denote by ξ_z the section $\xi_z(w) = \phi(z, w)$. As in the case $p = 1$ we can define a map $\xi : DET_2 \to \Gamma(DET_2^*)$ by $\xi(z, \lambda) = \overline{\lambda}\xi_z$; this is well-defined by

$$\phi(zt, w) = \phi(z, w)\overline{\omega_2(z_+, t)}.$$

Let again V be the space consisting of finite linear combinations of the vectors ξ_z, $z \in St_2$. A Hermitian form is defined as in the case $p = 1$ by $< \xi_z, \xi_w >= \phi(z, w)$. However, the proof of positivity is not valid in the present context. At the moment it is not known whether the Hermitian form is positive semidefinite in the case $p = 2$.

In fact, we are not so much interested in the positivity of the metric in $V(p = 2)$ since this space is not the correct choice if we want to construct highest weight representations of $\widehat{GL_2}$. The reason is that the space V is not invariant under the action of $\widehat{GL_2}$. We have to extend the space such that it will contain also sections of the type

$$e^{\operatorname{tr}\xi(F-\epsilon)} \cdot \psi, \qquad \psi \in V.$$

Spherical function for a highest weight representation of $\widehat{GL_1}$

Let us return for a moment to the case $p = 1$. The highest weight representation of $\widehat{GL_1}$ in $\Gamma(DET_1^*)$ can be defined in terms of the *spherical function* as follows. Consider the space R of formal linear combinations of the symbols $T(z)\psi_0$ with $z \in \widehat{GL_1}$. Define a conjugation (antiautomorphism) in $\widehat{GL_1}$ by

$$(g, q, \lambda)^* = (g^*, q^*, \overline{\lambda})$$

where the asterisk on the right-hand-side means Hermitian conjugation. Given a function $X : \widehat{GL_1} \to \mathbf{C}$ we can define a Hermitian form in R by setting

$$< T(z)\psi_0, T(z')\psi_0 >=< \psi_0, T(z^*z')\psi_0 >= X(z^*z') \tag{6.5.9}$$

How should we choose X? The answer follows from the requirement that ψ_0 should be the highest weight vector; this means that it is invariant under the subgroup B_+ represented by the triples $(g, a, 1)$, where

$$g = \begin{pmatrix} a & 0 \\ c & d \end{pmatrix} \in \widehat{GL_1}.$$

A vector in R is then a linear combination of the vectors $\psi_b = T(g(b), 1, 1)\psi_0$, where $g(b) = \left(\begin{smallmatrix} 1 & b \\ 0 & 1 \end{smallmatrix}\right)$. To compute the inner product of vectors ψ_b and $\psi_{b'}$ we write (we assume for the moment that $1 + b'b^*$ has an inverse)

$$\begin{aligned} g(b)^* g(b') &= \begin{pmatrix} 1 & b' \\ b^* & 1 + b^* b' \end{pmatrix} \\ &= \begin{pmatrix} (1 + b'b^*)^{-1} & b' \\ 0 & 1 + b^* b' \end{pmatrix} \begin{pmatrix} 1 & 0 \\ (1 + b^* b')^{-1} b^* & 1 \end{pmatrix}. \end{aligned}$$

Using the invariance property of the highest weight vector ψ_0 we reduce that

$$(6.5.10) \qquad <\psi_b, \psi_{b'}> = <\psi_0, T(1, 1 + b'b^*, 1)\psi_0> = \det(1 + b'b^*).$$

The right-hand side is well-defined even in the case when $1 + b'b^*$ does not have an inverse.

The formula (6.5.10) defines in fact the spherical function X. Namely, given an element (g, q, λ) in $\widehat{GL_1}$ we can write (assuming that d and the a-block $a_{g^{-1}}$ of g^{-1} have an inverse)

$$(g, q, \lambda) = (g_1, a_{g^{-1}}^{-1}, 1) \cdot (1, a_{g^{-1}} q, \lambda) \cdot (g_2, 1, 1),$$

where

$$g_1 = \begin{pmatrix} a_{g^{-1}}^{-1} & b \\ 0 & d \end{pmatrix} \quad g_2 = \begin{pmatrix} 1 & 0 \\ d^{-1} c & 1 \end{pmatrix}.$$

By the invariance property of the vacuum we obtain

$$\begin{aligned} X(g, q, \lambda) &= <\psi_0, T(g, q, \lambda)\psi_0> = <\psi_0, T(1, a(g^{-1} q, \lambda)\psi_0> \\ (6.5.11) \qquad &= \lambda^{-1} \det(a_{g^{-1}} q). \end{aligned}$$

The spherical function X is well-defined and continuous also at the points where d and $a_{g^{-1}}$ are not invertible.

The spherical function for $\widehat{GL}_2$

Let us now return to the case $p = 2$. The first problem is to define the conjugation in $\widehat{GL}_2$. In order to profit from the simple action $F \mapsto gFg^{-1}$ of GL_p on the complexified Grassmannian $\mathbf{C}Gr_p$ we shall here consider the extension of GL_p by the Abelian group $Map(\mathbf{C}Gr_p, \mathbf{C}^\times)$ instead of $Map(Gr_p, \mathbf{C}^\times)$; in fact, using our choice of the cocycle $\alpha(g, q; w)$ we observe that we can restrict to the subgroup of $Map(\mathbf{C}Gr_p, \mathbf{c}^\times)$ consisting of the exponentials $\lambda = \exp(\beta + \mathrm{tr}\xi(F - \epsilon))$ for $\beta \in \mathbf{C}$ and ξ a bounded operator in $H_+ \oplus H_-$ such that $\xi_{12}, \xi_{21} \in L_4$ and $\xi_{11} - 1, \xi_{22} + 1 \in L_2$; we shall denote this extension also by $\widehat{GL}_1$. We use the ansatz

$$(g, q, \lambda)^* = (g^*, q^*, \lambda^* \mu(g, q; \cdot)), \tag{6.5.12}$$

where $\lambda^*(F) \equiv \overline{\lambda(g^{-1}F^*g)}$ and $F \mapsto \mu(g, q; F)$ is a function on the *complexified Grassmannian* . We have to choose μ in such a way that

$$((g, q, \lambda)(g', q', \lambda'))^* = (g', q', \lambda')^*(g, q, \lambda)^*. \tag{6.5.13}$$

Using the multiplication rule in $\widehat{GL}_2$ we obtain the condition

$$\begin{aligned} &\mu(g, q; F)\mu(g', q'; g^*Fg^{*-1})\mu(gg', qq'; F)^{-1} \\ &\quad = \overline{\alpha}_1(g, q; g^{-1}F^*g)\overline{\alpha}_1(g', q'; g'^{-1}g^{-1}F^*gg') \\ &\quad \times \overline{\alpha}_1(gg', qq'; g'^{-1}g^{-1}F^*gg')^{-1}\alpha_1(g^*, q^*; F)^{-1}, \end{aligned} \tag{6.5.14}$$

where

$$\alpha_1(g, q; F) = e^{-\frac{1}{2}\mathrm{tr}q^{-1}bF_{21} + \mathrm{tr}(1 - q^{-1}a)}$$

is well-defined only in the subgroup $\widehat{GL}_1$ ($b, F_{21} \in L_2$ and $1 - q^{-1}a \in L_1$) but the product of α_1's in (6.5.14) extends to a continuous function for all $g \in GL_2$ and $F \in \mathbf{C}Gr_2$. Formally, we could write a solution of (6.5.14) as $\mu = \overline{\alpha}_1(g, q; g^{-1}F^*g)\alpha_1(g^*, q^*; F)^{-1}$ but the trouble is that this is well-defined only in the case $p = 1$. Similarly as before we have to introduce a regularizing factor. With a little experimentation the correct choice turns out to be

$$\mu(g, q; F) = \frac{\overline{\alpha}_1(g, q; g^{-1}F^*g)h(g^*Fg^{*-1})}{\alpha_1(g^*, q^*; F)h(F)}, \tag{6.5.15}$$

where $h(F) = \exp(-\frac{1}{4}\mathrm{tr}\epsilon(F - \epsilon))$.

Exercise 6.5.16. Prove that the right-hand side of (6.5.15) extends to a continuous function in the range of variables corresponding to the index $p = 2$.

With the conjugation at hand we shall define the inner product by $< T(z)\psi_0, T(z')\psi_0 >= X(z^* z')$ where $X : \widehat{GL_2} \to \mathbf{C}$ is the spherical function. Again we require that the vacuum is invariant under the subgroup B_+ of elements $(g, a, 1)$, where $g \in GL_2$ is a lower triangular operator $(b = 0)$. From this follows that

$$X(z_1^* z z_2) = X(z)$$

when z_1 and z_2 belong to the subgroup B_+. Almost all elements (g, q, λ) can be split as $z_1^* z z_2$, where $z_1, z_2 \in B_+$ and z is in the Abelian subgroup consisting of functions $\exp(\xi(F - \epsilon) + \eta)$. Using the same notation as in the case $p = 1$ above we can write

$$(g, q, \lambda) = (g_1, a_{g^{-1}}^{-1}, 1) \cdot (1, 1, \lambda') \cdot (g_2, 1, 1) \cdot (1, t, (\det{}_2 t^{-1})^{-1}) \tag{6.5.17}$$

where $t = a(g^{-1}q)$ and

$$\begin{aligned} \lambda'(F) =& \lambda(g^{-1}Fg)\det{}_2 t^{-1} \cdot e^{\frac{1}{2}\operatorname{tr} a_{g^{-1}} b F_{21}} \\ & \times e^{-\frac{1}{2}\operatorname{tr} q^{-1} b (g_2^{-1} F g_2)_{21}} e^{-\operatorname{tr} q^{-1} b d^{-1} c}. \end{aligned} \tag{6.5.18}$$

Note that the individual factors in (6.5.18) do not converge for $p = 2$ but only the product converges to a regular function when extending from $p = 1$ to $p = 2$.

Looking at the formula above we observe a problem: λ' has a singularity at the points where d and $a_{g^{-1}}$ are not invertible. However, by defining

$$X(1, 1, e^{\operatorname{tr}\xi(F-\epsilon)+\eta}) = e^{-\eta - 2\operatorname{tr}\xi^2} \tag{6.5.19}$$

the singularity will disappear from the expression for $X(1, 1, \lambda')$. In fact, writing $\lambda(F) = e^{\operatorname{tr}\xi(F-\epsilon)+\eta}$, we obtain by a straightforward but a bit tedious computation

$$\begin{aligned} X(g, q, \lambda) = \det t \cdot \exp\{ & \operatorname{tr}[2aq^{-1} - a_{g^{-1}} a q^{-1} a - 1 \\ & + 2(1 - a_{g^{-1}} a)\xi_{11} - 2c_{g^{-1}} a \xi_{12} \\ & -2a_{g^{-1}} b \xi_{21} - 2c_{g^{-1}} b \xi_{22}] - \eta \} \end{aligned}$$

This expression is manifestly free of singularities at the singular points of d and $a_{g^{-1}}$. Again, not all of the individual terms in the exponent are well-defined for $p = 2$ but the whole expression converges. Using the fact that $\xi_{11} - 1, \xi_{22} + 1 \in L_2$ and $b, c \in L_4$ it is easy to see that all traces involving the components of ξ converge. The rest, modulo the manifestly finite factor $\det_2 t\, e^{-\eta}$, is equal to

$$\begin{aligned}&\exp\left\{\mathrm{tr}[(t-1)+2aq^{-1}-a_{g^{-1}}aq^{-1}a-1]\right\}\\ &=\exp\left\{\mathrm{tr}[(q-a)a_{g^{-1}}+(aq^{-1}-1)+(1-aa_{g^{-1}})(aq^{-1}-1)]\right\}.\end{aligned}$$

The last term on the right-hand side in the exponent is finite since $1-aa_{g^{-1}}=bc_{g^{-1}}\in L_2$ and $aq^{-1}-1\in L_2$. The sum of the first two terms can be written as

$$(q-a)(a_{g^{-1}}-q^{-1})$$

which is finite.

By a direct computation we obtain that

$$\overline{X(z)}=X(z^*)$$

for all $z\in\widehat{GL}_1$. Now we have a Hermitian form $<\cdot,\cdot>$ in the space of formal linear combinations of the vectors $T(z)\psi_0$ defined by $<T(z)\psi_0, T(z')\psi_0>=X(z^*z')$. It is not a priori clear that this defines a Hermitian form in the space Γ of sections $\psi=\sum\alpha_i T(z_i)\psi_0$ of $\mathbf{C}DET_2^*$, $\psi_0(w)=\det_2 w_+$. We have to show that if ψ is zero as a section of $\mathbf{C}DET_2^*$ then the *formal* linear combination $\sum\alpha_i T(z_i)\psi_0$ in R is orthogonal to every vector in R, that is, to every vector $T(z')\psi_0$. Since the section ψ is the zero section if and only if $0=T(z'^*)\psi=\sum\alpha_i T(z'^*z_i)\psi_0$ it follows that the problem boils down to showing that

$$\sum\alpha_i X(z_i)=0$$

if $\sum\alpha_i T(z_i)\psi_0=0$ as a section of $\mathbf{C}DET_2^*$. Taking account of the invariance properties of the section ψ_0 we may without loss of generality assume that each $z_i=(g_i,1,\lambda_i)$ with $g_i=\left(\begin{smallmatrix}1&b_i\\0&1\end{smallmatrix}\right)$ and $\lambda_i(F)=\exp(\mathrm{tr}\xi_i(F-\epsilon)+\eta_i)$. Now $z_i=(g_i,1,1)\cdot(1,1,\lambda)$ and thus

$$\sum\alpha_i X(z_i)=\sum\alpha_i X(1,1,\lambda_i)=\sum\alpha_i e^{-\eta_i-2\mathrm{tr}\xi_i^2}. \tag{6.5.20}$$

On the other hand, at $w=\left(\begin{smallmatrix}1\\b\end{smallmatrix}\right)$

$$\begin{aligned}\psi(w)&=\sum\alpha_i e^{-\eta_i-\mathrm{tr}\xi_i(g_i^{-1}Fg_i-\epsilon)}e^{\mathrm{tr}b_i(\frac{1}{2}F_{21}-b)}\det_2(1-b_ib)\\ &=\sum\alpha_i\det(1-b_ib)\,e^{-\eta_i-\mathrm{tr}\tilde{\xi}_i(F-\epsilon)+\mathrm{tr}\epsilon(\xi_i-g_i\xi_ig_i^{-1})}\end{aligned}$$

(6.5.21)

where

$$\tilde{\xi}_i = g_i \xi_i g_i^{-1} - \begin{pmatrix} 0 & \frac{1}{2} b_i \\ 0 & 0 \end{pmatrix}.$$

Since the exponential functions with different parameters $\tilde{\xi}$ are linearly independent (with coefficients in the ring of polynomials) it follows from $\psi = 0$ that $\tilde{\xi}_i = const. = \tilde{\xi}$ for all i. Comparing (6.5.20) with (6.5.21) we observe that

$$\sum \alpha_i X(z_i) = e^{-2\mathrm{tr}\tilde{\xi}^2} \psi(w_+ = 1, w_- = 0) = 0$$

and therefore indeed a zero section is orthogonal to every vector in R. The Hermitian form in the space Γ is not positive definite. One might hope that it is positive semidefinite. However, from the results in Pickrell [1988] it follows that there is no invariant inner product in Γ. In fact, the group $\widehat{GL_1}$ does not have any *unitary* representations such that the Abelian ideal is represented nontrivially.

Exercise 6.5.22. Show that $< T(z)\psi_0, T(z)\psi_0 >$ is strictly positive for all $z \in \widehat{GL_1}$.

By construction the Hermitian form is invariant under all elements (g, q, λ) such that g, q are unitary and

$$|\lambda(F)|^{-2} = \mu(g, q; F)\alpha(g^*, q^*; gwq^{-1})\alpha(g, q; w). \tag{6.5.23}$$

Note that the right-hand-side is always real and positive; for unitary g, q it can be written as

$$|\alpha_1(g, q; F)|^2 \cdot h(F)h(gFg^{-1})^{-1}.$$

CHAPTER 7 THE VIRASORO ALGEBRA

7.0. Introduction

In this chapter we shall study the Lie algebra $Vect\, S^1$ of vector fields on a circle and some of its generalizations. The Lie algebra $Vect\, S^1$ has a central extension, the *Virasoro algebra* . The representation theory of the Virasoro algebra is closely related to the representation theory of affine Lie algebras. In fact, through the Sugawara construction, to be defined below, a highest weight representation of an affine Lie algebra carries always a highest weight representation of the Virasoro algebra. All the irreducible highest weight representations of the Virasoro algebra are known and they can be exponentiated to representations of associated infinite-dimensional Lie groups. The representation theory of the algebra of vector fields on a higher dimensional manifold is much less understood; we shall discuss the extensions of these algebras in Section 7.6.

The Virasoro algebra is important in physics mainly for the following two reasons. First, it gives the quantized energy momentum tensor in two-dimensional field theory. Consider the massless two-component Dirac field $\psi(x_0, x_1)$. The energy-momentum tensor of the Dirac field is

$$T_{\mu\nu} = \frac{1}{4}(\overline{\psi}\gamma_\mu \overleftrightarrow{\partial_\nu}\psi + \overline{\psi}\gamma_\nu \overleftrightarrow{\partial_\mu}\psi),$$

where $\overline{\psi}\overleftrightarrow{\partial_\mu}\phi = \overline{\psi}\partial_\mu\phi - (\partial_\mu\overline{\psi})\phi$. A Lorentzian metric is assumed and x_0 is the time coordinate. The γ matrices are defined by

$$\gamma_0 = \begin{pmatrix} 0 & 1 \\ 1 & 0 \end{pmatrix} \qquad \gamma_1 = \begin{pmatrix} 0 & 1 \\ -1 & 0 \end{pmatrix}.$$

The conjugate field is defined as $\overline{\psi} = \psi^*\gamma_0$. The field equations

$$\gamma^\mu \partial_\mu \psi = 0 \tag{7.0.1}$$

split into two one-component equations. Denoting $\psi = \begin{pmatrix} \psi_- \\ \psi_+ \end{pmatrix}$ and $x_\pm = x_0 \pm x_1$ the equations of motion become

$$\partial_+\psi_- = \partial_-\psi_+ = 0. \tag{7.0.2}$$

Correspondingly, the energy-momentum tensor splits into components $T_{\pm\pm}$. As a consequence of (7.0.2) $T_{-+} = T_{+-} = 0$ and the nonzero components are $T_{++} = \frac{1}{2}\psi_+^* \overleftrightarrow{\partial}_+ \psi_+$ and $T_{--} = \frac{1}{2}\psi_-^* \overleftrightarrow{\partial}_- \psi_-$. In the Hamiltonian formulation of the quantized Dirac theory (in one space dimension) one postulates the canonical anticommutation relations

$$[\overline{\psi_\pm}(x), \psi_\pm(y)]_+ = \delta(x-y)$$

and all other anticommutators are zero. Naively, applying these relations to the components of the energy-momentum tensor one gets

$$[T_{++}(x), T_{++}(y)] = T_{++}(x)(\partial_x - \partial_y)\delta(x-y).$$

Writing T_{++} in terms of its Fourier components L_n (assuming that the space is compactified to the circle) one obtains the commutation relations

(7.0.3) $$[L_n, L_m] = (n-m)L_{n+m}.$$

The components of T_{--} satisfy the same commutation relations and $[T_{--}, T_{++}] \equiv 0$. However, it turns out by a detailed analysis of the action of the operators in the fermionic Fock space that (7.0.3) is not valid; the correct form of the commutation relations involves a c-number on the right-hand side of (7.0.3), i.e., we obtain a central extension of the naive algebra (7.0.3). The relation to the algebra $Vect\, S^1$ is simple: The commutation relations of the operators L_n are precisely the commutation relations of the complex vector fields $ie^{in\phi}\frac{d}{d\phi}$.

The second application comes from statistical physics in two dimensions. Let us introduce the complex coordinates $z = x+iy$ and $\overline{z} = x-iy$ in the Euclidean plane. Denote the vector field $z^{n+1}\frac{d}{dz}$ by L_n and $\overline{z}^{n+1}\frac{d}{d\overline{z}}$ by $\overline{L}_n$. These represent infinitesimal conformal transformations of the plane. In a conformally invariant statistical system the generators L_n and $\overline{L}_n$ appear as coefficients of the stress-energy tensor,

$$T_{\mu\nu}dx^\mu dx^\nu = T(z)(dz)^2 + T(\overline{z})(d\overline{z})^2$$
$$T(z) = \sum z^{-n-2}L_n, \qquad T(\overline{z}) = \sum \overline{z}^{-n-2}\overline{L}_n.$$

We have assumed periodic boundary conditions for the physical system in order to apply the Fourier expansion. The components L_n and $\overline{L}_n$ satisfy the same algebra as in the field theory case, including a central term,

$$[L_n, L_m] = (n-m)L_{n+m} + \frac{c}{12}(n^3 - n)\delta_{n,-m}.$$

The center c is constant in an irreducible repesentation of the algebra. It turns out that in a unitary representation (where L_n is the Hermitian conjugate of L_{-n}) the "central charge" c can have only the values $c \geq 1$ or a discrete set of values between zero and one. It is a remarkable fact that the discrete values appear in certain models in statistical physics; to each c one can associate a finite number of fractional numbers, the allowed eigenvalues of the operator L_0 when acting on the vacuum vector, and these are precisely the critical exponents of the models.

Virasoro algebras were introduced in physics in the beginning of 1970s in connection of string models but it is relevant in any two-dimensional conformally invariant system. For references to the early litereture we refer to the review article by Scherk [1975] as well as to the string theory book by Green, Schwartz, and Witten [1987]. The author has also profited from the review article by Goddard and Olive [1986] and the original article by Goddard, Kent, and Olive [1985] which explains the construction of the representations with $o < c < 1$; the necessary condition for the existence of these representations was first proven in Friedan, Qiu, and Shenkar [1984].

Section 7.5, which deals with a generalization of the Virasoro algebra to a Riemann surface (meromorphic vector fields with a pair of poles) is based on Krichever and Novikov [1987]. The discussion on gravitational anomalies which forms the basis for the construction of extensions of the diffeomorphism groups in higher dimensions in Section 7.6 leans on the paper of Alvarez, Singer, and Zumino [1984]; see also Zoller [1986]. It remains to be seen whether is is possible to develop some kind of highest weight theory for these groups.

7.1. The Sugawara construction

Let l_n denote the smooth complex vector field $ie^{in\phi}\frac{d}{d\phi}$ on the circle S^1. The commutation relations are

$$[l_n, l_m] = (n-m)l_{n+m}. \tag{7.1.1}$$

In this section we want to show how a highest weight representation of an affine Kac-Moody algebra generates a representation of a central extension of the Lie algebra defined by the commutation relations (7.1.1).

Following the notation of Section 2.2, let g be a simple Lie algebra and $\widehat{\mathrm{g}}$ the associated affine algebra so that the central element together with the derivation d and the vectors T_a^n form a basis of $\widehat{\mathrm{g}}$, where $n \in \mathbf{Z}$ and $1 \leq a \leq dim\,\mathrm{g}$. We recall the commutation relations

$$[T_a^n, T_b^m] = \lambda_{ab}^c T_c^{n+m} + kn(T_a, T_b)\delta_{n,-m}.$$

We define a *normal ordering* in the enveloping algebra of $\widehat{\mathbf{g}}$ by setting

$$(7.1.2) \qquad :T_a^n T_b^m: \;:= \begin{cases} T_a^n T_b^m, & \text{for } n \leq m \\ \frac{1}{2}(T_a^n T_b^m + T_b^m T_a^n), & \text{for n=m} \\ T_b^m T_a^n, & \text{for } n > m \end{cases}$$

We shall work in an orthonormal basis $\{T_1, T_2, \ldots\}$ of $\mathbf{g}$,

$$\mathrm{tr}(ad\, T_a \cdot ad\, T_b) = (T_a, T_b) = \delta_{ab}.$$

Let V be a vector space which carries an integrable highest weight representation of the algebra $\widehat{\mathbf{g}}$ with the highest weight vector x. We can now define the operators

$$(7.1.3) \qquad L_n = \frac{1}{Q+2k} \sum_{a=1}^{dim\mathbf{g}} \sum_{k \in \mathbf{Z}} :T_a^k T_a^{n-k}:$$

acting in V. Here Q is equal to the eigenvalue of the Casimir operator in the adjoint representation,

$$\lambda_{ab}^c \lambda_{fc}^b = \delta_{af} Q.$$

With our normalization $Q = 1$. It is sometimes convenient to normalize the inner product differently from the Killing form and therefore we shall not insert $Q = 1$ in (7.1.3). The expression for L_n is the *Sugawara formula;* it was used in Sugawara [1968] for constructing the energy-momentum tensor in terms of currents. This type of idea was promoted also in Sommerfield [1968].

Denoting by θ the length of the longest root of the Lie algebra $\mathbf{g}$, we have a relation between the *dual Coxeter number* κ defined by

$$(7.1.4) \qquad \kappa = (\mathrm{rank}\, \mathbf{g})^{-1} \sum_{\alpha \in \Phi} \alpha^2/\theta^2$$

and Q, given by $Q = \theta^2 \kappa$.

Exercise 7.1.5. Show that $\kappa(A_\ell) = \ell + 1$, $\kappa(B_\ell) = 2\ell - 1$, $\kappa(C_\ell) = \ell + 1$ and $\kappa(D_\ell) = 2\ell - 2$.

Remember that $k = \frac{\theta^2}{2} x$, where $x \geq 0$ is the level of the representation and therefore $Q + 2k > 0$. The infinite sum in (7.1.3) reduces in fact to a finite sum when acting on any given vector v in V. Namely, in L_n there are only a finite number of terms containing T_a^m with $m < 0$ on the right and on the other hand $T_a^m v \neq 0$ only for a finite number of positive indices m.

PROPOSITION 7.1.6.

$$[L_n, L_m] = (n-m)L_{n+m} + \frac{c}{12}n(n^2-1)\delta_{n,-m},$$

where $c = \frac{2k\dim \mathfrak{g}}{Q+2k}$.

PROOF: The proof is by direct computation starting from the commutation relations of the operators T_a^n. We shall check here only the coefficient of the central terms and leave the rest to the reader. Let ψ be a highest weight vector in a highest weight representation of $\widehat{\mathfrak{g}}$. Then for any odd $n > 0$

$$\begin{aligned}
(\psi, [L_n, L_{-n}]\psi) &= (\psi, L_n L_{-n}\psi) \\
&= 4(Q+2k)^{-2}(\psi, \sum_{n/2\le m\le n} T_a^{n-m}T_a^m \sum_{-n/2\le j\le 0} T_b^{-n-j}T_b^j\psi) \\
&= 4(2k+Q)^{-2}\Big(\psi, \{ \sum_{n/2\le m\le n} mkT_a^{n-m}T_a^{m-n} \\
&\quad + \sum_{n/2\le m\le n}\ \sum_{-n/2\le j\le 0} \lambda_{ab}^c T_a^{n-m}T_c^{m-n-j}T_b^j\}\psi\Big) \\
&= 4(Q+2k)^{-2}k^2 \dim \mathfrak{g} \sum_{n/2\le m\le n} m(n-m) \\
&\quad + 4(Q+2k)^{-2} \sum_{n/2\le m\le n}\ \sum_{-n/2\le j\le 0} \Big(\psi, \{\lambda_{ab}^c\lambda_{ac}^f T_f^{-j}T_b^j \\
&\quad + \lambda_{ab}^c\lambda_{ab}^f T_c^{m-n-j}T_f^{n-m+j}\}\psi\Big) \\
&= 4k^2(Q+2k)^{-2}\dim \mathfrak{g}\frac{1}{12}(n^3-n) \\
&\quad + 4k(Q+2k)^{-2}\dim \mathfrak{g}\, Q\frac{1}{24}(n^3-n) \\
&= \frac{2k\dim \mathfrak{g}}{Q+2k}\frac{n^3-n}{12}.
\end{aligned}$$

The central charge c of the Virasoro algebra is ≥ 1. Namely,

$$\begin{aligned}
c - \operatorname{rank}\mathfrak{g} &= \frac{\theta^2 x}{\theta^2 x + \theta^2\kappa}\dim\mathfrak{g} - \operatorname{rank}\mathfrak{g} \\
&= \frac{x}{x+\kappa}\dim\mathfrak{g} - \operatorname{rank}\mathfrak{g} \geq \frac{1}{2}\dim\mathfrak{g} - \operatorname{rank}\mathfrak{g} \geq 0
\end{aligned}$$

Thus $c \geq \operatorname{rank}\mathfrak{g}$.

We leave as an exercise for the reader the proof of the following result:

PROPOSITION 7.1.7. *$[L_n, T_a^m] = -mT_a^{n+m}$ for all $n, m \in \mathbf{Z}$ and $1 \leq a \leq \dim \mathfrak{g}$.*

7.2. Embedding $Diff S^1/S^1$ in Gr_1

We denote by $Diff S^1$ the group of orientation preserving diffeomorphisms of the circle S^1. The derivative of an element $h \in Diff S^1$ is a real positive function on the circle. We can define a representation ρ_n of $Diff S^1$ in the Hilbert space $H = L_2(S^1)$ of square integrable complex valued functions on S^1 by setting

$$[\rho_n(h) \cdot \psi](\phi) = \psi(h^{-1}(\phi)) \cdot g'(\phi)^n, \tag{7.2.1}$$

where $n \in \mathbf{Z}/2$ and g is the inverse function of h. This representation corresponds to the natural action of diffeomorphisms on "differential forms" $\psi(\phi)(d\phi)^n$. The case $n = 1$ corresponds to true differential forms of degree 1, the case $n = -1$ to vector fields on the circle, the case $n = -1/2$ to spinors, etc. An L_2-basis for the n-forms is given by the vectors $e_k = e^{i(k+1)\phi}(d\phi)^n$, $k \in \mathbf{Z}$. We denote by ϵ the operator in H such that $\epsilon e_k = e_k$ for $k \geq 0$ and $\epsilon e_k = -e_k$ for $k < 0$. The integral kernel of the operator ϵ is

$$\begin{aligned} K(\theta, \phi) &= \sum_{k \geq 0} e^{ik(\theta-\phi)} - \sum_{k<0} e^{ik(\theta-\phi)} \\ &= 1 + i \cot \frac{1}{2}(\theta - \phi). \end{aligned} \tag{7.2.2}$$

The kernel for $\rho_n(h)$ is $g'(\theta)^n \delta(g(\theta) - \phi)$ and therefore the commutator $[\rho_n(h), \epsilon]$ is represented by the kernel

$$\begin{aligned} &\int \{\delta(g(\theta) - \eta) g'(\theta)^n K(\eta, \phi) - K(\theta - \eta)\delta(g(\eta) - \phi) g'(\eta)^n\} d\eta \\ &\qquad = g'(\theta)^n K(h^{-1}(\theta), \phi) - K(\theta, h(\phi)) h'(\phi)^n \end{aligned} \tag{7.2.3}$$

If $n = 1/2$ then the kernel is smooth, since $K(\theta, \phi) = 2i/(\theta - \phi) +$ a smooth function and so (7.2.3) can be written as

$$g'(\theta)^{1/2} \frac{2i}{g(\theta) - \phi} - h'(\phi)^{1/2} \frac{2i}{\theta - h(\phi)} + \text{smooth function}$$

which has a potential singularity at points (θ_o, ϕ_o) with $\theta_o = h(\phi_o)$. Expanding $h(\phi)$ as a Taylor series at $\phi = \phi_o$ shows however that the singularities cancel each other. It follows that the kernel of $[\rho_n(h), \epsilon]$ is smooth when $n = 1/2$ and therefore in this case the off-diagonal blocks (with respect to the decomposition $H = H_- \oplus H_+$ to eigenspaces of the operator ϵ) of $\rho_n(h)$ are Hilbert-Schmidt operators. In the general case we can write $\rho_n(h) = R(h) \cdot \rho_{1/2}(h)$, where $R(h)$ is the multiplication operator by the smooth function $g'(\theta)^{n-1/2}$. From the discussion in Chapter 6 it follows that $R(h)$ is in GL_1 and thus also $\rho_n(h) \in GL_1$. We have now constructed a homomorphism $Diff\, S^1 \to GL_1$. However, this map is continuous only with respect to the discrete topology of $Diff\, S^1$ but the composite map $Diff\, S^1 \to GL_1 \to GL_1/N = Gr_1$ is smooth [Pressley and Segal, 1986, Section 6.8]. Since the subgroup S^1 is mapped to diagonal matrices in GL_1 we have in fact a smooth mapping from $M = Diff\, S^1/S^1$ to Gr_1.

THEOREM 7.2.4. *The pull-back of the canonical two-form on Gr_1 with respect to the embedding $M \to Gr_1$ is the two-form on M given by*

$$curv(\ell_k, \ell_m) = \left[(6n^2 - 6n + 1)\frac{1}{6}k^3 - \frac{1}{6}k\right] \delta_{k,-m}.$$

PROOF: The vector field ℓ_k is represented in H by the matrix

$$\rho_n(\ell_k) = \sum_{j \in \mathbf{Z}} (j + kn) e_{k+j,j} \tag{7.2.5}$$

in the basis $\{e_j\}$. By the definition of the canonical two-form (= curvature of the canonical line bundle DET_1) on Gr_1 (see Section 6.3) the pull-back is computed by taking traces of products of the off-diagonal blocks,

$$\begin{aligned} curv(\ell_k, \ell_m) =& \delta_{k,-m} \sum_{j \geq 0, k+j<0} (j + kn)(j - m + mn) \\ &- \sum_{j \geq 0, j+m<0} (j + mn)(j - k + kn) \\ =& \left[(n^2 - n + \frac{1}{6})k^3 - \frac{1}{6}k\right] \delta_{m,-k}. \end{aligned}$$

The term linear in k can be shifted by a redefinition of zero-point energy. Namely, for each $N \in \mathbf{N}$ we can define a decomposition $H = H_+^N \oplus H_-^N$ such that the vectors e_i with $i \geq N$ form a basis of H_+^N. Defining the curvature with respect to this decomposition and using the standard formula $\mathrm{tr}(b_1c_2 - b_2c_1)$ for the curvature the induced two-form on M is

$$(7.2.6) \qquad F_{N,n}(\ell_k, \ell_m) = \left[(n^2 - n + \frac{1}{6})k^3 - (N^2 - N + \frac{1}{6})k\right] \delta_{k,-m}.$$

In particular when $N = n$ we get the simple expression

$$(7.2.7) \qquad F_{n,n}(\ell_k, \ell_m) = (n^2 - n + \tfrac{1}{6})(k^3 - k)\delta_{k,-m}.$$

Finally, we state without proof (for a proof see Fuks [1987]) the following theorem:

THEOREM 7.2.8. *Let $c'(\ell_n, \ell_m)$ be any two-cocycle on the Lie algebra (7.1.1). Then there is a linear form a such that the two-cocycle $c = c' + \delta a$ $\{i.e. c(x,y) = c'(x,y) + a([x,y])\}$ is of the form*

$$c(\ell_n, \ell_m) = const \times (n^3 - n)\delta_{n,-m}.$$

The cocycle c is characterized by the property that it is rotation invariant,

$$c(x, [\ell_0, y]) = -c([\ell_0, x], y)$$

for all elements x, y of the algebra of vector fields on S^1, and in addition the restriction of c to the subalgebra $\mathbf{sl}(2, \mathbf{R})$ spanned by $\ell_0, \ell_{\pm 1}$ vanishes.

7.3. Semi-infinite forms and representations of the Virasoro algebra

Let f_k be the linear form in the Hilbert space H defined by $f_k(e_i) = \delta_{-i,k}$. Recall the definition of semi-infinite forms from Section 6.2. We denote

$$(7.3.1) \qquad \psi_N = f_N \wedge f_{N-1} \wedge f_{N-2} \wedge \dots.$$

These should be understood as formal expressions and not as true differential forms; N is an arbitrary integer. We call ψ_N the *vacuum of index N*. More generally, we denote for each decreasing sequence $(i) = (i_1, i_2, \dots)$ of integers such that $-\mathbf{N} \setminus (i)$ and $\mathbf{N} \cap (i)$ are finite the forms

$$(7.3.2) \qquad \psi(i) = f_{i_1} \wedge f_{i_2} \wedge f_{i_3} \wedge \dots.$$

The integer $ind(i)$ which is the number of positive integers in (i) minus the number of missing nonpositive integers in (i) is called the *index* of the set (i). The Fock space $\mathcal{F}_N$ is the completion of the space of finite linear combinations of the vectors $\psi(i)$ with $ind(i) = N$ the inner product being defined such that the semi-infinite forms $\psi(i)$ form an orthonormal basis. The wedge product between a finite form and a semi-infinite form can be defined as in finite-dimensional differential calculus. We write once more the anticommutation relations for the creation operators $a_i^* = f_i\wedge$ and the annihilation operators $a_i = \imath(f_i)$ for each $i \in \mathbf{Z}$,

$$a_i a_j^* + a_j^* a_i = \delta_{i,j} \; a_i a_j + a_j a_i = 0, \;\; a_i^* a_j^* + a_j^* a_i^* = 0. \tag{7.3.4}$$

In the case $N = 0$ the vacuum is the "Dirac sea": all the positive energy levels are empty and the negative energy levels are all filled. In general, $a_i \psi_N = 0$ for $i > N$ and $a_i^* \psi = 0$ for $i \leq N$.

Set $E_{ij} = a_i a^* j - 1$ when $i = j > N$ and $E_{ij} = a_i a_j^*$ otherwise. Then $E_{ij}\psi_N = 0$ for $i, j \leq N$ and $i > N$. Let us denote by V_N the subspace of $\mathcal{F}_N$ consisting of finite linear combinations of the vectors $\psi(i)$. The operators E_{ij} map V_N into itself and the commutation relations are

$$[E_{ij}, E_{kl}] = \delta_{jk} E_{il} - \delta_{il} E_{kj} + c(E_{ij}, E_{kl}), \tag{7.3.5}$$

where the *Kac-Peterson cocycle* c is

$$c(E_{ij}, E_{kl}) = \begin{cases} 1 \text{ for } j = k \leq N, \; i = l > N \\ -1 \text{ for } j = k > N, \; i = l \leq N \\ 0, \text{ otherwise} \end{cases} \tag{7.3.6}$$

If $x \in V$ then $E_{k+m,k}x = 0$ when $|k|$ is large enough and it follows that the operators

$$\ell_k = \sum_{j \in \mathbf{Z}} (j + kn) E_{k+j,j}$$

are well-defined in V_N. They cannot be extended to $\mathcal{F}_N$. From (7.3.6) we obtain the central term of the Virasoro algebra,

$$c(\ell_k, \ell_m) = [(n^2 - n + \tfrac{1}{6})k^3 - (N^2 - N + \tfrac{1}{6}k)]\delta_{k,-m}. \tag{7.3.7}$$

Here we recognize again the 2-form computed in the previous section. Note that the central term here differs from the expression obtained through the Sugawara construction; however, the difference is not very essential because by a redefinition $\ell_0 \mapsto \ell_0 + a$, where a is a constant, the term linear in k can be shifted arbitrarily.

The unitarity conditions $\ell_n^* = \ell_{-n}$ are fulfilled if and only if $n = 1/2$; this follows directly from the relations $E_{ij}^* = E_{ji}$.

The relation between the computation of the central term c in semi-infinite cohomology and the computation of the curvature in Section 7.2 is very simple. Using the embedding $M \subset Gr_1$ we can pull back the sections of the canonical line bundle DET_1 over Gr_1 to sections in a line bundle over M. In particular, to each semi-infinite form $\psi(i)$ of index=0 there corresponds a section over M which is the pull-back of the section ψ_S defined in Section 6.2, for $S = (i)$. Since both the vectors ψ_S and $\psi(i)$ form an orthonormal basis this correspondence *defines a unitary isomorphism* between Γ_1 and $\mathcal{F}_0$. One easily sees that the isomorphism is equivariant with respect to the action of $\mathbf{gl}_1$ in Γ_1 and $\mathcal{F}_0$.

In order to describe the correspondence in the case of index$\neq 0$ we have to generalize the discussion in 6.2, where we constructed the holomorphic sections ψ_S only over the *connected component* of the Grassmannian described by matrices $g = \begin{pmatrix} a & b \\ c & d \end{pmatrix}$ such that the Fredholm index of a is equal to zero. The set of planes W obtained by the action $g \cdot H_+$ by operators with $\operatorname{ind} a = N$ is characterized by the property

$$N = \operatorname{ind} W = \dim(W \cap H_-) - \dim(W^\perp \cap H_+).$$

A typical representative of the sector Gr_1^N is the plane spanned by the vectors $e_{-N}, e_{-N+1}, e_{-N+2}, \ldots$. Any other plane in the same sector is obtained by an action by an element of GL_1 in the connected component of the identity. For each $S \subset \mathbf{Z}$ such that $S \cap -\mathbf{N}$ and $\mathbf{N} \setminus S$ are finite we define the plane H_S spanned by the vectors e_i, $i \in S$. If $N = \operatorname{ind} S$ then $H_S \in Gr_1^N$.

For each S of index N we can define a complex function ψ_S in the set St_1^N of admissible basis of elements of Gr_1^N by

$$\psi_S(w) = \det w_S$$

where w_S is the matrix obtained by choosing the rows labelled by the numbers $i \in S$ from the matrix (w_{ij}), $w_j = \sum w_{ij} e_i$. Clearly $\psi_S(wt) = \psi_S(w) \det t$ for any matrix t with determinant, so ψ_S is a section of a line bundle over Gr_1^N, namely, of the dual determinant bundle DET_1^*. The curvature form on Gr_1^N is defined by the formula

$$\omega(X, Y) = \frac{1}{8} \operatorname{tr} \epsilon_N [[\epsilon_N, X], [\epsilon_N, Y]],$$

where $\epsilon_N e_i = e_i$ for $i \geq -N$ and $\epsilon_N e_i = -e_i$ for $i < -N$.

We can now define an isomorphism between the space of holomorphic sections over Gr_1^N, spanned by the ψ_S's, and the Fock space $\mathcal{F}_N$ simply by mapping the element $\psi(i) \in \mathcal{F}_N$ to the section ψ_S, $S = (i)$. The curvature on M [formula (7.2.6)], described by the cocycle c of the Virasoro algebra, is then the pull-back of the curvature on Gr_1^N with respect to the embedding described in Section 7.2.

7.4. Representations of the Virasoro algebra with central charge $c < 1$.

Let V be a vector space carrying an irreducible representation of the Virasoro algebra

$$[L_n, L_m] = (n-m)L_{n+m} + \frac{c}{12}n(n^2-1)\delta_{n,-m}$$

containing the highest weight vector v,

$$L_n v = 0, \text{ for } n > 0, \; L_0 v = h v_0. \tag{7.4.1}$$

The representation is said to be unitary if there is an inner product $(\cdot,\cdot)$ in V such that

$$(L_n v, v') = (v, L_{-n} v') \; \forall n \in \mathbf{Z}, \; v, v' \in V. \tag{7.4.2}$$

Because L_n is supposed to represent the complex vector field ℓ_n on the circle and the complex conjugate of ℓ_n is $-\ell_{-n}$ the unitarity relation simply means that real vector fields are represented by anti-Hermitian operators. Clearly in a unitary representation the highest weight h has to be real. The central charge c has to be a non-negative real number. This follows from

$$\begin{aligned} 0 \le \|L_{-n}v\|^2 = (v, L_n L_{-n} v) &= \left(v, \left[2nL_0 + \frac{c}{12}n(n^2-1)\right] v\right) \\ &= n\left[2h + \frac{c}{12}(n^2-1)\right] \text{ for } n > 0 \end{aligned}$$

where we have normalized $\|v\| = 1$. In all of the examples studied so far we have had $c \ge 1$. From the above computation follows also that $h \ge 0$ (put $n = 1$). Since $[L_0, L_{-n}] = nL_{-n}$, all other eigenvalues of L_0 must be bigger than h. The eigenvalue h has multiplicity =1. Namely, from the commutation relations follows that any vector in $\mathcal{U}v$ ($\mathcal{U}$ is the enveloping algebra of the Virasoro algebra) is a linear combination of the vectors

$$|n_1, n_2, \ldots n_k >= L_{-n_1} \ldots L_{-n_k} v, \tag{7.4.3}$$

where $0 < n_1 \leq n_2 \leq \cdots \leq n_k$ and the corresponding eigenvalue (the energy) of L_0 is $h + n_1 + \cdots + n_k$. The dimension of the eigenspace belonging to the eigenvalue $m > h$ is smaller or equal to the number of different partitions of $m - h$ as a sum of positive integers (partitions differing by the order of the integers are counted only once). The dimension can be strictly smaller because the vectors (7.4.3) are in general not linearly independent.

The inner product between the vectors (7.4.3) can in principle be evaluated using the unitarity condition and the commutation relations,

$$< n_1, \ldots, n_k \mid n'_1, \ldots, n'_l >= (v, L_{n_k} \ldots L_{n_1} L_{n'_1} \ldots L_{n'_l} v).$$

The L_n's with a positive index are shifted to the right using the commutation relations. By (7.4.1) we are left with an expression involving only L_n's with negative index; but by (7.4.2) all the inner products $(v, L_{-n} v')$ vanish for $n > 0$ and so the result will be of the form *const.*$\times (v, v)$, where the constant is some polynomial of c and h. The inner product can be nonzero only if the L_0-eigenvalues are equal, $n_1 + \ldots n_k = n'_1 + \cdots + n'_l$. Let $M(c, h; m)$ be the matrix formed from the inner products of the vectors (7.4.3) belonging to energy m. The existence of a unitary irreducible representation with the characteristics (c, h) is equivalent with the condition that the matrices $M(c, h; m)$ are all positive semidefinite for $m = h, h+1, h+2, \ldots$. In these considerations the *Kac determinant formula* for $\det M(c, h; m)$ plays a central role:

$$\begin{aligned} \det M(c,h;m) &= \prod_{k=1}^{m} \eta_k(c,h)^{\pi(m-k)} \\ \eta_k(c,h) &= \prod_{p,q \in \mathbf{Z}, pq=k} (h - h(p,q,c)), \qquad (7.4.4) \\ h(p,q,c) &= \frac{[(z+3)p - (z+2)q]^2 - 1}{4(z+2)(z+3)}, \end{aligned}$$

where we have written the central charge c in the form

$$c = 1 - \frac{6}{(z+2)(z+3)}.$$

Exercise 7.4.5. Show that for any fixed $c > 1$ the matrices $M(c, h; m)$ are positive definite when h is large enough.

Since the determinants $\det M(c, h; m)$ are positive when $c > 1$ and $h \geq 0$ it follows from the result of the exercise that $M(c, h; m)$ is positive semidefinite for all $c \geq 1$ and $h \geq 0$. One can show that in the range $0 < c < 1$ a necessary condition for the positive semidefiniteness of the matrices is that c corresponds to the values $z = 0, 1, 2, \dots$ and $h = h(p, q, c)$ with $p = 1, 2, \dots, z+1$, $q = 1, 2, \dots p$. This is in fact also a sufficient condition for the existence of unitary irreducible representations in the range as the following explicit construction shows. In addition to these representations the only remaining unitary highest weight representation is the trivial representation corresponding to $c = h = 0$.

Let $\mathbf{g}$ be a finite-dimensional semisimple Lie algebra, $\mathbf{g} = \mathbf{g}_1 + \dots \mathbf{g}_r$, where each $\mathbf{g}_i$ is simple. We can construct a Virasoro algebra associated to $\mathbf{g}$ from a highest weight representation of $\hat{\mathbf{g}} = \hat{\mathbf{g}}_1 + \cdots + \hat{\mathbf{g}}_r$ simply by taking a sum of the Virasoro algebras,

$$L_n(\mathbf{g}) = L_n(\mathbf{g}_1) + \cdots + L_n(\mathbf{g}_r). \tag{7.4.6}$$

Since $\mathbf{g}_i$ commutes with $\mathbf{g}_j$ for $i \neq j$ it is clear that the terms in (7.4.6) commute with each other and the central charge of the total Virasoro algebra is the sum of the central charges of the subalgebras. Note that the central charge k_i of the affine Kac-Moody algebra $\hat{\mathbf{g}}_i$ may in general depend on i.

Let $\{T_a\}_{a=1}^{\dim \mathbf{g}}$ be a basis of $\mathbf{g}$ such that the T_a's with $1 \leq a \leq \dim \mathbf{h}$ form a basis for a subalgebra $\mathbf{h} \subset \mathbf{g}$. We assume that the commutation relations of $\hat{\mathbf{h}}$ are of the form

$$[T_a^n, T_b^m] = \lambda_{ab}^c T_c^{n+m} + kn\delta_{n,-m}(T_a, T_b). \tag{7.4.7}$$

This is not the most general situation. For example, if $\mathbf{h}$ is a sum of several simple pieces $\mathbf{h}_i$ then we might have a different value for the central charge k in (7.4.7) for each $\mathbf{h}_i$. We shall also assume that the eigenvalues of the Casimir of $\mathbf{h}$ in the adjoint representation of the subalgebra $\mathbf{h}$ are all equal $= Q_h$; this is of course automatic if $\mathbf{h}$ is simple. In this case we define the Virasoro generators for $\mathbf{h}$ by

$$L_n^h = \frac{1}{Q_h + 2k} \sum_{a=1}^{dim\, \mathbf{h}} \sum_{i \in \mathbf{Z}} : T_a^i T_a^{n-i} : \tag{7.4.8}$$

Let us denote

$$K_n = L_n - L_n^h$$

where the L_n's are the Virasoro generators constructed from $\mathbf{g}$.

THEOREM 7.4.9. *The K_n's satisfy the Virasoro commutation relations with the central charge $c = c_g - c_h$.*

PROOF: We know from Proposition 7.1.7 that

$$[L_n, T_a^m] = -mT_a^{n+m} \text{ for } a = 1, 2, \ldots \dim \mathbf{g}$$

(7.4.10)

$$[L_n^h, T_a^m] = -mT_a^{m+n} \text{ for } a = 1, 2, \ldots \dim \mathbf{h}.$$

By a subtraction it follows that

$$[K_n, T_a^m] = 0, \text{ for } a = 1, 2, \ldots \dim \mathbf{h}.$$

Since the L_n^h's are constructed from the T_a^m with $1 \leq a \leq \dim \mathbf{h}$ it follows from the above that

(7.4.11) $$[K_n, L_m^h] = 0 \ \forall n, m \in \mathbf{Z}.$$

The Virasoro generators $L_n = L_n^h + K_n$ split into two commuting pieces and therefore

$$\begin{aligned}[L_n, L_m] &= [L_n^h, L_m^h] + [K_n, K_m] \\ &= (n-m)L_{n+m}^h + \frac{c_h}{12}n(n^2-1)\delta_{n,-m} + [K_n, K_m].\end{aligned}$$

Comparing the left-hand side of this with the Virasoro commutation relations of the operators L_n we can conclude that the K_n's satisfy the Virasoro commutation relations with $c = c_g - c_h$.

Let us next consider the example where $\mathbf{g} = A_1 \oplus A_1$ [the Lie algebra of $SU(2) \times SU(2)$] and $\mathbf{h}$ is the diagonal A_1 algebra consisting of pairs (x, x). Let k_1 be the central charge for the first A_1 algebra and k_2 the central charge for the second A_1. With our conventions each k is a non-negative integer times $\frac{1}{4}$ (the squared length of a root with our present normalizations is equal to $1/2$). The central charge for the algebra spanned by the K_n's is now

$$c = \frac{6k_1}{1+2k_1} + \frac{6k_2}{1+2k_2} - \frac{6(k_1+k_2)}{1+2(k_1+k_2)}.$$

Denoting $k_2 = \frac{1}{4}z$, in the case $k_1 = 1/4$ the value of c is

$$c = 1 - \frac{6}{(z+2)(z+3)}, \quad z = 0, 1, 2, \ldots.$$

By this example we have constructed all representations of the Virasoro algebra in the discrete series $0 < c < 1$. A second application of Theorem 7.4.9 is the quantum equivalence theorem (so called because in quantum field theory in 1+1 space-time dimensions the Virasoro algebra represents the Fourier modes of the energy-momentum tensor):

THEOREM 7.4.12. *If* $c = c_g - c_h = 0$ *then* $L_n^g = L_n^h$ *for all* n.

PROOF: The only unitary highest weight representation with $c = 0$ is the trivial representation with $h = 0$. The representation of the Virasoro algebra K_n is unitary and the spectrum of K_0 is bounded from below by the construction,

$$K_0 = L_0 - L_0^h = \sum_{a=\dim \mathbf{h}+1}^{\dim \mathbf{g}} \frac{1}{Q_a + 2k_a}(T_a^0 T_a^0 + 2\sum_{n>0} T_a^{-n} T_a^n),$$

where we have included the possibility that $\mathbf{g} \ominus \mathbf{h}$ could contain pieces corresponding to different values of the central charge k. Taking account the unitarity relations $(T_a^n)^* = T_a^{-n}$ we observe that K_0 is a positive operator and thus, due to the vanishing of the central charge, we have $K_n = 0$ for all n.

In string theory (more about strings in Chapter 9) an important example of the quantum equivalence result is the case when $\mathbf{g}$ is simple and $\mathbf{h}$ is the Cartan subalgebra of $\mathbf{g}$. Then the values of k for $\mathbf{g}$ and $\mathbf{h}$ are equal and $\dim \mathbf{h} = \operatorname{rank} \mathbf{g}$. Now

$$c(K) = \frac{2k \dim \mathbf{g}}{Q + 2k} - \operatorname{rank} \mathbf{g} = \frac{x \dim \mathbf{g}}{\kappa + x} - \operatorname{rank} \mathbf{g}.$$

Using the result of Exercise 7.1.5 one sees that for the classical simple Lie algebras the only cases when $c(K)$ vanishes are A_ℓ and D_ℓ with the level $x = 1$. In addition to these the level one representations of the exceptional simply laced algebras E_6, E_7 and E_8 lead to the vanishing of the K_n's.

7.5. Riemann surfaces and generalizations of Virasoro algebras

Let M be a two-dimensional compact oriented manifold without boundary (a Riemann surface) and $x_\pm$ two distinct points on M. We shall study the Lie algebra L of holomorphic vector fields on $M \setminus \{x_\pm\}$. Consider first the example $M = S^2$. Choose x_- as the south pole and x_+ the north pole and map the punctured sphere to the complex plane such that x_- is the origin and x_+ is the point at infinity. The Virasoro algebra without the central term is generated by the holomorphic vector fields $z^n \frac{d}{dz}$ and an arbitrary holomorphic vector field in $\mathbf{C} \setminus 0$ is in the completion (with respect to a suitable Frechet topology) of the linear span of the basis fields. For this reason L (or its central extensions) can be viewed as a generalization of the Virasoro algebra.

A Riemann surface is characterized topologically by its genus. Any Riemann (oriented) surface is obtained from the unit sphere by adding a certain number g of handles; this number is the genus. For example, the torus is obtained by adding one handle. A basis in L can be obtained as follows. Fix a local holomorphic coordinate $z_\pm$ at the points $x_\pm$ such that $z_-(x_-) = 0 = z_+(x_+)$. For each $i \in g/2 + \mathbf{Z}$ there is a unique holomorphic vector field e_i such that at $x_\pm$

$$e_i = a_i^\pm z_\pm^{\pm i+1-3g/2}[1 + O(z_\pm)]\frac{d}{dz_\pm}, \text{with } a_i^+ = 1. \tag{7.5.1}$$

Let us denote $g_0 = 3g/2$. The representation (7.5.1) is a consequence of the Riemann-Roch theorem: The sum of multiplicities of the zeros minus the sum of degrees of the poles is equal to $2 - 2g_0$. The Lie algebra L is not graded like the Virasoro algebra but it is "almost graded". By counting the multiplicities of zeros and poles for the commutator we get from (7.5.1)

$$[e_n, e_m] = \sum_{i=n+m-g_0}^{n+m+g_0} c_{nm}^i e_i, \tag{7.5.2}$$

where the structure constants c_{nm}^i are some complex numbers. For a fixed g_0, Krichever and Novikov call a Lie algebra satisfying commutation relations of the type (7.5.2) a g_0 graded algebra.

Let γ be a small circle around the point x_- . ("Small" means that the local coordinate z_- is defined in a neighborhood of γ.) Let f and g be two vector fields on the punctured Riemann surface. We define

$$\chi(f, g) = \frac{1}{24\pi i}\int_\gamma f'' g' dz_-. \tag{7.5.3}$$

By a simple computation one proves that this is a two-cocycle for the Lie algebra L, i.e., $\chi([f,g],h)+\chi([g,h],f)+\chi([h,f],g) \equiv 0$. Thus (7.5.3) defines a central extension $\hat{L}$ of L. It requires some more work to show that this cocycle is essentially unique (see Krichever and Novikov [1987], Fuks [1987]), that is, any other two-cocycle is obtained from χ by multiplying by a constant and adding a coboundary $\beta([f,g])$. The path γ must be noncontractible but is otherwise arbitrary. For example, when $g = 0$ we obtain (when choosing γ as the unit circle around the origin in the complex plane)

$$\chi(e_n, e_m) = \frac{1}{12}(n^3 - n)\delta_{n,-m}.$$

We can construct projective Fock space representations for the algebra L essentially in the same manner as for the Virasoro algebra. Let $H(N)$ be the linear space of holomorphic forms $f(z)(dz)^N$ of degree N on the surface $M \setminus \{x_\pm\}$. Again by the Riemann-Roch theorem for each $i \in g_0 + \mathbf{Z}$ there is a unique form f_j of degree N such that

$$f_j = b_j^{\pm} z_{\pm}^{\pm j - g/2 + N(g+1)}[1 + O(z_\pm)](dz_\pm)^N, \text{with } b_j^+ = 1. \tag{7.5.4}$$

Computing the action of the basis vector fields e_i to the basic differential forms one obtains

$$e_i \cdot f_j = \sum_{k=i+j-g_0}^{i+j+g_0} A_{kj}^{(i)} f_k \text{ with } A_{i+j-g_0,j}^{(i)} = j - g/2 + N\left(i - \frac{g}{2} + 1\right). \tag{7.5.5}$$

for some complex matrix $A^{(i)}$. Thus the representation of the Lie algebra L in $H(N)$ is g_0 graded. Using this fact we can transport the representation of L in $H(N)$ to a representation in the Fock space just like in the case of the Virasoro algebra. The vacuum is the semi-infinite form

$$\psi_0 = f_N \wedge f_{N-1} \wedge f_{N-2} \wedge \cdots, \qquad \begin{cases} N \in \mathbf{Z} \text{ if } g \text{ is even} \\ N \in 1/2 + \mathbf{Z} \text{ if } g \text{ is odd} \end{cases}$$

The action of the Lie algebra $\mathbf{gl}(\infty)$ in the Fock space $\mathcal{F}_N$ is defined as before. We recall that $E_{nm}\psi_N = 0$ for $n > N$ and $E_{nm}\psi_N = 0$ for $n, m \leq N$. The operators e_i are not strictly speaking in $\mathbf{gl}(\infty)$ but they have nevertheless a well-defined action in the subspace $V_N \subset \mathcal{F}_N$ consisting of all finite linear combinations of the basis vectors $\psi(i)$ of index N. (In the case $N \in \frac{1}{2} + \mathbf{Z}$ the index is $-1/2$ plus the number of positive i_ν's minus the number of missing negative i_ν's.) By (7.5.5) the embedding of L to the general linear Lie algebra is

$$e_i \mapsto \sum_{j \in g_0 + \mathbf{Z}} \sum_{k=i+j-g_0}^{i+j+g_0} A_{kj}^{(i)} E_{kj} \tag{7.5.6}$$

and thus $e_n \psi(i)$ indeed is a finite linear combination of the basis vectors.

THEOREM 7.5.7. *The commutation relations of the operators representing the basic vector fields e_i in $\mathcal{F}_N$ are*

$$[e_n, e_m] = \sum_{i=n+m-g_0}^{n+m+g_0} c_{nm}^i e_i + c(e_n, e_m)$$

where the central term c satisfies $c(e_n, e_m) = 0$ *for* $|n+m| > 3g$. *In addition,* $e_n\psi_N = 0$ *for* $n > N + g_0$.

PROOF: 1) $c(e_n, e_m) = 0$ for $|n+m| > 3g$: By (7.5.6) we have

$$c(e_n, e_m) = \sum_{j,k\in g_0+\mathbf{Z}} \sum_{p=n+j-g_0}^{n+j+g_0} \sum_{q=m+k-g_0}^{m+k+g_0} A_{pj}^{(n)} A_{qk}^{(m)} c(E_{pj}, E_{qk}).$$

Using the explicit form of the Kac-Peterson cocycle given in (7.3.6) we observe that the range of the indices j, k, p, q is empty except when $|n+m| \leq 2g_0$.

2) The second assertion is a consequence of $E_{nm}\psi_N = 0$ for $n \geq m$ and of (7.5.6).

There is no nice explicit formula for the cocycle like in the case of the Virasoro algebra. However, by a simple computation from (7.5.1) one has

$$\chi(e_n, e_{-n+2g_0}) = \frac{1}{12}[(n-g_0)^3 - (n-g_0)].$$

7.6. Extensions of *Diff* S^n and diffeomorphism anomalies

In Chapter 5 we saw that the Schwinger terms in the commutation relations of the infinitesimal gauge transformations were related to the curvature of the parameter space $\mathcal{A}/\mathcal{G}$ of the chiral Dirac operator coupled to vector potentials which in turn is related to the chiral anomaly of the Dirac determinant. In Chapter 5 we kept the metric and connection of the space-time manifold M fixed. Here we shall consider the case when the metric and connection in M are varied. The manifold M is assumed to be even dimensional.

Let $\mathcal{M}$ be the space of Riemannian metrics on M. In an analogy with the gauge potential case it would be natural to think of the determinant of the chiral Dirac operator as a section of a complex line bundle over the moduli space $\mathcal{M}/Diff\,M$. However, the quotient space is not a smooth manifold because the group *Diff* M does not act freely in $\mathcal{M}$ and the isotropy subgroup depends on the point in $\mathcal{M}$. In the case of vector potentials this problem was avoided by restricting to the subgroup of the group of gauge transformations consisting of *based* maps $f: M \to G$. In the present case this is not sufficient: even after fixing a point $p \in M$ and restricting to the diffeomorphisms leaving p fixed the quotient space can have singularities. Let w be a frame at the point p. Let $Diff_0\, M$ be the group of diffeomorphism which leave both the point p and the frame w invariant. It can be shown that $\mathcal{M}/Diff_0\, M$ is a smooth manifold.

Let $\psi(g)$ be the determinant of the chiral Dirac operator $D_+ = (\gamma^\mu \nabla_\mu)_+$ with respect to some fixed regularization. The covariant derivative ∇ is defined by the Levi-Cività connection of the metric g. The determinant of the full Dirac operator is $\mathit{Diff}_0\, M$ invariant but $\psi(g)$ is not since the determinant can be computed only with respect to fixed operator T from the space of negative chirality spinors to positive chirality spinors. The variation of ψ with respect to $\mathit{Diff}_0\, M$ is called the *diffeomorphism anomaly* of the Dirac operator. Explicit formulas for the anomaly have been computed. Instead of going into the rather involved computations of the determinants we shall give the cohomological principle how the anomaly can be found.

In Section 4.1 we used the transgression method to derive various cocycles for the action of $\mathcal{G}$ on $\mathcal{A}$. Starting from a gauge invariant differential form $c^{0,2n} = c^{0,2n}(A)$ of degree $2n$ we obtain a one cocycle $c^{1,2n-2}$ (differential form degree $2n-2$) such that $dc^{1,2n-1} = \delta c^{0,2n-1}$, where $c^{0,2n-1}$ is some local form such that $dc^{0,2n-1} = c^{0,2n}$. If A is a vector potential in a nontrivial principal bundle over M then the forms are only locally defined on M (they depend on the chosen local trivialization) but they are always globally defined as forms in the total space of P. The cocycle $c^{1,2n-2}$ is the gauge anomaly in $2n-2$ space-time dimensions. The gauge variation of $c^{1,2n-2}$ is a closed form and can be written as an exterior derivative of a form $c^{2,2n-3}$ of degree $2n-3$. This latter form is a cocycle of degree 2; the corresponding Lie algebra cocycle is the commutator anomaly of the Lie algebra of infinitesimal gauge transformations.

In the gauge potential case the form $c^{o,2n}$ is essentially the nth Chern class. These exist in all even dimensions. In the case of metrics we can use the *Pontrjagin classes* . They exist only in dimension $4n$. The reason is algebraic. Given a metric g the bundle of frames of the tangent bundle TM reduces naturally to the bundle of orthonormal frames FM. The structure group is then the orthogonal group $O(N)$, where $N = dim\, M$. To define characteristic classes we need invariant polynomials in the Lie algebra of $O(N)$. One can show that a complete set of invariant polynomials is obtained by expanding $\det(1 + \frac{\lambda}{2\pi i} T)$ in powers of λ. The Lie algebra of $O(N)$ consists of all real antisymmetric matrices and therefore we can write

$$\det\left(1 + \frac{\lambda}{2\pi i} T\right) = \sum_{k=0}^{N/2} P_{2k}(T)\lambda^{2k}, \tag{7.6.1}$$

where P_{2k} is a polynomial of degree $2k$ in the matrix elements of T. The nth Pontrjagin class of a real vector bundle with structure group $O(N)$

is by definition the de Rham cohomology class on the base manifold represented by the differential form $p_n = P_{2n}(R)$, where R is the curvature form of an arbitrary connection in the vector bundle; p_n is a differential form of degreee $4n$. The first two Pontrjagin classes are

$$p_1 = \frac{1}{2(2\pi i)^2} R^j_k \wedge R^k_j$$

$$p_2 = -\frac{1}{4!(2\pi i)^4}(10R^i_j \wedge R^j_k \wedge R^k_l \wedge R^l_i - -6R^i_j \wedge R^j_i \wedge R^k_l \wedge R^l_k).$$

We can now apply the same method as before. Compute the Pontrjagin classes of the tangent bundle TM using the Levi-Cività connection. Locally we may write (and globally in the total space of FM) p_n as an exterior derivative of a differential form $c^{0,4n-1}$ which is a functional of the metric g. Let X be an element of the Lie algebra of *Diff* M, i.e., a vector field on M. X acts on $c^{0,4n-1}$ through the standard action of *Diff* M on metrics. We define a one-cocycle $c^{1,4n-1}$ by $c^{1,4n-1}(g;X) = X \cdot c^{0,4n-1}(g)$. Since the action of diffeomorphisms commutes with the exterior differentiation we can write $c^{1,4n-1} = dc^{1,4n-2}$. Applying the coboundary operator to this one- cocycle we obtain a two-cocycle

$$c^{2,4n-2}(g;X,Y) = X\cdot c^{1,4n-2}(g;Y) - Y\cdot c^{1,4n-2}(g;X) - c^{1,4n-2}(g;[X,Y]).$$

This $4n-2$ form is closed and we may write $c^{2,4n-2} = dc^{2,4n-3}$. The cocycle $c^{1,4n-2}$ gives the infinitesimal form of the *diffeomorphism anomaly* in $4n-2$ dimensions: One has $X \cdot \psi = \alpha(g;X)\psi(g)$, where

$$\alpha(g;X) = 2\pi i \int_M c^{1,4n-2}(g;X),$$

when $\dim M = 4n-2$. As we have said, the anomaly depends on the regularization but changing the regularization cannot change the Lie algebra cohomology class of $\alpha(g;X)$; the difference is a coboundary, $(\alpha' - \alpha)(g;X) = X \cdot \beta(g)$ for some function β in $\mathcal{M}$.

The integral of $c^{2,4n-3}$ over a $4n-3$ dimensional manifold S gives the commutator anomaly (Schwinger terms) in the Hamiltonian approach. We shall illustrate this by an example. The first Pontrjagin form can be locally written as $p_1(g) = \frac{-1}{8\pi^2} d\mathrm{tr}(\Gamma \wedge d\Gamma + \frac{2}{3}\Gamma^3) = c^{0,3}(g)$, where Γ is the Levi-Cività connection of a metric g. We have written the connection form as a matrix $(\Gamma_\mu)^\beta_\alpha = \Gamma^\beta_{\mu\alpha}$ with respect to some local coordinate system. The effect of a coordinate transformation when acting on the 3-form $c^{0,3}$ can be split into the usual action of a change of coordinates in the differential form indices and to the gauge action on Γ corresponding to a change of local frame in the tangent bundle. We shall be interested only on the integrals of the differential forms over some manifold and therefore we may discard the pull-back action of diffeomorphisms on differential forms and concentrate only on the gauge action

$$\Gamma'_\mu = \Phi\Gamma_\mu\Phi^{-1} - d\Phi\Phi^{-1}$$

where $\Phi_{\beta\alpha} = (\frac{\partial \phi_\beta}{\partial x_\alpha})$ is the derivative of a diffeomorphism ϕ. The gauge variation of $c^{0,3}$ in the direction of a vector field X is then the exterior derivative of the form $c^{1,2}(g;X) = \frac{-1}{8\pi^2}\mathrm{tr}\Gamma \wedge dX'$, where X' denotes the Jacobian matrix $(\frac{\partial X_\beta}{\partial x_\alpha})$. The infinitesimal diffeomorphism anomaly on a two-dimensional manifold M is

$$\omega(g;X) = \frac{1}{4\pi i}\int_M \Gamma^\alpha_{\mu\beta}\partial_\nu(\partial_\alpha X_\beta)dx_\mu dx_\nu \tag{7.6.2}$$

in terms of a fixed coordinate system (x_1, x_2); a compact 2-manifold is always a quotient of a polygon D in the plane by a discrete group acting on the boundary [Bers, 1981], and therefore we may use for example the Cartesian coordinates in the plane. Of course, not any vector field in D gives a vector field on M but one has to impose invariance with respect to the discrete group action.

Using the transgression formula we obtain $c^{1,2}(g;X,Y) = \frac{-1}{8\pi^2}\mathrm{tr}X'\,dY'$ and the corresponding infinitesimal two-cocycle is obtained by integrating this over a one-dimensional space. In one dimension we can write the vector fields in terms of the Fourier modes $\ell_n = ie^{in\phi}\frac{d}{d\phi}$ and the two-cocycle is then

$$c^{1,2}(\ell_n, \ell_m) = n^3\delta_{n,-m}.$$

This is up to a coboundary the central term (7.1.6) of the Virasoro algebra with the choice $c/12 = 1$. Note that the two-cocycle does not depend on the metric g; this is not so in higher dimensions.

In dimension three there are no Schwinger terms for Diff_0 because there are no gravitational anomalies in dimension four.

Exercise 7.6.3. Compute the commutator anomaly for the Lie algebra of vector fields on S^5 by transgression from p_2.

CHAPTER 8 THE BOSON FERMION CORRESPONDENCE

8.0. Introduction

In 1932 Louis de Broglie suggested that photons could be constructed from pairs of neutrinos [de Broglie, 1932]. Both are massless particles (except for some recent unconfirmed experiments according to which the neutrino could have a very small mass) and are electrically neutral. The main difference between free photons and free neutrinos is that the former obeys Bose statistics and the second Fermi statistics. The spin of a photon is 1 and the spin of a neutrino is 1/2 and therefore kinematically it should be possible to think of a photon as a neutrino pair. However, this old formulation of boson fermion equivalence has not been very fruitful in particle physics. Instead, there has been a lot of work on certain quantum field theory models in 1+1 space-time dimensions which admit equivalent formulations in terms of bosonic or fermionic fields.

A prototype for the boson fermion relation is the fermionic Thirring model together with the bosonic sine-Gordon model. The former is described by the field equations

$$\gamma^{\mu}\partial_{\mu}\psi + ig(\overline{\psi}\psi)\psi = 0,$$

where the Dirac field ψ on 1+1 dimensions has two space-time components in addition to a possible internal symmetry label a corresponding to a finite-dimensional representation of a symmetry group G. The sine-Gordon model is described by the equation

$$\partial_t\partial_x\phi(x,t) = \beta sin\, \phi(x,t)$$

in the case of one real scalar field ϕ. This model has a generalization involving the symmetry group G. The equivalence between the sine-Gordon and Thirring models was pointed out in Coleman [1975], and soon afterwards there appeared more detailed discussions on the Fermi-Bose correspondence in these models and related models; see, e.g., Mandelstam [1974], Halpern [1975], Kogut and Susskind [1975]. There are mathematically more precise treatments, of which we specifically want to mention Frenkel [1981], Carey and Hannabuss [1987], and Carey and Ruijsenaars [1987] because they relate the problem to the representation theory of loop algebras.

As stressed in Frenkel [1981], basically the quantum equivalence of the Thirring and sine-Gordon models is nothing else than the equivalence of two different realizations of the same highest weight representations of affine Lie algebras, namely, the (bosonic) vertex operator construction and the (fermionic) spinor realization. Both models can be described (in the Hamiltonian formulation) in terms of currents; the components of the currents are the generators of an affine Lie algebra based on G. In order to be more specific consider the example of $G = SO(2N)$ and the fermions transforming to the (real) defining representation of G. Assuming that the space is the circle S^1 the commutation relations of (real) fermions are assumed to be

$$[\psi_n^i, \psi_m^j]_+ = -\delta_{i,-j}\delta_{n,-m} \tag{8.0.1}$$

where $i \in \{\pm 1, \ldots, \pm N\}$ is the internal index and $n \in \mathbf{Z}$ is the Fourier index. The Fock vacuum v_0 for this system is defined by the relations $\psi_n^i v_0 = 0$ for $n > 0$ and $\psi_0^i v_0 = 0$ for $i > 0$. The hermiticity relations are assumed to be $(\psi_n^i)^* = -\psi_{-n}^{-i}$.

A normal ordering for products of fermion operators is defined by

$$: \psi_n^i \psi_m^j := \begin{cases} \psi_n^i \psi_m^j, & m > n \\ \psi_n^i \psi_m^j - \psi_m^j \psi_n^i, & n = m \\ -\psi_m^j \psi_n^i, & n > m \end{cases}$$

Define now the fermion bilinears (currents)

$$a_{ij}(n) = -a_{ji}(n) = \frac{1}{2} \sum_{m \in \mathbf{Z}} : \psi_m^i \psi_{n-m}^j : . \tag{8.0.2}$$

By a straightforward computation one obtains the current algebra commutation relations

$$\begin{aligned} [a_{ij}(n), a_{kl}(m)] &= \delta_{i,-k} a_{jl}(n+m) + \delta_{j,-k} a_{li}(n+m) + \delta_{j,-l} a_{ik}(n+m) \\ &\quad + \delta_{i,-l} a_{lj}(n+m) + (a_{ij}, a_{kl}) n \delta_{n,-m}, \end{aligned} \tag{8.0.3}$$

where $(a_{ij}, a_{kl}) = \delta_{i,-k}\delta_{j,-l} - \delta_{j,-k}\delta_{i,-l}$ is the invariant bilinear form on $SO(2N)$. We recognize these as the commutation relations of the affine Lie algebra related to the Lie algebra of $SO(2N)$. The representation of the affine Lie algebra $\widehat{o(2N)}$ in the Fock space is a sum of two highest weight representations. The irreducible components are the even and odd subspaces with respect to the grading by the degree of the polynomials of creation operators.

One can construct the same representation of the Lie algebra $\widehat{o(2N)}$ starting from the canonical commutation relations

$$(8.0.4) \qquad [\phi_n^i, \phi_m^j] = \delta_{ij}\delta_{n,-m}, \qquad 1 \le i, j \le N, n, m \in \mathbf{Z}.$$

This is achieved by the so-called vertex operator construction, to be discussed in Section 8.2. Note that the number of the components of the Bose field is equal to the rank of the group $SO(2N)$; this is a general feature of the vertex operator construction, valid for all G.

Vertex operators were first invented by physicists to construct scattering amplitudes in string models. We are not going to discuss the string scattering amplitudes in this book (but some other aspects of string models in the next chapter); we refer to the review article by Scherk [1975] for references to the early literature on string theory and the article by Goddard and Olive [1986] for a more updated presentation of vertex operators. The vertex operator construction of highest weight representations of affine algebras was first found in Frenkel and Kac [1980], and Kac, Kazhdan, Lepowsky, and Wilson [1981] (see also Segal [1981]) for the Lie algebras of the type A, D, and E and the generalization to arbitrary affine algebras was completed in Goddard, Nahm, Schwimmer, and Olive [1986] and Bernard and Thierry-Mieg [1987].

Any loop algebra can be thought of as a subalgebra of the Lie algebra $\mathbf{gl}_1$ through the gauge action on Dirac field in one space dimension, as was discussed in Chapter 6. In general, the representation of the corresponding affine algebra in the fermionic Fock space is not irreducible [except in the case $G = SU(N)$ acting on the Dirac field through the defining representation] but is in any case a direct sum of finitely many irreducible representations. For this reason, if we can understand the boson fermion relation on the basis of representation theory of the Lie algebra $\mathbf{gl}_1$, we have a general picture of the relation for an arbitrary (simple) symmetry group G. Of course, if one needs detailed information about the energy spectrum, partition function, etc. more work is required. In this chapter we shall concentrate on the general characteristics of the boson fermion equivalence and we refer to the articles by Goddard and Olive [1986], Bernard and Thierry-Mieg [1987], and Carey and Ruijsenaars [1987] for more refined analysis.

The author has profited most from the article by Kac and Peterson [1986]. Similar discussion can be found also in Date, Jimbo, Kashiwara, and Miwa [1983].

8.1. Representations of gl(∞)

We shall study highest weight representations of the infinite-dimensional Lie algebra **gl**(∞) generated by elements e_{ij}, $i, j \in \mathbf{Z}$, with commutation relations

$$[e_{ij}, e_{kl}] = \delta_{jk} e_{il} - \delta_{il} e_{kj}. \tag{8.1.1}$$

Let $\{\lambda_i\}_{i\in\mathbf{Z}}$ be a collection of (complex) numbers. The highest weight modules L_λ are defined as for the finite-dimensional algebras **gl**(n) [or for the simple algebras **sl**(n)]: First one constructs the Verma module V_λ as the quotient $\mathcal{U}(\mathbf{gl}(\infty))/\mathbf{g}_+$ where $\mathbf{g}_+$ consists of the vectors e_{ij} with $i < j$ and of the vectors $e_{ii} - \lambda_i$. The irreducible module L_λ is then equal to V_λ/W, where W is a maximal submodule of the Verma module not containing the highest weight vector $v_\lambda = 1 + \mathcal{U}(\mathbf{gl}(\infty))\mathbf{g}_+$. The highest weight vector in L_λ is characterized by the properties

$$e_{ij} v_\lambda = 0, \text{ for } i < j \qquad e_{ii} v_\lambda = \lambda_i v_\lambda.$$

Next we shall assume that all the λ_i's are integers, $\lambda_i \geq \lambda_{i+1}$ for all $i \in \mathbf{Z}$, $\lambda_i = 0$ for i large and positive, and λ_i constant for i large negative.

For each integer k define the subspace $L_{\lambda,k}$ to consist of linear combinations of the vectors $e_{i_1,j_1} e_{i_2,j_2} \cdots e_{i_n,j_n} v_\lambda$ with $(i_1 + \cdots + i_n) - (j_1 + \cdots + j_n) = k$. From $e_{ij} v_\lambda = 0$ for $i < j$ follows that $L_{\lambda,k} = 0$ when $k < 0$. From our assumption $\lambda_i - \lambda_j \neq 0$ only for a finite number of indices i, j follows furthermore that all the subspaces $L_{\lambda,k}$ are finite-dimensional and we can define the formal power series

$$\dim_q L_\lambda = \sum_{k\geq 0} (\dim L_{\lambda,k}) q^k \tag{8.1.2}$$

which is called the q-dimension of the module L_λ. Vectors in the space $L_{\lambda,k}$ are said to have *energy* k.

The Lie algebra **gl**(∞) can be thought of as an infinite simple algebra. The Cartan subalgebra is generated by the elements e_{ii} and a system of simple roots correspond to the set of generators $e_{i,i+1}$. An invariant bilinear form on **gl**(∞) is defined by $(x, y) = \operatorname{tr} xy$. The infinite Cartan matrix has nonzero elements $A_{ii} = 2$, $A_{i,i+1} = A_{i+1,i} = 1$. With a little effort one can extend the character formulas in Section 2.5 to the case of **gl**(∞); in fact, the formula in Theorem 2.5.13 can be directly applied to this case:

$$\dim_q L_\lambda = \prod_{i<j} \frac{1 - q^{\lambda_i - \lambda_j + j - i}}{1 - q^{j-i}}. \tag{8.1.3}$$

Denote by ω_m the weight λ such that $\lambda_i = 0$ for $i > m$ and $\lambda_i = 1$ for $i \leq m$. The set Λ_+ of positive weights we subject to the conditions above is then $\{\sum_i n_i\omega_i \mid n_i \in \mathbf{N}$ and $n_i = 0$ for almost all $i\}$. The *basic representation* of $\mathbf{gl}(\infty)$ corresponds by definition to the weight ω_0.

Exercise 8.1.4. Show that $\dim_q L_{\omega_m} = \phi(q)^{-1}$, where

$$\phi(q) = \prod_{i\geq 1}(1 - q^i).$$

The *fundamental representations* ω_m are related to the semi-infinite constructions in Section 7.3. The vacuum $\psi_m = f_m \wedge f_{m-1} \wedge \ldots$ is the highest weight vector v_{ω_m} when the algebra $\mathbf{gl}(\infty)$ is realized as

$$e_{ij} = a_i^* a_j.$$

As before, $\{f_i\}_{i\in\mathbf{Z}}$ is a basis in the dual H^* of the Hilbert space H. The space $\mathcal{U}(\mathbf{gl}(\infty))\psi_m$ is the carrier space for the representation L_{ω_m}. This identification gives a positive definite inner product for all the fundamental representations. In fact, it can be shown that for any $\lambda \in \Lambda_+$ there is a unique positive definite inner product in the representation space such that $(v_\lambda, v_\lambda) = 1$ and the adjoint of e_{ij} is e_{ji} for all i, j. The proof is obtained by an appropriate modification of Proposition 2.4.11 in Kac [1985].

8.2. The principal Heisenberg subalgebra

We shall study further the representation ω_m of $\mathbf{gl}(\infty)$ in the Fock space $\mathcal{F}_m$ corresponding to the Fock vacuum ψ_m. Following the discussion in Section 7.3 we introduce the operators $E_{ii} = a_i^* a_i - 1$ for $i \leq m$ and $E_{ij} = a_i^* a_j$ otherwise.

For each $k \in \mathbf{Z}$ we define the operator

$$h_k = \sum_{i\in\mathbf{Z}} E_{i,i+k}.$$

They satisfy the Heisenberg commutation relations

$$[h_j, h_k] = j\delta_{j,-k}. \tag{8.2.1}$$

The generators h_k, $k \neq 0$, together with the unit element 1 define the *principal Heisenberg subalgebra* $\mathbf{h}$ of $\mathbf{gl}(\infty)$. The vacuum is annihilated by all elements h_k with $k > 0$. Acting by the algebra $\mathcal{U}(\mathbf{h})$ on the vacuum we thus obtain a bosonic Fock space $\mathcal{F}_m^B \subset \mathcal{F}_m$. Note that the subspace $\mathcal{F}_{m,k}$ of vectors of energy k is the eigenspace of the operator

$$H = \sum_k E_{kk}.$$

The action of the element h_k changes the energy eigenvalue by k. It follows that $\mathcal{F}_{m,k}^B$ is spanned by the vectors $h_{i_1}^{n_1} h_{i_2}^{n_2} \dots h_{i_s}^{n_s} \psi_m$ with $i_1 \leq i_2 \leq \dots \leq i_s < 0$ and $i_1 n_1 + \dots i_s n_s = k$ and therefore the dimension of $\mathcal{F}_{m,k}^B$ is equal to the number of partitions of k into positive integers; but this is equal to the coefficient of q^k in the expansion of $\phi(q)^{-1}$. Thus $\dim\mathcal{F}_{m,k}^B = \dim\mathcal{F}_{m,k}$ and so $\mathcal{F}_m = \mathcal{F}_m^B$. This equality is the basis of the Fermi-Bose equivalence in quantum field theory in 1+1 space-time dimensions.

Let $\mathcal{B}_m$ be the space of polynomials with complex coefficients in the variables $x_1, x_2, \dots$. The space $\mathcal{B}_m$ carries a representation of the algebra $\mathbf{h}$ when we associate to h_k for negative k the operator kx_k and for positive values of k the operator $\frac{\partial}{\partial x_k}$. We want to clarify further the relation between the realization of the Heisenberg algebra in the standard bosonic Fock space $\mathcal{B}_m$ and in the fermionic Fock space $\mathcal{F}_m$. Note that the value of the index m does not really appear in the definition of $\mathcal{B}_m$ but we want to keep it to remind us that we are considering the bosonic realization together with the fermionic realization in $\mathcal{F}_m$.

The polynomial 1 clearly plays the role of ψ_m since $h_k \cdot 1 = 0$ for all $k > 0$. The isomorphism σ between $\mathcal{F}_m$ and $\mathcal{B}_m$ is given by the map

$$h_{k_1} h_{k_2} \dots h_{k_s} \psi_m \mapsto x_{k_1} x_{k_2} \dots x_{k_s}. \tag{8.2.2}$$

The grading $\mathcal{B}_m = \bigoplus_{k \geq 0} \mathcal{B}_{m,k}$ transported by σ from the grading by energy levels in $\mathcal{F}_m$ has then the property that the degree of the monomial $x_{k_1} \dots x_{k_s}$ is equal to $k_1 + \dots k_s$. The Hermitian inner product in $\mathcal{B}_m$ is defined by $(1,1) = 1$ and the condition that h_k is the adjoint of h_{-k}. The monomials form an orthogonal basis and

$$(x_1^{n_1} \dots x_s^{n_s}, x_1^{n_1} \dots x_s^{n_s}) = \prod_{1 \leq i \leq s} n_i! \, i^{-n_i}.$$

It will be convenient to define the *formal completions* $\overline{\mathcal{F}}_m$ and $\overline{\mathcal{B}}_m$ of $\mathcal{F}_m$ and $\mathcal{B}_m$ to consist of all formal power series in their respective generators (i.e., infinite polynomials in the variables x_i in the case of $\overline{\mathcal{B}}_m$ and infinite polynomials of a_i, a_i^* acting on the vacuum in the case of $\overline{\mathcal{F}}_m$). We define also

$$\mathcal{F} = \bigoplus_{m \in \mathbf{Z}} \mathcal{F}_m, \; \mathcal{B} = \bigoplus_{m \in \mathbf{Z}} \mathcal{B}_m$$

and the corresponding formal completions $\overline{\mathcal{F}}$ and $\overline{\mathcal{B}}$. Then the series

$$X(z)=\sum_{j\in\mathbf{Z}} z^j a_j \qquad X^*(z)=\sum_{j\in\mathbf{Z}} z^j a_j^*,$$

where $z\in\mathbf{C}^\times$, can be considered as linear operators from $\mathcal{F}$ to $\overline{\mathcal{F}}$. We want to describe explicitly the operators $\sigma X(z)\sigma^{-1}$ and $\sigma X^*(z)\sigma^{-1}$ from $\mathcal{B}$ to $\overline{\mathcal{B}}$. Since the action of $X(z)$ changes the vacuum level by $+1$ and the action of $X^*(z)$ by -1, the transported operators in the bosonic Fock space are maps $\overline{\mathcal{B}}_m \mapsto \overline{\mathcal{B}}_{m\pm1}$ respectively. Let $\#$ be the shift operator in $\overline{\mathcal{B}}_m$ which increases the index m by 1.

Let us define *the vertex operator*

$$\Gamma(z)=\exp\left(\sum_{j>0} z^j x_j\right)\cdot\exp\left(-\sum_{j>0}\frac{z^{-j}}{j}\frac{\partial}{\partial x_j}\right) \tag{8.2.3}$$

as linear maps from $\mathcal{B}$ to $\overline{\mathcal{B}}$.

THEOREM 8.2.4. *The fermionic operators $X(z)$ and $X^*(z)$ are related to the bosonic vertex operator $\Gamma(z)$ by*

$$\sigma X(z)\sigma^{-1}=z^{m+1}\#\Gamma(z),\ \sigma X^*(z)\sigma^{-1}=z^{-m-1}\#^{-1}\Gamma(z)^*$$

in the subspace $\mathcal{B}_m$.

PROOF: Note first that the adjoint $\Gamma(z)^*$ of the vertex operator is obtained simply by reversing all the signs in front of x_i and $\frac{\partial}{\partial x_i}$; this follows from the fact that ix_i is the adjoint of $\frac{\partial}{\partial x_i}$.

The commutation relations

$$[h_j,X(z)]=z^jX(z),\ [h_j,X^*(z)]=-z^jX^*(z)$$

imply the corresponding relations in the bosonic Fock space,

$$\left[\frac{\partial}{\partial x_n},\sigma X(z)\sigma^{-1}\right]=z^n\sigma X(z)\sigma^{-1},\ \ [nx_n,\sigma X(z)\sigma^{-1}]=z^{-n}\sigma X(z)\sigma^{-1}$$

and similarly for $X^*(z)$. Denote by T the operator

$$e^{\sum_{j>0}\frac{z^{-j}}{j}\frac{\partial}{\partial x_j}}.$$

Then $[nx_n,T]=-z^{-n}T$ and therefore $V=\sigma X(z)\sigma^{-1}T$ commutes with the operators x_n; it follows that V can be expressed as a formal power series in the variables x_n. On the other hand, T commutes with the partial derivatives and therefore V obeys the same commutation relations with the partial derivatives as the operator $\sigma X(z)\sigma^{-1}$; but these commutation relations tell us that V must be of the form

$$C(z)e^{\sum_{j>0} z^j x_j}$$

where $C(z)$ is some constant depending only on z. Thus $\sigma X(z)\sigma^{-1} = C(z)\#\Gamma(z)$. Comparing the action of both sides to the vacuum vector 1 in $\mathcal{B}_m$ we conclude that $C(z) = z^{m+1}$. The case of $X^*(z)$ is proven in a similar way.

The generators e_{ij} of $\mathbf{gl}(\infty)$ can be transported to linear operators in $\mathcal{B}_m$ by the isomorphism σ. For brevity, denote the transported operators also by e_{ij}. We define a vertex operator depending on two complex variables z, w by

$$\Gamma(z,w) = \exp\left[\sum_{j>0}(z^j - w^j)x_j\right] \cdot \exp\left(-\sum_{j>0}\frac{z^{-j} - w^{-j}}{j}\frac{\partial}{\partial x_j}\right).$$

THEOREM 8.2.5. *The generators of* $\mathbf{gl}(\infty)$ *in the bosonic Fock space are obtained from the expansion* ($|w| < |z|$)

$$\sum_{i,j\in\mathbf{Z}} z^i w^{-j} e_{ij} = \frac{1}{1 - w/z}[(w/z)^m \Gamma(z,w) - 1].$$

PROOF: First, directly from the definitions, we have

$$\sum_{i,j} z^i w^{-j} e_{ij} = \sigma X(z) X^*(w)\sigma^{-1}.$$

But the claim is then a simple consequence of the identity

$$e^{\alpha\frac{\partial}{\partial x}} e^{\beta x} = e^{\alpha\beta} e^{\beta x} e^{\alpha\frac{\partial}{\partial x}}.$$

Exercise 8.2.6. Verify the identity used in the above proof.

We complete the discussion of this section by determining the image of the basis vectors $f_{i_1} \wedge f_{i_2} \wedge \ldots$ under the map $\sigma : \mathcal{F}_m \to \mathcal{B}_m$. Define the polynomials $S_k(x_1, x_2, \ldots)$ by the formula

$$e^{\sum_{k>0} z^k x_k} = \sum_{k\geq 0} z^k S_k(x). \tag{8.2.7}$$

We set $S_k = 0$ for $k < 0$. Expanding the right-hand side in (8.2.7) we get $S_0 = 1$ and

$$S_k = \sum_{s>0} \sum_{n_1+2n_2+\cdots+sn_s=k} \prod_{1\leq i\leq s} \frac{x_i^{n_i}}{n_i!}, \quad k > 0. \tag{8.2.8}$$

The first three polynomials are $S_1 = x_1$, $S_2 = \frac{1}{2}x_1^2 + x_2$, $S_3 = \frac{1}{6}x_1^3 + x_1x_2 + x_3$.

Let $\lambda = \{\lambda_1 \geq \lambda_2 \geq \cdots \geq \lambda_n \geq 0\}$ be a sequence of integers; the set of all finite sequences of integers satisfying the inequalities is denoted by P. The *Schur polynomials* are defined by

$$S(\lambda)(x) = \det(S_{\lambda_i+j-i}(x))_{i,j}. \tag{8.2.9}$$

8.3. Properties of the Schur polynomials

The Schur polynomials are related to characters of irreducible finite-dimensional representations of the groups $GL(n)$. Let $L(\lambda)$ be the representation space for such a representation T with highest weight $\lambda = (\lambda_1, \ldots, \lambda_n)$. Let $g(\epsilon) = diag(\epsilon_1, \ldots, \epsilon_n) \in GL(n)$ and denote

$$x_j = \frac{\epsilon_1^j + \cdots + \epsilon_n^j}{j}.$$

Then $\mathrm{tr}_{L(\lambda)}T(g(\epsilon)) = S_\lambda(x)$; see Macdonald [1979] for a proof.

The Schur polynomials give an explicit realization of the boson fermion correspondence defined in the last section. Namely, we have the following Theorem.

THEOREM 8.3.1. *Let $v = f_{i_1} \wedge f_{i_2} \wedge \cdots \in \mathcal{F}_m$. Then $\sigma(v) \in \mathcal{B}_m$ is the polynomial $S_\lambda(x)$ with $\lambda_k = i_k + m + k - 1$.*

PROOF: Note first that $i_k = -k + 1 + m$ when $k >> 0$ and therefore only a finite number of the λ_k's are non-zero. Denote

$$F(x_1, x_2, \ldots) = (\psi_m, e^{\sum_{j\geq 1} x_j h_j} v).$$

From (8.2.7) it follows that the matrix A corresponding to the action of $\exp(\sum_{j>0} x_j h_j)$ in the vector space $H = \bigoplus_{j\in\mathbf{Z}} \mathbf{C}e_j$ is

$$A_{ij} = S_{j-i}(x), \; i, j \in \mathbf{Z}$$

and therefore in the fermionic Fock space (which is the semi-infinite tensor algebra of H) we have

$$\begin{aligned} F(x) &= (\psi_m, Af_{i_1} \wedge Af_{i_2} \wedge \dots) \\ &= \det(A_{i_\mu, i})_{\mu \geq 1, j \leq m} = \det(S_{j-i}(x))_{i \in \{m, m-1, \dots\}}^{j \in \{i_1, i_2, \dots\}}. \end{aligned} \tag{8.3.2}$$

In the bosonic Fock space h_j is represented by $\frac{\partial}{\partial x_j}$ when $j > 0$ and thus, denoting $\sigma(v)$ by G,

$$F(y) = \left(\exp \sum_{j \geq 1} y_j \frac{\partial}{\partial x_j}\right) G|_{x=0} = G(x+y)|_{x=0} = G(y).$$

Comparing this with (8.3.2) completes the proof.

From the proof of (8.3.1) we obtain the following corollary:

COROLLARY 8.3.3. *Let λ be as in (8.3.1). Then*

$$S_\lambda(x) = (\psi_m, (\exp \sum_{j>0} x_j h_j) a_{i_1}^* \dots a_{i_p}^* a_{k_1} \dots a_{k_q} \psi_m),$$

where $i_1 > \dots > i_p$ are those integers in (i) which are bigger than m and $\{k_1 > \dots > k_q\}$ is the set $\{i \ni (i) \mid i \leq m\}$.

PROOF: There is nothing more than to remember the definition of the creation and annihilation operators from which follows

$$f_{i_1} \wedge f_{i_2} \wedge \dots = a_{i_1}^* \dots a_{i_p}^* a_{k_1} \dots a_{k_q} \psi_m.$$

Since the semi-infinite forms $\psi(i)$ form on orthonormal basis of the Fock space $\mathcal{F}_m$ we conclude from (8.3.1) that the Schur polynomials $S(\lambda)$ form an orthonormal basis in $\mathcal{B}_m$.

Let $A \in GL(\infty)$ (i.e., A is an invertible matrix and $A_{ij} - \delta_{ij}$ is zero for almost all i, j). Let us denote by ρ_m the representation of $GL(\infty)$ in $\mathcal{F}_m$ induced by the natural representation in H. By the complete antisymmetry of the wedge product we have

$$\rho_m(A) f_{i_1} \wedge f_{i_2} \wedge \dots = \sum_{j_1 > j_2 > \dots} \det(A_{i_1 i_2 \dots}^{j_1 j_2 \dots}) f_{j_1} \wedge f_{j_2} \wedge \dots,$$

where $(A_{i_1 i_2 \dots}^{j_1 j_2 \dots})$ is the matrix formed from A by taking only the elements at positions (j_μ, i_ν). Using (8.3.1) we deduce that the matrix elements of A in the bosonic realization are given by

$$\rho_m(A) S(\lambda) = \sum_{\mu \in P} \det(A_{\lambda_1+m, \lambda_2+m-1, \dots}^{\mu_1+m, \mu_2+m-1, \dots}) S(\mu).$$

CHAPTER 9 HOLOMORPHIC ASPECTS OF STRING THEORY

9.0. Introduction

A string is a piecewise smooth map of the interval to a manifold M. A closed string is a map of the circle S^1 into M. In string theory the strings replace the points of the manifold M as fundamental objects. The enormous amount of work done on quantized string models in physics has been motivated by the hope that the quantum string theory would produce a finite quantized theory of gravity, free of the divergences of the ordinary quantized Einstein theory of gravitation. So far the proof is missing but work is continuing. It has been proposed that some kind of string theory would be the unified theory of all fundamental interactions in physics. However, the fundamental principles of string theory have not been clearly formulated and up to now there are no predictions of string theory which could be experimentally tested. The great interest in string models cannot be explained only by the (rather meager) physical results, but there are very interesting mathematical problems which require an exciting blend of algebraic and geometric methods.

The path of a string on the background manifold M is a two-dimensional surface. The classical action of a string $X(\sigma, \tau)$ is postulated to be

$$S(X) = \int_R \sqrt{g} g^{ab} \partial_a X^\mu \partial_b X^\nu \eta_{\mu\nu} d\sigma d\tau.$$

Here g_{ab} is the metric tensor of the parameter space R of the surface, g^{ab} is its inverse, $g = |\det(g_{ab})|$ and $\eta_{\mu\nu}$ is the metric of the background space. The action has the remarkable property that it is invariant under both the diffeomorphisms of R and the conformal rescalings $g_{ab} \mapsto \rho \cdot g_{ab}$; the diffeomorphisms act on both g_{ab} and $X(\sigma, \tau)$ in the natural way. In fact, the inclusion of the parameter space metric g_{ab} is a consequence of the requirement that the action should not depend on the chosen parametrization of the string "world sheet". Note that we have now increased the dynamical degrees of freedom: in addition to the string coordinates we have the metric g_{ab}. The "space-time" metric $\eta_{\mu\nu}$ is assumed to be fixed.

Let $\mathcal{R}$ be the space of metrics on a given R. Since the action $S(X, g)$ is invariant under the group $\mathit{Diff}\, R$ and the conformal group $C(R)$ one would expect that $\mathcal{T} = \mathcal{R}/[C(R) \times \mathit{Diff}\ R]$ would be the relevant parameter space for the string world sheet geometries. However, it turns out that the quantization of the string breaks in general the classical symmetries. There are a few cases where one gets a conformally and reparametrization invariant quantum theory. For example, if M is the flat Minkowski space $\mathbf{R}^{d-1,1}$ in d dimensions then the invariance is preserved when $d = 26$. If the space M is curved the condition will be different; we shall consider the case of a string moving on a group manifold in Section 9.5.

A string moving in flat space-time can be thought of as a collection of d real scalar fields on R. Therefore a string theory is a particular case of conformally invariant field theory in two dimensions. A purely bosonic string theory is not considered to be a very realistic description of nature. There are so-called superstring theories which are hoped to give a better description of the fundamental interactions. They involve fermionic fields on R in addition to the bosonic components. The most promising candidate at the moment is a theory in 10 space-time dimensions with a symmetry group $E_8 \times E_8$. We are not going to discuss the superstring theories any further in this book but refer the interested reader to the vast literature on the subject, to which Green, Schwarz and Witten [1987] is a good introduction.

The parameter space R is assumed to be a smooth compact oriented manifold of dimension two, i.e., a Riemann surface. The space $\mathcal{T}$ is the *Riemann moduli space* of the surface. If R has no boundary then it is topologically characterized by the genus g (number of handles of the surface) and $\mathcal{T}$ has dimension $6g-6$ for $g > 1$, $\dim\mathcal{T} = 2$ for $g = 1$ and $\mathcal{T}$ is one point when $g = 0$. One can write $\mathcal{T} = \mathcal{T}_0/\Gamma$ where $\mathcal{T}_0 = \mathcal{R}/\mathit{Diff}_0 R$ is the *Teichmüller space* of R; $\mathit{Diff}_0\, R$ is the connected component of identity of $\mathit{Diff}\, R$, and $\Gamma = \mathit{Diff}\, R/\mathit{Diff}_0\, R$ is the modular group. The Teichmüller space is, in contrast to the Riemann moduli space, a smooth manifold and therefore one usually first formulates a string theory on $\mathcal{T}_0$ and then imposes the requirement of modular invariance on basic physical quantities.

The Teichmüller spaces are complex manifolds with a preferred metric, the *Weil-Peterson metric.* Furthermore, they are Kähler manifolds [Ahlfors, 1961]. The complex structure plays a central role in the quantization of string theories. In the Lagrangian formulation of the quantized string theory the basic objects are the Teichmüller spaces and various holomorphic vector bundles over $\mathcal{T}_0$. (Here one meets again the determinant bundle of the Dirac operator, this time parametrized by metrics.) There is also a Hamiltonian formulation of string theory which we shall discuss in this chapter. The precise relation between string theories in the respective Hamiltonian and Lagrangian formulations is not well understood. We shall consider some aspects of string theories in the

Hamiltonian approach as an application of the theory of current algebras (Virasoro and Kac-Moody). It is not surprising that the holomorphic structures are of importance also here.

In order to construct the Fock space of a bosonic string one has to define a normal ordering for the components of the quantized string operator. The normal ordering is not invariant under reparametrizations of the string (we shall consider closed strings) except under the global rotations of the circle S^1 parametrizing the string. Thus one is naturally led to consider a bundle of Fock spaces parametrized by elements of $\mathit{Diff}\, S^1/S^1$. This bundle is holomorphic. The Fock bundle does not have a reparametrization invariant vacuum and therefore one has to introduce a *ghost field*. The ghost "lives" in a holomorphic line bundle over $\mathit{Diff}\, S^1/S^1$. The curvature of this line bundle is opposite (in dimension 26, in case of a flat background and in some special cases for string on a group manifold) to the curvature of the Fock bundle and therefore the curvature of the tensor product vanishes; the curvature is the obstruction for constructing the vacuum.

The group $\mathit{Diff}\, S^1$ is relevant also in moduli space problems related to the Lagrangian formulation of string theory. Namely, $\mathit{Diff}\, S^1$ acts "infinitesimally transitively" in the moduli space of Riemann surfaces together with a local parameter; there is a more precise statement according to which the tangent space of the moduli space (including a local parameter on R) at an arbitrary point is isomorphic with $\mathit{Vect}S^1/D_0$ where D_0 is the Lie algebra of vector fields on S^1 (boundary of the parameter domain) which can be extended to holomorphic vector fields in the interior of S^1. We shall return to this kind of moduli problems in Section 12.2. One can say that at least infinitesimally $\mathit{Diff}\, S^1$ is the "universal moduli space" for Riemann surfaces (together with a local parameter).

The construction of the Kähler structure of $\mathit{Diff}\, S^1$ and the discussion on strings in Sections 9.1 and 9.2 is based on Bowick and Rajeev [1987] (see also Kirillov [1987]). In Section 9.3 we have used Pilch and Warner [1987]. The discussion in Section 9.4 relies mainly on Frenkel, Garland, and Zuckerman [1986] and in Section 9.5 we follow Mickelsson [1987b].

9.1. The Kähler structure of $\mathit{Diff}\, S^1/S^1$

We denote by $\mathit{Diff}\, S^1$ the group of orientation preserving diffeomorphisms of the circle. We shall first define a complex structure on the manifold $M = \mathit{Diff}\, S^1/S^1$. The topology and differentiable structure on M can be defined as an *inverse limit* of Hilbert manifolds (defined by Sobolev norms in the space of smooth mappings); we do not need the technical details of the construction; see Omori [1973]; Michor [1980].

An almost complex structure on a manifold is a smooth distribution $\{J_x\}$ of linear operators J_x in the tangent spaces T_x such that $J_x^2 = -1$ at each point x on the manifold. As before, we identify the tangent space of *Diff* S^1 at the identity with the space of smooth vector fields on the circle. The canonical projection $\mathit{Diff}\, S^1 \to \mathit{Diff}\, S^1/S^1$ sends the elements $\ell_n = ie^{in\phi}\frac{d}{d\phi}$, $n \neq 0$, to a basis in the complexified tangent space $T_o^{\mathbf{C}}$ of M at the "origin" o (= the class represented by the identity diffeomorphism). We can define an almost complex structure on M by setting

$$J_o \sum_{n\neq 0} v_n \ell_n = -i \sum_{n \neq 0} \frac{n}{|n|} v_n \ell_n, \tag{9.1.1}$$

At an arbitrary point $[h] \in M$, $h \in \mathit{Diff}\, S^1$, the operator $J_{[h]}$ is defined by a left translation s_h

$$J_{[h]} v = s_h \circ J_o \circ s_h^{-1} v, \quad v \in T_{[h]}^{\mathbf{C}} M.$$

If h' differs from h by a multiplication by a rotation r from the right then $s_h \circ J_o \circ s_h^{-1} = s_{h'} \circ J_o \circ s_{h'}^{-1}$ since obviously r commutes with the sign operator in (9.1.1). Thus the distribution J is well-defined on M.

A complex vector field X is *holomorphic*, or of type (1,0), if $J_x X_x = -iX_x$ at each point x. Similarly a vector field is of type $(0,1)$ (antiholomorphic) if $J_x X_x = iX_x$. Any vector field can be uniquely written as a sum $X = X^{(1,0)} + X^{(0,1)}$ of a holomorphic and an antiholomorphic vector field. If one can define an open covering $\{U_\alpha\}$ of the manifold and on each set U_α a basis $(X_1, X_2, \dots)$ for the space of holomorphic vector fields such that the commutators $[X_i, X_j]$ are all linear combinations of the elements in the basis then the almost complex structure is *integrable.*The integrability implies that the manifold is *a complex manifold* in the sense that one can choose an atlas of complex coordinates in such a way that the coordinate transformations are holomorphic functions and that the vector fields $\frac{\partial}{\partial z_i}$ determined by complex coordinates z_i are of type $(1,0)$ [Kobayashi and Nomizu II, 1969]. The splitting of vector fields to (1,0) and (0,1) components defines a splitting

$$d = \partial + \overline{\partial}$$

of the exterior differentiation to a (1,0) component ∂ and to a (0,1) component $\overline{\partial}$.

Let $\tilde{\ell}_n$ denote the left invariant vector field on *Diff* S^1 which is equal to ℓ_n at the identity; we denote by the same symbol the vector field on M obtained through the canonical projection to the quotient. By definition, the vector fields $\tilde{\ell}_n$ are holomorphic for $n > 0$ and they span at each point of M the eigenspace $T_x^{(1,0)} \subset T_x^{\mathbf{C}}M$ which corresponds to the eigenvalue $-i$ of J_x. Since $[\ell_n, \ell_m] = (n-m)\ell_{n+m}$, the commutator of two basis elements with a positive index is again a vector field of the same type; thus the almost complex structure on M is integrable.

Since the vector fields $\tilde{\ell}_n$ span the tangent space $T_x^{\mathbf{C}}M$ at an arbitrary point x we can define a 2-form ω on M by setting

$$\omega(\tilde{\ell}_n, \tilde{\ell}_m) = i(am^3 + bm)\delta_{m,-n} \tag{9.1.2}$$

where $a, b \in \mathbf{C}$ are arbitrary constants. This form is closed,

$$\begin{aligned} d\omega(\tilde{\ell}_n, \tilde{\ell}_m, \tilde{\ell}_p) &= \omega([\tilde{\ell}_n, \tilde{\ell}_m], \tilde{\ell}_p) + \omega([\tilde{\ell}_m, \tilde{\ell}_p], \tilde{\ell}_n) + \omega([\tilde{\ell}_p, \tilde{\ell}_n], \tilde{\ell}_m) \\ &= (n-m)\omega(\tilde{\ell}_{n+m}, \tilde{\ell}_p) + (m-p)\omega(\tilde{\ell}_{m+p}, \tilde{\ell}_n) + (p-n)\omega(\tilde{\ell}_{p+n}, \tilde{\ell}_m) = 0. \end{aligned}$$

The form ω is *compatible with the complex structure* J in the sense that

$$\omega(JX, JY) = \omega(X, Y) \tag{9.1.3}$$

for arbitrary vector fields X, Y.

PROPOSITION 9.1.4. *If $a > 0$ and $a + b > 0$ or $a = 0$ and $b > 0$ then $(X,Y) \mapsto \omega(X, JY)$ defines a positive definite inner product on each of the tangent spaces T_xM.*

PROOF: A real tangent vector at a point $x \in M$ can be written as a linear combination

$$v = \sum_{n>0} \alpha_n \tilde{\ell}_n + \sum_{n<0} \bar{\alpha}_{-n} \tilde{\ell}_n,$$

evaluated at the point x, where $\alpha_n \in \mathbf{C}$. Then

$$\omega(v, J_x v) = \sum_{n>0} 2|\alpha_n|^2 (an^3 + bn).$$

A closed 2-form ω which is compatible with the complex structure and for which the real symmetric bilinear form $g(X,Y) = \omega(X, JY)$ is positive definite is called a *Kähler structure.* Thus (M, ω) is a *Kähler manifold* when a and b are in the allowed domain.

Let us study the case $a > 0$, $a + b > 0$ in more detail. First we want to compute the Riemann curvature tensor. To do that we need a formula for the covariant derivative corresponding to the metric g defined above. We use the formula

$$\begin{aligned} 2g(Z,\nabla_X Y) =& X\cdot g(Z,Y)+Y\cdot g(X,Z)-Z\cdot g(X,Y)+g([X,Y],Z) \\ &+g([X,Z],Y)+g([Y,Z],X) \end{aligned} \tag{9.1.5}$$

The Christoffel symbols in the basis $\{\tilde{\ell}_n\}$ are then $\Gamma^k_{ij}=g^{kl}\Gamma_{ijk}$, where $\Gamma_{ijk}=g(\tilde{\ell}_k,\nabla_{\tilde{\ell}_i}\tilde{\ell}_j)$, $g_{ij}=g(\tilde{\ell}_i,\tilde{\ell}_j)$ and (g^{ij}) is the inverse of the matrix (g_{ij}). Inserting the expression $g_{mn}=\delta_{n,-m}(a|m|^3+b|m|)$ into (9.1.5) [the first three terms in (9.1.5) drop away because $g(\tilde{\ell}_n,\tilde{\ell}_m)$ is a constant function] we get

$$\Gamma^k_{ij}=\frac{1}{2}\delta_{i+j,k}\left[i-j-(k+i)f(j)/f(k)-(k+j)f(i)/f(k)\right], \tag{9.1.6}$$

where $f(n)=a|n|^3+b|n|$. The metric is invariant under the action of the group $K=\mathit{Diff}\,S^1$ and therefore also the Riemannian connection Γ. In the notation of Section 3.6 (the principal bundle P is now the frame bundle of M) the subalgebra $\mathbf{m}\subset\mathbf{k}$ is spanned by the elements ℓ_n, $n\neq 0$, and $\Lambda(\ell_n)$ is the infinite matrix with elements $(a_{ij})=\Gamma^i_{nj}$ when $n\neq 0$. From Proposition 3.6.5 we obtain the following expression for the curvature tensor at the origin:

$$R^l_{ijk}=[\Lambda(\ell_i),\Lambda(\ell_j)]_{lk}+\Lambda([\ell_i,\ell_j]_{\mathbf{m}})_{lk}+\lambda([\ell_i,\ell_j]_{\mathbf{l}})_{lk},$$

where $i,j,k,l\neq 0$, $\mathbf{l}$ is the one-dimensional Lie algebra spanned by ℓ_0 and $\lambda(\ell_0)_{lk}=k\delta_{lk}$ is the matrix of ℓ_0 in the adjoint representation. After some algebra one obtains

$$\begin{aligned} R^l_{ijk}=&\left\{\theta(i-k)\,(i+k)^2\frac{a(i-k)^3+b(i-k)}{ak^3+bk}\right. \\ &\left.+(k+2i)^2\frac{ak^3+bk}{a(k+i)^3+b(k+i)}-2ik\right\}\delta_{i,-j}\delta_{l,-k} \end{aligned} \tag{9.1.7}$$

where $\theta(n)=1$ for $n\geq 0$ and $\theta(n)=0$ otherwise. The *Ricci form* in the Kähler sense is by definition the 2-form

$$(Ric)_{ij}=\sum_{k>0}R^{-k}_{ijk}=\left(\frac{26}{12}i^3-\frac{1}{6}i\right)\delta_{i,-j} \tag{9.1.8}$$

Note that the result is independent of the values of the parameters a,b. The complex line bundle L with the curvature equal to Ric is called the *canonical line bundle* over M. In fact, from Theorem 7.2.4 we have an explicit construction of this bundle as a pull-back of the determinant bundle over Gr_1 by setting the parameter n equal to -1.

9.2. The Fock space of the bosonic string

The classical bosonic string theory is defined by fixing the phase space (X, ω), where X is the space of smooth loops in a real vector space $\mathbf{R}^{d-1,1}$ with a Lorentzian inner product of signature $(d-1,1)$ and a 2-form ω on X defined by

$$\omega(u,v) = \frac{1}{2\pi}\int_{S^1} \eta_{\mu\nu} u^{\mu} \frac{d}{d\phi} v^{\nu}\, d\phi, \tag{9.2.1}$$

where $\eta_{\mu\mu} = +1$, $1 \leq \mu \leq d-1$, $\eta_{dd} = -1$ and $\eta_{\mu\nu} = 0$ otherwise. The form ω is clearly closed (it has constant coefficients) but it is not quite symplectic since it has null-directions: $\omega(u,v) = 0$ for all $v \in X$ when u is a constant loop. We have identified the tangent space of X with itself at any point $x \in X$ because X is a vector space.

We recall that to any smooth function f on a symplectic manifold there corresponds a unique vector field u_f (the Hamiltonian vector field associated with f) such that $\omega(v, u_f) = v \cdot f$ for any vector field v. The Poisson bracket $\{f,g\}$ of a pair of functions f, g is then defined to be the function $\omega(u_f, u_g) = u_f \cdot g = -u_g \cdot f$. The map $f \mapsto u_f$ is a Lie algebra homomorphism: $[u_f, u_g] = u_{\{f,g\}}$; see, e.g., Abraham and Marsden [1978].

Since ω is degenerate not all functions f on X have a vector field u_f such that $\omega(u_f, v) = v \cdot f \ \forall v$ and if u_f exists it is fixed only up to an addition of a vector field along the null-directions of ω. A necessary and sufficient condition for the existence of u_f is that the derivative $v \cdot f$ vanishes when v is a constant loop; this means that f is really a function on the quotient $X/(\text{constant loops})$. This latter manifold can be identified as the space of *based loops* $x(\phi)$ which take a prescribed value, say 0, at the end points $\phi = 0, \phi = 2\pi$. Denoting by $\Omega\mathbf{R}^{d-1,1}$ the space of based loops we obtain a symplectic form on $\Omega\mathbf{R}^{d-1,1}$ simply by restricting ω to the submanifold.

For each $n \in \mathbf{Z}$ we define on X the function

$$\lambda_n(x) = \frac{1}{4\pi i}\int_{S^1} e^{-in\phi} x'^2\, d\phi, \tag{9.2.2}$$

where $x' = \frac{d}{d\phi}x(\phi)$ and $u^2 = u \cdot u = \eta_{\mu\nu}u^{\mu}u^{\nu}$. The associated (complex) Hamiltonian vector fields U_n are given by

$$[U_n(x)](\phi) = ie^{-in\phi}x'(\phi). \tag{9.2.3}$$

The Poisson brackets of the λ_n's amomg themselves are given by

$$\{\lambda_n, \lambda_m\} = (n-m)\lambda_{n+m}. \tag{9.2.4}$$

Thus the set $\{\lambda_n\}$ generates the Lie algebra of $Diff\, S^1$.

Exercise 9.2.5. Show that the Hamiltonian flows [= the local 1-parameter groups generated by the Hamiltonian vector fields] defined by the real functions $\lambda_n + \lambda_{-n}$ and $i\lambda_n - i\lambda_{-n}$ actually correspond to the natural action of $Diff\, S^1$ in the loop space X.

We fix a complex structure J on the manifold $\Omega\mathbf{R}^{d-1,1}$. Define the Fourier modes of a loop u by

$$u(\phi) = \frac{1}{\sqrt{2\pi}} \sum_n u_n e^{in\phi}.$$

We set

$$(Ju)_n = \begin{cases} -iu_n \text{ for } n > 0 \\ iu_n \text{ for } n < 0. \end{cases}$$

The pair (J, ω) makes $\Omega\mathbf{R}^{d-1,1}$ into a "pseudo-Kähler" manifold. The corresponding pseudo-Riemannian metric is

$$g(u,v) = \omega(u, Jv) = \frac{1}{2\pi} \sum_n |n| u_n \cdot v_{-n}. \tag{9.2.5}$$

Note that $u_{-n} = \overline{x}_n$ for all real loops and therefore g is real.

We can now apply the method of geometric quantization to the system (X, ω). The first step is to construct a complex line bundle E over X with a Hermitian fiber product and a compatible connection such that the curvature is equal to ω. In our case there is no problem because X is contractible as a vector space and all bundles over X are trivial. The connection is determined by giving the covariant derivative of the global section $\psi(x) = 1$; $\nabla_u \cdot 1 = -i\theta(u)$. One easily checks that $d\theta = \omega$ for the one-form

$$\theta_x(u) = -i \sum_{n=1}^{\infty} n x_n u_{-n}.$$

If α, β are two elements in the complex line over $x \in X$ the inner product can be written as

$$(\alpha, \beta)_x = \alpha \overline{\beta} e^{-K(x)}, \tag{9.2.6}$$

where K is some real valued function on X. The metric is compatible with the connection θ if

$$u \cdot (\psi, \psi') = (\nabla_{u(1,0)} \psi, \psi') + (\psi, \nabla_{u(0,1)} \psi')$$

for all sections ψ, ψ' and for any complex vector field u. This is equivalent to the condition

$$-u \cdot K = -i\theta(u^{(1,0)}) + \overline{i\theta(u^{(0,1)})} \tag{9.2.7}$$

A solution of this is given by

$$K(x) = \sum_{n=1}^{\infty} n x_n \cdot x_{-n} \tag{9.2.8}$$

where we have used the Fourier decomposition. Because of $x_n = \overline{x}_{-n}$ the function K is real valued. The next step in the geometric quantization program is to choose a *polarization.* The effect of a polarization is to divide the coordinates of the phase space (at least locally) to space coordinates and momenta and to consider sections of the line bundle which depend only on the space coordinates. The reason for doing this is that otherwise the representation of the algebra of observables in the space of sections of the line bundle would be highly reducible. If the phase space of the physical system happens to be just a cotangent bundle $M = T^*Q$ of a configuration space Q then there is a natural polarization which splits the tangent space of M at an arbitrary point $(q,p) \in T^*Q$ to the space T_qQ of tangent vectors along Q and the vertical space $V_{(q,p)}$ of tangent vectors along the fiber of the cotangent bundle. The wave functions in the Schrödinger picture are then functions $\psi(q)$ on the base Q of the cotangent bundle, i.e., functions on M such that the derivatives of ψ vanish along the fiber directions.

A complex structure on a manifold defines a natural polarization. The complexified tangent space is divided into type $(1,0)$ and type $(0,1)$ tangent vectors. The wave functions in the Schrödinger picture are sections of the complex line bundle with vanishing derivatives along the $(0,1)$ directions; these are by definition the *holomorphic* sections. We have a natural indefinite inner product in the space of polynomial functions $\psi(x_1, x_2, \ldots, x_n)$ (a polynomial is holomorphic precisely when it depends only on the coordinates x_k, $k > 0$):

$$< \psi, \psi' >= \int \psi \overline{\psi'} e^{-K} \prod_{n=1}^{\infty} dx_n \overline{x}_n, \tag{9.2.9}$$

where the Gaussian normalized integral has been used. A basis is given by the monomials

$$\prod_{n>0} \sqrt{n} \prod_{\mu_n=1}^{d} \frac{(\sqrt{n} x_n^{\mu_n})^{N_{n,\mu_n}}}{\sqrt{N_{n,\mu_n}! \, n}}.$$

All these vectors are orthogonal to each others and their norms are ± 1.

Following the principles of geometric quantization we want to quantize the classical functions λ_n by associating to them the operators

$$L_n = -i\nabla_{U_n} + \lambda_n$$

acting on the space of square-integrable sections of our line bundle. Let us first consider the case $n \geq 0$. When U_n acts on a holomorphic polynomial $\psi(x_1, x_2, \dots)$ it produces another holomorphic polynomial (in other words, the vector field U_n, $n \geq 0$, preserves the polarization defined by the complex structure). By a simple computation,

$$(9.2.10) \qquad L_n = -\sum kx_k \cdot \frac{\partial}{\partial x_{k-n}} - \frac{1}{2}\sum_{k=1}^{n-1} k(n-k)x_k \cdot x_{n-k}, \; n \geq 0.$$

All the operators L_n with $n \geq 0$ map the space of holomorphic polynomials V into itself and they satisfy the commutation relations $[L_n, L_m] = (n-m)L_{n+m}$. However, we cannot use the same formula (9.2.10) to define the operators L_n for $n < 0$ because these would not leave the space V invariant. The second term in (9.2.10) is somewhat reminiscent (for $n < 0$) of the kinetic energy term $g_{ij}p^ip^j$ in the usual Schrödinger quantization; the nonholomorphic coordinate x_{-k}, $k > 0$, corresponds to the momentum p^k. Thus we quantize the quadratic expression by replacing the coordinates kx_{-k} by the operators $\frac{\partial}{\partial x_k}$. In addition we project out the nonholomorphic part of the vector field and obtain

$$(9.2.11) \qquad L_n = -\sum_{k>0} kx_k \cdot \frac{\partial}{\partial x_{k-n}} - \frac{1}{2}\sum_{k=1}^{-n-1} \frac{\partial}{\partial x_k} \cdot \frac{\partial}{\partial x_{-n+k}}, \; n < 0.$$

The derivation of the operators L_{-n} is not very satisfactory but it (almost) works: The commutation relations are

$$(9.2.12) \qquad [L_n, L_m] = (n-m)L_{n+m} + \frac{d}{12}n^3\delta_{n,-m}.$$

Except for the scalar term the operators satisfy the same algebra as the classical functions λ_n. They also satisfy the hermiticity relations $L_n^* = L_{-n}$ which guarantee that the real vector fields on the circle are represented by symmetric operators.

Since the function λ_0 is the integral of the square of the derivative of $x(\phi)$, the operator $H = -L_0$ is called the Hamiltonian of the free string. The operators L_n are the generators of diffeomorphisms of S^1 acting in the (indefinite) Fock space V of the string. The *vacuum vector* $\psi_0(x) = 1$ satisfies $L_n\psi = 0$ for all $n \leq 0$. If $\psi(x)$ is a homogeneous polynomial of degree k then $H\psi = k\psi$. Thus 0 is the lowest eigenvalue of the Hamiltonian. The vacuum is not invariant under $Diff\, S^1$ since $L_n\psi_0 \neq 0$ for $n > 0$. In addition, the L_n's do not commute with the Hamiltonian H, and so $Diff\, S^1$ is clearly not an invariance group of the quantum system. However, if the string $x(\phi)$ should really describe physical particles it is natural to require that the theory is invariant under reparametrizarions of the circle S^1. How to recover reparametrization invariance is the subject of the next section.

9.3. Reparametrization invariance in string theory

The complex structure of $\Omega\mathbf{R}^{d-1,1}$ is invariant under the rotations of the circle generated by λ_0 since $\{\lambda_0, \lambda_n\} = -n\lambda_n$.; S^1 is the maximal subgroup of $Diff\, S^1$ which leaves invariant the complex structure. Therefore the complex structures obtained by acting by $Diff\, S^1$ on the original complex structure are parametrized by points on $M = Diff\, S^1/S^1$. Each complex structure J defines a Fock space V_J as explained above. We obtain a vector bundle E with fibers V_J isomorphic to the standard Fock space V, and base M. In order to construct a string theory invariant under $Diff\, S^1$ we shall consider the space $\Gamma(E)$ of sections of the vector bundle E. Note that E is topologically trivial since M is contractible and so a section can be viewed as an ordinary function on M with values in V. The group $Diff\, S^1$ acts in $\Gamma(E)$ in two ways. First, it acts on the base M in a natural way and secondly it acts projectively in the target space V through the generators L_n.

To set up a string theory in E we try to construct a vacuum vector which is invariant under $Diff\, S^1$. However, this is not possible because of the commutation relations (9.2.12). (If ψ is invariant then $L_n\psi = 0$ for all n and so $\frac{d}{12}n^3\psi = 0$.) What we can do at the moment is to define the *vacuum line bundle* Vac over M. The vacuum in V is the unique complex line in V which is annihilated by all the *annihilation operators*

$$a_n^\mu = \frac{1}{n}\frac{\partial}{\partial x_n^\mu},\; n > 0.$$

With respect to the indefinite metric in V the adjoint of a_n^μ is the *creation operator* $a_n^{*\,\mu} = x_n^\mu$ and their only nonzero commutation relations are $[a_n^\mu, a_n^{*\,\mu}] = \frac{1}{n}$. In the following we shall suppress the space-time indices $\mu, \nu, \ldots$ whenever possible. Let now $h \in Diff\, S^1$ and let A be the matrix defined by

$$(9.3.1) \qquad A_{mn} = \frac{1}{2\pi} \int_{S^1} e^{i(m\phi - nh^{-1}(\phi))} d\phi.$$

We want to write the transformation laws of the creation and annihilation operators in terms of the matrix A. In principle one could try to exponentiate the infinitesimal projective representation, given by the operators L_n, of the Lie algebra of $Diff\, S^1$ to a projective representation T of $Diff\, S^1$ in the vector space V. Then the transformation laws could be written as

$$(9.3.2) \qquad a'_n = T(h)a_nT(h)^{-1},\ a_n^{*\prime} = T(h)a_n^*T(h)^{-1}.$$

This is not easy; however, there is a simpler way. First, the natural action of h on an element $x \in \Omega\mathbf{R}^{d-1,1}$, $x'(\phi) = x(h^{-1}(\phi))$ gives for the coefficients

$$x'_n = \sum_{nm} A_{nm}x_m.$$

From this we get immediately the transformation laws for creation and annihilation operators when we remember that the coordinates x_{-n}, $n > 0$, were quantized as the derivatives $\frac{1}{n}\frac{\partial}{\partial x_n} = a_n$. Thus

$$(9.3.3) \qquad \begin{aligned} a'_n &= \sum_{m>0} A_{-n,m}a_m^* + \sum_{m<0} A_{-n,m}a_{-m} \\ a_n^{*\prime} &= \sum_{m>0} A_{nm}a_m^* + \sum_{m<0} A_{nm}a_{-m},\ n > 0. \end{aligned}$$

From (9.3.2) we conclude that the commutation relations are preserved under the action of the diffeomorphism h; this implies the following quadratic relations for the matrix elements of $A = \begin{pmatrix} a & b \\ c & d \end{pmatrix}$:

$$(9.3.4) \qquad aPb^t - bP^ta^t = 0,\ aPa^* - bP^tb^* = P,$$

where we have taken into account the relations $\overline{A}_{nm} = A_{-n,-m}$ which follow from (9.3.1), and P is the infinite matrix labelled by elements of $\mathbf{N} \times (-\mathbf{N})$ with $P_{n,-m} = \frac{1}{n}\delta_{nm}$, $n, m > 0$.

Let $\gamma_{nm} = \sqrt{\frac{n}{m}}(\overline{a}^{-1}\overline{b})_{n,-m}$, $m, n > 0$. From the first of the relations in (9.3.4) it follows that the matrix γ is symmetric (multiply by a^{-1} from the left and by a^{t-1} from the right and take the complex conjugate). The matrix is well-defined: By the second of the relations in (9.3.4) the matrix a is invertible.

LEMMA 9.3.5. *The matrix γ is a Hilbert-Schmidt operator.*

PROOF: b is a Hilbert-Schmidt operator as a consequence of the discussion in Section 7.2. ($\mathit{Diff}\, S^1$ is embedded in GL_1) and the Hibert-Schmidt operators form an ideal in the space of all bounded operators.

If the diffeomorphism h is just a rotation of S^1 then A is a multiplication operator by a constant phase and so $b = 0$ and $\gamma = 0$. Thus $h \mapsto \gamma$ can be viewed as a function on M. For any element of M, represented by a diffeomorphism h, we can now define the vector $\Phi(\gamma)$ in the Fock space V by

$$\Phi(\gamma) = \exp\left(-\frac{1}{2}\sum_{m,n=1}^{\infty} \gamma_{mn} a^*_m \cdot a^*_n\right)\psi_0, \tag{9.3.6}$$

where ψ_0 is the vacuum represented by the polynomial $\psi_0(x) = 1$.

Exercise 9.3.7. Show that $(\Phi(\gamma), \Phi(\gamma)) = [\det(1 - \overline{\gamma}\gamma)]^{-\frac{d}{2}}$.

The vector $\Phi(\gamma)$ is a vacuum vector for the transformed operators $a'_n, a^{*\prime}_n$. Namely,

$$\begin{aligned} a'_n\Phi(\gamma) &= \left(\sum_{m>0} \bar{b}_{n,-m} a^*_m + \sum_{m>0} \bar{a}_{nm} a_m\right)\Phi(\gamma) \\ &= \left(\sum_{m>0} \bar{b}_{n,-m} a^*_m - \sum_{k,m>0} \frac{1}{m}\bar{a}_{nm}\gamma_{mk} a^*_k\right)\Phi(\gamma) = 0. \end{aligned}$$

We have now defined a complex line bundle Vac over M; the line corresponding to $h \bmod S^1 \in M$ is $\mathbf{C}\Phi(\gamma)$. This is in fact a *holomorphic line bundle.* We know from Chapter 7 that the embedding of M into the Grassmannian Gr_1 defined by the mapping $h \mapsto A$ is holomorphic; on the other hand γ is obviously a holomorphic coordinate on the Grassmannian and therefore $\gamma(h)$ is a holomorphic function which implies that $\Phi(g)$ depends analytically on h.

The formula in Exercise 9.3.7 defines a positive definite inner product in the fibers of Vac. We now recall a few facts about complex Hermitian vector bundles; for more details see Chern [1979]. *A Hermitian structure* in a complex vector bundle E over a manifold M is a smooth distribution of positive definite Hermitian inner products H_x, $x \in M$ in the fibers. This means that if ψ, ψ' is a pair of (local) sections then $x \mapsto H_x(\psi(x), \psi'(x)) \equiv H(\psi, \psi')$ is a (local) smooth function which depends linearly on ψ, antilinearly on ψ' and $H(\psi, \psi) > 0$ for $\psi \neq 0$. In addition, $H(\psi, \psi') = \overline{H(\psi', \psi)}$. If $\{s_1, s_2, \ldots, s_N\}$ is a $C^\infty(M, \mathbf{C})$ basis for local sections in E we can define the Hermitian matrix function

$H_{ij} = H(s_i, s_j)$. Suppose that E has a connection and let X be a vector field on M. We can define the connection forms ω^i_j by

$$\nabla_X s_i = \omega^j_i(X) s_j. \tag{9.3.8}$$

The connection is called *admissible* if the inner product of a pair of elements in a fiber remains invariant under parallel transport. In particular, the infinitesimal change of $H(s_i, s_j)$ to the direction of the vector field X is

$$\mathcal{L}_X H_{ij} - H_{kj}\omega^k_i(X) - H_{ik}\overline{\omega^k_j(X)}$$

so that ω is admissible if and only if

$$dH_{ij} = H_{kj}\omega^k_i + H_{ik}\overline{\omega}^k_j \tag{9.3.9}$$

or using the matrix notation

$$dH = \omega H + H\omega^*. \tag{9.3.10}$$

Taking the exterior derivative from (9.3.9) one obtains a Hermiticity condition for the curvature form $\Omega = d\omega - \omega \wedge \omega$,

$$\Omega H + H\Omega^* = 0. \tag{9.3.11}$$

In particular, in an orthonormal basis one has $\omega = -\omega^*$ and $\Omega = -\Omega^*$.

When M is a complex manifold there is an admissible connection ω of type (1,0),

$$\omega = \partial H \cdot H^{-1}. \tag{9.3.12}$$

Exercise 9.3.13. Show that the admissible (1,0) connection in a Hermitian vector bundle E is uniquely defined.

Writing again $d = \partial + \overline{\partial}$ one gets for the curvature the formula

$$\Omega = \overline{\partial}\partial H \cdot H^{-1} + \partial H \cdot H^{-1} \wedge \overline{\partial} H \cdot H^{-1}. \tag{9.3.14}$$

When the dimension $N = 1$ the matrix H is a positive scalar function and (9.3.14) can be written as

$$\Omega = \overline{\partial}\partial \log H. \tag{9.3.15}$$

We shall apply this formula to the Hermitian metric in Exercise 9.3.7 on Vac. Using the formula in 9.3.7 for $H = (\Phi(\gamma), \Phi(\gamma))$ we obtain first

$$\Omega = \frac{d}{2}\mathrm{tr}(d\overline{\gamma} \wedge d\gamma). \tag{9.3.16}$$

From (9.3.1) we get

$$dA_{nm}(\ell_p) = m\delta_{n-m,-p},\ \ d\overline{A}_{nm}(\ell_p) = -m\delta_{n-m,p}.$$

This gives

$$d\gamma_{nm}(\ell_p) = \sqrt{nm}\delta_{n+m,-p},\ d\overline{\gamma}_{nm}(\ell_p) = \sqrt{nm}\delta_{n+m,p} \tag{9.3.17}$$

and inserting in (9.3.16) we obtain

$$\Omega(\ell_p,\ell_q) = \frac{d}{12}(p^3 - p)\delta_{p,-q}. \tag{9.3.18}$$

As we observed in the beginning of this section there cannot be any vacuum in the Fock bundle E over M which is invariant under diffeomorphisms of the circle. The nonzero curvature of the line bundle Vac is just one more manifestation of this fact. In order to solve this problem we have to introduce *ghosts.* This means that we have to modify the vector bundle E by tensoring with a line bundle L^* over M. The line bundle must have opposite curvature as compared to Vac so that the curvature of the tensor product bundle $Vac \otimes L^*$ vanishes. At this stage we could choose any line bundle with curvature $-\Omega$. However, there is a reason, to which we shall return in the next section, that the line bundle should be the dual of the canonical line bundle over M. The curvature $(-\frac{26}{12}p^3 + \frac{1}{6}p)\delta_{p,-q}$ essentially cancels (9.3.18) when the dimension d of the space-time is equal to 26. The second term $\frac{1}{6}p$ does not quite match the term $-p$ in (9.3.18) but this can be taken care of by the following observation. The projective action of *Diff* S^1 in $\hat{E} = E \otimes L^*$ consists of the following pieces: First, *Diff* S^1 acts projectively in the fibers of the (topologically trivial) vector bundle E such that the central term is given by (9.2.12). Secondly, the group acts naturally in the base space M of E. Finally, it acts projectively in the bundle L^* such that the central term is equal to the curvature. The central term in $\hat{E}$ will be equal to zero if we redefine the action of L_0 by an additive constant equal to $-\frac{1}{6}$.

The shifting of the value of the operator L_0 is not as unnatural as it may seem to be; the definition of the Virasoro algebra involved some choices anyway which are related to the choice of normal ordering in the old quantization of the bosonic string [Green, Schwarz, and Witten, 1987]. We can now explicitly construct the vacuum section in the modified Fock bundle $\hat{E}$. Let ξ be a section of the dual bundle Vac^* defined by

$$\xi_\gamma(z) = (z, \Phi(\gamma)),\ z \in Vac_\gamma. \tag{9.3.19}$$

Assuming $d = 26$ we can think of ξ as a section of the dual canonical line bundle L^*. The connection in the dual bundle is defined by $\nabla_X \xi = -\omega(X)\xi$ which gives an action of the Virasoro algebra in the space of sections of L^*, $\ell_n \psi = \nabla_{\ell_n} \psi$. The vacuum vector in $\hat{E}$ is then

$$\psi_0 = \Phi \otimes \xi. \tag{9.3.20}$$

The covariant derivative of ψ_0 vanishes to each direction ℓ_n and thus ψ_0 is indeed invariant under diffeomorphisms. It is clearly nonvanishing at each point $\gamma \in M$.

9.4. The BRST complex

Let $\mathbf{g}$ be any Lie algebra and V a $\mathbf{g}$ module. We have defined in Section 3.6 the cohomology groups $H^n(\mathbf{g}, V)$ for $\mathbf{g}$ with values in the vector space V. Let $\{x_i\}$ be a basis of $\mathbf{g}$ and define the structure constants

$$[x_i, x_j] = \lambda_{ij}^k x_k.$$

Let $\{f_i\}$ be the dual basis in $\mathbf{g}^*$, $f_i(x_j) = \delta_{ij}$. An element of $C^n(\mathbf{g}, V)$ is a linear combination of elements

$$v\, f_{i_1} \wedge f_{i_2} \wedge \ldots f_{i_n},\ v \in V.$$

As before in Section 7.2 we define the creation operators $a_i^* = f_i \wedge$ and annihilation operators $a_i = \imath(x_i)$. The coboundary operator δ can be written as

$$\delta = \sum_i a_i^* x_i - \frac{1}{2} \sum_{ijk} \lambda_{ij}^k a_i^* a_j^* a_k. \tag{9.4.1}$$

The first term on the right gives the second term in (4.1.13) and the second term here is equal to the first term in (4.1.13). In physics literature the coboundary operator δ is called the BRST operator and the a_i^*'s are "ghosts" and the a_i's are "antighosts". The formula (9.4.1) works equally well for infinite-dimensional Lie algebras as in the finite-dimensional situation. However, we want to apply this to the case of semi-infinite cohomology. There is then an additional problem due to the fact that the infinite sum in (9.4.1) when applied to a semi-infinite form does not in general converge; this will be taken care of using a normal ordering prescription when $\mathbf{g}$ is the Virasoro algebra. Let us start from the anomalous commutation relations

$$[\ell_n, \ell_m] = (n-m)\ell_{n+m} + \frac{n^3 - n}{12} \delta_{n,-m} \cdot c, \tag{9.4.2}$$

where c is the central element, and denote as before by f_n the elements of the dual basis. In an irreducible representation c is represented by multiplication by a constant. We have fixed the ratio of the coefficient of the cubic term to the coefficient of the linear term n to conform with the usual notation; remember that the linear term can be chosen arbitrarily by a shift $\ell_0 \mapsto \ell_0 + \alpha$, but once we have made the choice we must keep it consistently throughout. Let

$$\psi_N = f_N \wedge f_{N-1} \wedge \dots$$

be the vacuum of level N. The Virasoro algebra acts in the space Λ of semi-infinite forms through the operators

$$\rho(\ell_n) = (n-k) : a_{n+k} a_k^* : \tag{9.4.3}$$

where the normal ordering is defined by

$$\begin{cases} : a_k a_k^* := a_k a_k^* - 1 \text{ for } k > N \\ : a_k^* a_k := -a_k a_k^* \text{ for } k \leq N \end{cases}$$

and otherwise :: is the identity operation. When $n \neq 0$ the action of $\rho(\ell_n)$ is the same as the coadjoint action

$$\rho(\ell_n) f_{i_1} \wedge f_{i_2} \wedge \dots = \sum_{\mu} f_{i_1} \wedge f_{i_2} \dots \wedge (i_\mu - n) f_{n+i_\mu} \wedge \dots$$

but the coadjoint action of ℓ_0 diverges. We get the commutation relations as a special case of the computations in Section 7.3,

$$[\rho(\ell_n), \rho(\ell_m)] = (n-m)\rho(\ell_{n+m}) + \left[-\frac{26}{12}n^3 + (N^2 + N + \frac{1}{6})n\right] \delta_{n,-m}. \tag{9.4.4}$$

Thus the operators $\rho(\ell_n)$ form a representation of the Virasoro algebra with central charge $c = -\frac{26}{12}$ if and only if the vacuum level $N = 1$ or $N = -2$. We shall now assume that N is fixed to either of these two values. It is possible to work with a different vacuum level but then we should change the coefficient of n in (9.4.2). Let L_n denote the operator representing ℓ_n in V; they should satisfy the same commutation relations as the operators ℓ_n. Of course $[L_n, \rho(\ell_m)] = 0$ since they act in different spaces.

We define the normal ordered BRST operator by

$$Q = a_n^* L_n - \frac{1}{2}(n-m) : a_n^* a_m^* a_{n+m} : \tag{9.4.5}$$

The normal ordering takes care that the action of Q on any state in the semi-infinite cohomology is well-defined. Note that this definition differs from the expression which we would get from (4.1.13) in two respects: First there is the normal ordering in (9.4.5). Second, we have not introduced any ghost or antighost for the central element of the Virasoro algebra. In fact, the normal ordering introduces the necessary extra terms such that the usual rules of the game are valid:

PROPOSITION 9.4.6. *If* $c = 26$ *then* $Q^2 = 0$.

PROOF: By a careful application of the anticommutation relations the square of the second term A_2 in (9.4.5) is seen to be

$$A_2^2 = \sum_{n>0} \frac{26}{12}(-n^3 + n)a_n^* a_{-n}^*.$$

Ordinarily, for finite-dimensional algebras, $A_2^2 = 0$ but the nonzero contribution is due to the normal ordering prescribtion. The square of the first term is

$$\begin{aligned} A_1^2 &= L_n L_m a_n^* a_m^* = \frac{1}{2}[L_n, L_m] a_n^* a_m^* \\ &= \frac{1}{2}(n-m)L_{n+m} a_n^* a_m^* + \frac{c}{24}(n^3 - n)a_n^* a_{-n}^*. \end{aligned}$$

Finally,

$$A_1 A_2 + A_2 A_1 = -\frac{1}{2}(n-m)L_{n+m}$$

so that $(A_1 + A_2)^2 = 0$.

In fact, the value $c = 26$ can be (working backwards the proof above) obtained from the requirement that $Q^2 = 0$. One could criticize the computations above due to the fact that we did not really define Q as the coboundary operator in the sense of Lie algebra cohomology of Section 4.1 since we left out the ghost α^* and antighost α corresponding to the center. However, one can easely check that adding these missing pieces, i.e., defining

$$Q' = Q + \alpha^* c - \frac{1}{2}\frac{1}{12}(n^3 - n)a_n^* a_{-n}^* \alpha$$

would be disastrous; one has

$$Q'^2 = -\frac{26}{12}(n^3 - n)\frac{1}{2}a_n^* a_{-n}^*$$

and so Q' would not be nilpotent for any value of c.

Next we let Λ denote the space of semi-infinite forms with coefficients in the vector space V. We define a grading $\Lambda = \sum_{n\in\mathbf{Z}} \Lambda^n$ of Λ such that Λ^n is spanned by forms $v\,\ell_{i_1} \wedge \ell_{i_2} \wedge \ldots$ with $n =$ the number of non-negative indices i_p - the number of *missing* negative indices i_p. It is clear that Q maps Λ^n into Λ^{n+1} and we can define the cohomology groups as in the finite-dimensional case: The kernel of Q in C^n is denoted by Z^n, the image $Q\Lambda^{n-1}$ by B^n and $H^n(\mathbf{g}, V) = Z^n/B^n$. We shall define the *relative cohomology groups* H^n_{rel} with respect to the subalgebra spanned by ℓ_0 by taking into account only those forms which are annihilated by $\rho(\ell_0) + L_0$ and which do not contain the factor f_0.

We shall study in particular the case when V is the Fock space of the bosonic string moving in the background space $\mathbf{R}^{25,1}$. According to (9.2.12) the central term in the commutation relations of the Virasoro algebra is $\frac{26}{12}n^3\delta_{n,-m}$; by adding a constant $\frac{13}{12}$ to the formula defining the operator L_0 we obtain the central term $\frac{26}{12}(n^3 - n)\delta_{n,-m}$ to conform with the current conventions. We introduce a grading in V in the following way. The states in V are obtained by multiplying the vacuum by monomials $(x_1^{\mu_1})^{n_1}(x_2^{\mu_2})^{n_2}\ldots(x_p^{\mu_p})^{n_p}$ in the variables x_n^{μ}, $n > 0$ and $1 \leq \mu \leq 26$. The grading of the monomial is defined by $n_1 + 2n_2 + \ldots pn_p$ and is equal to the eigenvalue of the operator L_0 when applied to this state. Let us denote by V^n the subspace consisting of vectors of degree n. It is clear that all the subspaces V^n are finite-dimensional. The *formal character* of V is defined by

$$ch(V) = \sum_n q^n \dim V^n.$$

If the space-time dimension were $d = 1$ then we would have only one set of variables x_n and the dimension of V^n would be equal to the number of ways one can decompose n as a sum $n_1 + 2n_2 + \cdots + pn_p$; this is equal to the coefficient of of q^n in the expansion of the function $\phi(q)^{-1}$,

$$\phi(q) = \prod_{k>0}(1 - q^k).$$

In the d dimensional case the character is then $ch(V) = \sum_n q^n p_{(d)}(n)$ where $p_{(d)}(n)$ is the coefficient of q^n in the expansion of $\phi(q)^{-d}$.

In Section 9.2 we did not discuss the motion of the "center of mass" of the string. We considered only the based loops in $\mathbf{R}^{d-1,1}$ and therefore the Fourier modes x_0^{μ} did not appear in the formulas. We shall now add the operators k^{μ} representing the center of mass motion. They are assumed to commute among themselves and with the operators $x_n^\mu, \frac{\partial}{\partial x_n^\mu}$ and therefore in an irreducible representation of the string operator algebra we may assume that k^μ is represented by a multiplication by a constant p^μ.

The operators L_n will be also modified accordingly. A compact way to write the modified operators is the following. Define $\beta_n^\mu = x_n^\mu$ for $n > 0$, $\beta_0^\mu = k^\mu$ and $\beta_{-n}^\mu = n\frac{\partial}{\partial x_n^\mu}$ for $n < 0$. We have then the commutation relations

$$[\beta_n^{mu}, \beta_m^\nu] = m\delta_{\mu\nu}\delta_{n,-m}$$

and the Virasoro algebra is defined by

$$L_n = \frac{1}{2}\sum_{n\in\mathbf{Z}} :\beta_m \cdot \beta_{n-m}:$$

where the inner product is the indefinite product of $\mathbf{R}^{d-1,1}$ and the normal ordering is defined by

$$(9.4.7) \qquad :\beta_n\beta_m := \begin{cases} \beta_n\beta_m, & \text{for } n \neq -m,\ n = -m > 0 \\ \beta_m\beta_n, & \text{for } n = -m < 0 \end{cases}$$

The commutation relations remain unchanged after the insertion of the zero modes. We denote by $V(p)$ the modified Fock space. The vacuum vector in V is an eigenvector of L_0 corresponding to the eigenvalue 0 but in $V(p)$ the eigenvalue is shifted to $\frac{1}{2}p \cdot p$.

The character of the Virasoro algebra in $V(p)$ is computed as in the case of V but if we still want to define the grading by the eigenvalues of the energy operator L_0 then we have to include the factor $q^{\frac{1}{2}p\cdot p}$,

$$(9.4.8) \qquad ch(V(p)) = q^{\frac{1}{2}p\cdot p}\phi(q)^{-d}.$$

The representation of the Virasoro algebra in $V(p)$ is reducible. We shall next analyze the structure of $V(p)$ in terms of the Verma modules. First we write

$$V(p) = V(p') \otimes V(p'')$$

where p' denotes the first 25 components of p and $p'' = p^{26}$. Let us assume that $p' \neq 0$ and $p'' \neq 0$. The tensor product decomposition follows from the fact that the operators β_n^μ with different upper indices commute with each other. The value of the central element c in $V(p'')$ is equal to 1 and the energy of the vacuum is $-\frac{1}{2}(p'')^2$. On the other hand the Verma module $M(h, 1)$ is irreducible for $h < 0$. Comparing the characters of $V(p'')$ and $M(-\frac{1}{2}(p'')^2, 1)$ we conclude that the modules are isomorphic. The value of c in $V(p')$ is 25 and the energy of the vacuum is $\frac{1}{2}(p')^2$. The character is

$$ch(V(p')) = q^{\frac{1}{2}(p')^2}\phi^{-25}.$$

The character of the irreducible module $M(\frac{1}{2}(p')^2 + k, 25)$ is $q^{\frac{1}{2}(p')^2+k} \times \phi(q)^{-1}$ and therefore

$$ch(V(p')) = \sum_{k\geq 0} p_{(24)}(k) ch(M(\frac{1}{2}(p')^2 + k, 25).$$

We conclude that

$$(9.4.9) \qquad V(p) = M(-\tfrac{1}{2}(p'')^2, 1) \otimes \sum_{k\geq 0} p_{(24)}(k) M(\tfrac{1}{2}(p')^2 + k, 25).$$

We have assumed that both p' and p'' are nonzero. However, it is sufficient to assume that $p \neq 0$.

Exercise 9.4.10. Show by an appropriate splitting of $\mathbb{R}^{25,1}$ to complementary subspaces of dimension 25 and 1 that the result (9.4.9) is valid for all $p \neq 0$.

We define a positive definite subspace of $V(p)$ by

$$T = \bigoplus_{k\geq 0} p_{(24)}(k)\{u \otimes v_0 \mid u \in M(\tfrac{1}{2}(p')^2 + k, 25),\ L_n u = 0,\ n > 0\}$$

where $v_0 \in M(-\frac{1}{2}(p'')^2, 1)$ is the vacuum vector. Clearly T is annihilated by all L_n's for $n > 0$. The *physical subspace* P of $V(p)$ is defined to consist of all vectors v with $L_n v = \delta_{0,n} v$.

With a little more work one can prove the following theorem [Frenkel, Garland, and Zuckerman, 1986]:

THEOREM 9.4.11. *Assume $p \neq 0$. The cohomology groups $H^n_{rel}(\mathbf{g}, V(p))$ vanish for $n \neq 0$ and the dimension of $H^0_{rel}(\mathbf{g}, V(p))$ is equal to $p_{(24)}(1 - \frac{1}{2}(p)^2)$ if $1 - \frac{1}{2}(p)^2$ is an integer and is zero otherwise.*

The space of semi-infinite forms has a nondegenerate Hermitian form $(\cdot,\cdot)$ which is fixed by the Hermitian Fock space inner product in $V(p)$ and by the requirement that $(\omega, \omega) = 1$ where $\omega = f_{-1} \wedge f_{-2} \wedge \ldots$ and that a_n is the adjoint of a^*_n.

The physical subspace P is closely related to the cohomology group H^0_{rel}. Define a map $\kappa : V(p) \to C^0(\mathbf{g}, V(p))$ by

$$\kappa(v) = v f_{-1} \wedge f_{-2} \wedge \ldots.$$

Directly from the definition of Q we observe that $Q\kappa(v) = 0$ and therefore we get a well-defined map $V(p) \to H^0_{rel}(\mathbf{g}, V(p))$. Suppose that $\kappa(v) = Q\psi$ for some $\psi \in C^{-1}$, $v \in P$. Then

$$\kappa(v) = \sum_{k>0} L_{-k}\phi_k$$

for some $\phi_k \in C^0$ and therefore

$$(\kappa(v), \kappa(v)) = \sum_{k>0} (\phi_k, L_k\kappa(v)) = 0$$

which implies that v has zero norm. Denote by $radP$ the zero norm vectors in P. We have just shown that $\kappa^{-1}(B^0) \subset radP$. On the other hand, $T \cap radP = 0$, $\dim P \cap T = p_{(24)}(1 - \frac{1}{2}(p)^2)$ and therefore by dimensional grounds, using Theorem 9.4.11, we conclude that $T \cap P$, $P/radP$ and $H^0_{rel}(\mathbf{g}, V(p))$ are isomorphic vector spaces. Furthermore, all spaces have positive definite metric and the common dimension (in the integral case) is equal to $p_{(24)}(1 - \frac{1}{2}(p)^2)$ when $p \neq 0$.

The space $P/radP$ is postulated to describe the physical states of the bosonic string in 26 dimensions. The construction described above has a close resemblance with the construction of the space of physical states in quantum electrodynamics; the counterpart in QED is known as the Gupta-Bleuler quantization of the electromagnetic field [Itzykson and Zuber, 1980].

9.5. Strings on a group manifold

Let G be a connected, simply connected compact simple Lie group. In this section we want to describe the quantized system associated to a classical system which has as the phase space the space of loops on the group manifold G. The loop group LG can be written as a product $G \times \Omega G$ (as a manifold) where ΩG is the space of *based loops*, i.e., loops f such that $f(0) = 1$. The diffeomorphism is given by the map $LG \to G \times \Omega G$ which sends f to $(f(0), f(0)^{-1}f(\cdot))$. This is not a group homomorphism to the direct product; the composition in $G \times \Omega G$ is

$$(a, f) \cdot (a', f') = (aa', a'^{-1}fa'f'). \tag{9.5.1}$$

Thus LG can be viewed as a semidirect product of G and ΩG. The group ΩG is a symplectic manifold; the left-invariant symplectic form is

$$\omega(X, Y) = \frac{\theta^2}{4\pi} \int_{S^1} \left(X, \frac{d}{d\phi} Y \right) \tag{9.5.2}$$

where we have used the notation of Chapter 4. The group G is not a symplectic manifold; if one extends (9.5.1) to LG it becomes degenerate. However, G/T is a symplectic manifold when $T \subset G$ is a Cartan subgroup. Therefore, we shall take LG/T as the classical phase space; this is a bit similar to the situation we had in the case of a string moving in the Minkowski space, where we had to divide out the translational degrees of freedom in order to obtain a true symplectic phase space. We shall now describe the symplectic structure on G/T. After doing that we could define a symplectic form on LG/T as composed of the form (9.5.2) on ΩG and the symplectic form on G/T; in fact, we shall proceed later in a different way but it is instructive to look first at the finite-dimensional case in order to understand the infinite-dimensional setting.

The symplectic form on G/T can be described as follows. First, pick up a vector ξ in the Lie algebra $\mathbf{h}$ of T such that a vector $x \in \mathbf{g}$ commutes with ξ if and only if $x \in \mathbf{h}$ [Humphreys, 1980]. Then the orbit $\{ad_g \xi \mid g \in G\}$ can be identified with G/T. A homogeneous (left-invariant) 2-form on G/T is uniquely defined by its value at the point ξ; we set

$$\omega_\xi(x,y) = \frac{\theta^2}{4\pi}(\xi, [x,y]). \tag{9.5.3}$$

The tangent space of G/T at ξ can be identified as the quotient space $\mathbf{g}/\mathbf{h}$. On the other hand, $[x,y]$ is orthogonal to the vector ξ if either x or y is in $\mathbf{h}$; thus (9.5.3) is well-defined. It is also nondegenerate: If $x \notin \mathbf{h}$ then $[\xi, x] \neq 0$. But $(\xi, [x,y]) = ([\xi,x], y)$ and we can choose y such that the right-hand-side is nonzero.

Exercise 9.5.4. Show that the homogeneous form defined by (9.5.3) is closed.

For example, if $G = SU(2)$ then G/T is the unit sphere S^2 and ω is just the volume form on S^2 divided by 4π. The condition that the commutant of ξ is exactly $\mathbf{h}$ is equivalent to the condition that ξ is not contained in any of the hyperplanes in $\mathbf{h}$ which are orthogonal to the root vectors (we identify $\mathbf{h}$ with $\mathbf{h}*$ using the inner product in $\mathbf{h}$). If $g \in G$ is any element which maps $\mathbf{h}$ onto itself under the adjoint action (i.e., g is an element of the Weyl group) then one can show that the forms defined by the vectors ξ and $ad_g\xi$ are in the same de Rham cohomology class. In addition, the form ω is integral if and only if $\xi = h_\lambda$ for some integral weight $\lambda \in \mathbf{h}^*$; see Bott [1957]. Since each integral weight is in a Weyl group orbit through precisely one positive integral weight we observe that the cohomology classes of the 2-forms ω are in one-to-one correspondence with the equivalence classes of finite-dimensional irreducible representations of the Lie algebra $\mathbf{g}$.

We shall now return to the degenerate form (9.5.2) on LG. Each element X of the Lie algebra $L\mathfrak{g}$ generates a vector field on LG through the left action of LG on itself. Since the (degenerate) symplectic form is left-invariant we may expect that the vector fields X are Hamiltonian. This is true for the finite-dimensional manifolds G/T. For example, in the case of the unit sphere $S^2 = SU(2)/S^1$ the coordinate functions x_k generate the rotations, $\{x_i, x_j\} = \epsilon_{ijk}x_k$. However, this is not quite the case for the loop group LG. Before proceeding, let us look at the action of the right- invariant vector fields R_X on LG. The left-invariant vector field L_X is related to R_X through the adjoint action, $R_X(g) = g^{-1}L_X g$, and therefore

$$\begin{aligned} \omega(R_X, R_Y) &= \frac{\theta^2}{4\pi}\int_{S^1}(g^{-1}Xg, \frac{d}{d\phi}(g^{-1}Yg)) \\ &= \frac{\theta^2}{4\pi}\int(X, Y') + \frac{\theta^2}{4\pi}\int([X,Y], dgg^{-1}). \end{aligned} \tag{9.5.5}$$

Let us define for each $X \in L\mathfrak{g}$ the function

$$F_X(g) = \frac{\theta^2}{4\pi}\int(X, dgg^{-1}) \tag{9.5.6}$$

on LG. By a simple computation,

$$R_X \cdot F_Y = \omega(R_X, R_Y) \tag{9.5.7}$$

from which we obtain the Poisson brackets

$$\{F_X, F_Y\} = F_{[X,Y]} + \frac{\theta^2}{4\pi}\int(X, Y'). \tag{9.5.8}$$

Note that the action of a right-invariant vector field $R_X(g) = X.g$ on any function $F(g)$ is given by $R_X F = \frac{d}{dt}F(e^{tX}g)|_{t=0}$. Thus the right-invariant vector fields R_X are Hamiltonian, but the Poisson algebra has an "anomaly" term. If we had chosen the right-invariant form

$$\omega_R(L_X, L_Y) = \frac{\theta^2}{4\pi}\int[(X, Y') + ([X,Y], dgg^{-1})] \tag{9.5.9}$$

as the symplectic form then the vector fields L_X would be Hamiltonian.

Exercise 9.5.10. Show that the left-invariant fields L_X are not Hamiltonian with respect to the form ω.

The group $Diff\, S^1$ of orientation preserving diffeomorphisms of the circle acts naturally on the loop group. The infinitesimal generators $v\frac{d}{d\phi}$ act as Hamiltonian vector fields. To see this define the functions

$$G_v(g) = \frac{\theta^2}{8\pi}\int v(\phi)(g'g^{-1}, g'g^{-1})d\phi. \tag{9.5.11}$$

Let X_v be the vector field on LG which has the value $-vg^{-1}g'$ at the point g (that is, after left translating the tangent vector to the identity one obtains the Lie algebra element $-vg^{-1}g'$). We show that X_v is the Hamiltonian vector field associated to the function G_v. Let Y be a left-invariant vector field on LG. Then

$$Y.G_v = \frac{\theta^2}{8\pi}\int 2v(Y', g^{-1}g') = -\omega(X_v, Y)$$

which proves the assertion, because the left-invariant vector fields span at each point the tangent space of LG. Since apparently X_v is the tangent vector at $g \in LG$ generated by the infinitesimal action of $v\frac{d}{d\phi}$ the functions G_v are indeed generators of $Diff\, S^1$ acting on LG. The commutation relations are

$$\begin{aligned} \{G_u, G_v\} &= X_u \cdot G_v \\ &= \frac{\theta^2}{8\pi}\int (v'u - u'v)(dgg^{-1}, dgg^{-1}) = G_{v'u-u'v} \end{aligned} \tag{9.5.12}$$

which are compatible with the commutation relations of the vector fields $v\frac{d}{d\phi}$ on the circle. In particular, extending (9.5.12) to complex vector fields and defining $\ell_n = G_v$ when $v = ie^{in\phi}$ we obtain once again the familiar commutation relations

$$\{G_n, G_m\} = (n-m)G_{n+m}. \tag{9.5.13}$$

Let $\{T_1, T_2, \ldots, T_N\}$ be an orthonormal basis of $\mathbf{g}$. We use the basis $T_a^n = e^{in\phi}T_a$ for the loops in the complexified Lie algebra $\mathbf{g}_{\mathbf{C}}$ and set

$$F_a^n = \frac{\theta^2}{4\pi}\int (T_a^n, dgg^{-1}).$$

We can write

$$dgg^{-1} = \sum_{a,n} \frac{1}{\sqrt{2\pi}} T_a^n (\frac{1}{\sqrt{2\pi}} T_a^n, dgg^{-1}) = \frac{2}{\theta^2}\sum T_a^n F_a^n$$

and inserting this to (9.5.11) we obtain

$$G_n = \frac{1}{\theta^2} \sum_{a,k} F_a^k F_a^{n-k}. \tag{9.5.14}$$

This is the classical counterpart of the Sugawara construction. Next we want to quantize the classical system described by the form ω in the phase space LG/T. We shall in fact consider the wave functions of the quantum system as given on the bigger space LG; they must satisfy a certain set of constraints discussed below which guarantee that they are effectively defined on $G/T \times \Omega G$. The first step is to construct a complex line bundle over LG with a connection such that the curvature is equal to $k\omega$, where k is some fixed integer. This we have essentially done already in Section 4.2; the only thing which was missing in Section 4.2 is the factoring out of the subgroup T. Let us fix a positive integral weight λ of T. We define a complex line bundle $E^{k,\lambda}$ over LG/T as follows. A section of $E^{k,\lambda}$ is by definition a complex function ψ on $\widehat{LG}$ such that

(1) $\psi(f\mu) = (\mu)^k \psi(f)$ for all μ in the center of $\widehat{LG}$
(2) $\psi(fe^{ih}) = \psi(f)e^{i\lambda(h)}$ for all $h \in \mathbf{h}$.

Let ∇_a^n denote the covariant derivative in the direction of the right-invariant vector field corresponding to the Lie algebra element T_a^n. We could try to define the dynamics of the free string on the group manifold G by the "covariant Laplacian" operator

$$\Delta' = \sum_{n,a} \nabla_a^n \nabla_a^{-n}. \tag{9.5.15}$$

However, this definition involves an infinite summation and it is a priori clear how and where this should converge. The solution to this problem is similar to what we did earlier in the case of a string in a Minkowski space. We define a polarization with the help of a complex structure and we restrict the space of Schrödinger wave functions to holomorphic sections of the bundle $E^{k,\lambda}$.

To define the complex structure we use the *right action* of $\widehat{LG}$ on itself in the same way as was done in Section 3.3 for $S^2 = SU(2)/S^1$. We denote by Ξ_a^n the covariant derivative in the direction of the left-invariant vector field corresponding to the Lie algebra element T_a^n. Since right and left translations commute we have $[\nabla_a^n, \Xi_b^m] = 0$. A section ψ of $E^{k,\lambda}$ is by definition holomorphic if

(1) $\Xi_a^n \psi = 0$ for $n > 0$
(2) $\Xi_\alpha^0 \psi = 0$ for any positive root α of $\mathbf{g}$.

Of course, the space $\Gamma^{k,\lambda}$ of holomorphic sections may contain only the zero section. However, it has been shown in Pressley and Segal [1986] that if (k, λ) is an integral dominant weight of the affine algebra $\widehat{\mathbf{g}}$ then the space of holomorphic sections carries the irreducible highest weight representation of $\widehat{LG}$ with highest weight (k, λ). (The choice of the complex structure on LG/T is here opposite to Pressley and Segal; they use antidominant weights instead of dominant weights.) We shall discuss the proof and the explicit construction of the highest weight vector in the next chapter.

We take now as the space of physical states the space $\Gamma^{k,\lambda}$. We can use the normal ordering defined by the complex structure to make sense of the Schrödinger operator (9.5.15). The normal ordering is

$$:\nabla_a^n \nabla_b^m: \;:= \begin{cases} \nabla_a^n \nabla_b^m, & n < m \\ \nabla_b^m \nabla_a^n, & n > m \\ \frac{1}{2}(\nabla_a^n \nabla_b^n + \nabla_b^n \nabla_a^n), & n = m. \end{cases} \tag{9.5.16}$$

We define the normal ordered Schrödinger operator by

$$\Delta = \frac{1}{Q + k\theta^2} \sum_{a,n} :\nabla_a^n \nabla_a^{-n}: \tag{9.5.17}$$

where Q is the value of the Casimir operator $\sum T_a T_a$ in the adjoint representation of $\mathbf{g}$; it is related to the normalization of the T_a's by

$$\mathrm{tr}(ad\, T_a \cdot ad\, T_b) = Q\, \delta_{ab}.$$

We have chosen the T_a's to be normalized with respect to the Killing form on $\mathbf{g}$ and therefore $Q = 1$ in our case. The action of Δ is well-defined in a highest weight representation of $\widehat{LG}$ since the operators ∇_a^n annihilate the vacuum (highest weight vector) for $n > 0$ and therefore the action of these operators on an arbitrary state give a nonzero result only for a finite number of positive indices n. In fact, we can define the operators

$$L_n = \frac{1}{Q + k\theta^2} \sum_{a,k} :\nabla_a^k \nabla_a^{n-k}: \tag{9.5.18}$$

for any integer n; this is of course once again the Sugawara construction of the Virasoro algebra with the value of the central term equal to $\frac{c}{12} = k\, dim\mathbf{g}/(k + \kappa)$, where $\kappa = Q/\theta^2$ is the dual Coxeter number of the Lie algebra $\mathbf{g}$. Since the operators L_n commute with the operators Ξ_a^m the space $\Gamma^{k,\lambda}$ of holomorphic sections is invariant under the action of the Virasoro algebra.

From (9.5.17) and (9.5.18) we conclude that the vacuum ψ satisfies

$$\Delta\psi = \frac{2Q_\lambda}{Q+k\theta^2}\psi, \; L_n\psi = 0, \; \text{for}\, n > 0, \tag{9.5.19}$$

where Q_λ is the value of the Casimir operator of $\mathbf{g}$ for the highest weight vector.

We can define a second Virasoro algebra by using the left-invariant vector fields instead of right-invariant ones,

$$\overline{L}_n = \frac{1}{Q+k\theta^2}\sum : \Xi_a^k \Xi_a^{n-k} : . \tag{9.5.20}$$

Since the left-invariant vector fields corresponding to the Lie algebra generators T_a^n satisfy the same algebra as the right-invariant fields also the central term for the $\overline{L}_n$ algebra is the same as for the L_n algebra. However, the operators $\overline{L}_n$ do not act in $\Gamma^{k,\lambda}$, i.e., they do not leave invariant the complex structure defined by the operators Ξ.

Exercise 9.5.21. Check the commutation relations

$$[\overline{L}_n, L_m] = 0, \; [L_n, \nabla_a^m] = m\nabla_a^{m+n}, \; [\overline{L}_n, \Xi_a^m] = m\Xi_a^{n+m}.$$

To make things more complicated we note that there is still one more Virasoro algebra which is of interest to us. Namely, the natural action of *Diff* S^1 on LG can be lifted to a true action (without a central term) to the total space of $\widehat{LG}$. A diffeomorphism $h : S^1 \to S^1$ can be extended a diffeomorphism $\tilde{h} : D \to D$ by $\tilde{h}(\phi, r) = (h(\phi), r)$. Then h acts in $\widehat{LG} = (DG \times S^1)/\mathcal{G}$ by

$$h.(f,\mu) = (f \circ \tilde{h}, \mu). \tag{9.5.22}$$

In fact, it is straightforward to show that the equivalence class of the right-hand side does not depend on how h was extended to the disc D. The infinitesimal generators ℓ_n of the diffeomorphism group are then represented by linear operators W_n on the space of smooth sections of any associated bundle to the principal bundle $\widehat{LG}$ over LG. Since there is no projective factor in the global action (9.5.22) the operators close a Virasoro algebra with central term equal to zero. Since W_n corresponds to the natural action of $\ell_n = ie^{in\phi}\frac{d}{d\phi}$ on loops we have the commutation relations

$$[W_n, \Xi_a^m] = m\Xi_a^{m+n}. \tag{9.5.23}$$

From these relations we observe that the complex structure of LG/T is not invariant under the natural action of *Diff* S^1: Since the operators W_n when commuted with the Ξ's mix the elements with positive indices and those with negative indices a section annihilated with all the Ξ_a^n 's, $n > 0$, is in general mapped to a section not satisfying the holomorphicity conditions. The stability subgroup of the holomorphic structure is the group S^1 of global rotations. Thus the different holomorphic structures obtained through the action of *Diff* S^1 are parametrized by points on $M = \mathit{Diff}\, S^1/S^1$.

We have now essentially the same situation as in Section 9.3. Over each point $h \in M$ we can define the "Fock space" consisting of all holomorphic sections of the corresponding bundle $\Gamma^{k,\lambda}$; this gives a vector bundle over M with fiber $E^{k,\lambda}$. Since M is contractible the bundle $\mathcal{F}$ can be written as direct product $M \times E^{k,\lambda}$. The complex structure on LG/T, i.e., the splitting of Lg to positive and negative frequencies, depends on the point $h \in M$ through the natural action of *Diff* S^1 on loops and for this reason the vacuum ray will depend also on the base point. We shall next analyze the vacuum structure in $\mathcal{F}$.

As shown in Goodman and Wallach [1985] any Hermitian highest weight representation of the Virasoro algebra can be exponentiated to give a projective representation of the group *Diff* S^1. In particular, the representation by the operators L_n can be exponentiated; let us denote this by $D(h)$. We have

$$D(h_1)D(h_2) = D(h_1 h_2)e^{\epsilon(h_1,h_2)}$$

where e^{ϵ} is the projective factor; we do not need an explicit description of ϵ. From the commutation relations above (Exercise 9.5.21) we know that the adjoint action of $D(h)$ on the operators ∇_a^n is the same as the effect of the natural action of h and therefore the Schrödinger operator over the point h is equal to

$$\Delta_h = D(h)\Delta D(h)^{-1}. \tag{9.5.24}$$

The vacuum at $h \in M$ is the ray in $E^{k,\lambda}$ characterized by $\Delta_h \psi = \frac{2Q_\lambda}{Q+k\theta^2}\psi$; it is annihilated by the operators $D(h)L_n D(h)^{-1}$ for $n > 0$. It is convenient to think of the sections of the trivial bundle $\mathcal{F}$ as equivariant functions $\psi : \mathit{Diff}\, S^1 \to E^{k,\lambda}$ satisfying $\psi(hr(\phi)) = \psi(h)e^{i\alpha\phi}$ where α is the vacuum eigenvalue of Δ and $r(\phi)$ is a rotation of the circle by the angle ϕ.

The projective action of *Diff* S^1 in $\mathcal{F}$ consists of two parts. An element $h \in \mathit{Diff}\, S^1$ acts on the base manifold through left multiplication and h acts in the fiber through the representation D,

$$(h \cdot \psi)(h_1) = D(h)\psi(h^{-1}h_1). \tag{9.5.25}$$

As in Section 9.3 one cannot construct a *Diff* S^1 invariant string theory in the space $\Gamma(\mathcal{F})$ of sections of the bundle $\mathcal{F}$.

PROPOSITION 9.5.26. *There are no nonzero Diff S^1 invariant vectors in $\Gamma(\mathcal{F})$.*

PROOF: The infinitesimal action of *Diff* S^1 is given by

$$\rho(\ell_n)\cdot\psi = \mathcal{L}_n\psi + L_n\psi$$

where $\mathcal{L}_n$ is the Lie derivative in the direction of ℓ_n. Since

$$[\ell_n, \ell_m] = (n-m)\ell_{n+m}$$

we get

$$[\rho(\ell_n), \rho(\ell_m)] = (n-m)\rho(\ell_{n+m}) + \frac{c}{12}n(n^2-1)\delta_{n,-m}. \tag{9.5.27}$$

Thus for a vector satisfying $\rho(\ell_n)\psi = 0$ for all $n > 0$ one has $c\psi = 0$; but here $c \neq 0$.

The cure to the disease is the same as in Section 9.3. We introduce the ghost field as a section of the dual L^* of the canonical line bundle L over M and define the modified bundle $B = L^* \otimes \mathcal{F}$ over M. This bundle is topologically *and* holomorphically trivial. The group *Diff* S^1 acts on the tensor product without a central term:

THEOREM 9.5.28. *The curvature of B is equal to*

$$curv(\ell_n, \ell_m) = \frac{c-26}{12}(n^3 - n)\delta_{n,-m}.$$

In particular, for $c = 26$ there is a Diff S^1 invariant vacuum section in B given by $\psi(h) = \xi(h)D(h)\psi_0$, where ξ is a phase factor and ψ_0 is the vacuum in $\Gamma^{k,\lambda}$ (at h=1).

PROOF: As in Section 9.3 the curvature of $\mathcal{F}$ is given by the central term of the Virasoro algebra; on the other hand, the curvature of the product bundle B is equal to the sum of curvatures of the factors which gives immediately the curvature formula above. Let $V(h;\ell_n)$ denote the value of the vector potential in L^* [corresponding to the curvature (7.2.7) for $n = -1$] to the direction of the vector field ℓ_n. Let first

$\psi(h) = D(h)\psi_0$. The action of a generator ℓ_n on ψ is

$$\begin{aligned}
&\mathcal{L}_n\psi + L_n\psi + V(h;\ell_n)\psi \\
&\quad = \frac{d}{dt}D(e^{-t\ell_n}h)\psi_0|_{t=0} + L_n\psi + V(h;\ell_n)\psi \\
&\quad = \frac{d}{dt}D(e^{-t\ell_n})D(h)e^{\epsilon(\exp -t\ell_n, h)}\psi_0|_{t=0} + L_n\psi + V(h;\ell_n)\psi \\
&\quad = \left[-L_n + \frac{d}{dt}\epsilon(e^{-t\ell_n}, h)|_{t=0}\right] D(h)\psi_0 + L_n\psi + V(h;\ell_n)\psi \\
&\quad = \left[V(h;\ell_n) - \frac{d}{dt}\epsilon(e^{t\ell_n}, h)|_{t=0}\right]\psi(h).
\end{aligned}$$

Thus the section $D(h)\psi_0$ is covariantly constant up to a phase. But since the curvature vanishes it is always possible to modify the section by a phase factor ξ such that the new section is covariantly constant.

CHAPTER 10 THE NONLINEAR σ MODEL

10.0. Introduction

The nonlinear (principal) σ model has been for a long time a theoretical laboratory to test different approaches for quantizing classical field theories. Here we shall discuss as an application of the current algebra representation theory a construction of the quantized σ model.

We shall first study the σ model in two space-time dimensions; in the last section we discuss some features of the $3+1$ dimensional case. We have already met the σ model in Section 4.4 in context of the spin statistics connection. In Section 10.3 we shall make some complementary observations about the construction of the $3+1$ dimensional quantum σ model in terms of geometric quantization.

In the Lagrangian formulation the $1+1$ dimensional σ model is described by

$$L = \Lambda \int_M \mathrm{tr}\,(\partial_\mu g)(\partial^\mu g^{-1}) + \pi q_4 \int_B \mathrm{tr}\,(dgg^{-1})^3, \tag{10.0.1}$$

where Λ is some real constant, g is a map from the two-dimensional space-time (which is supposed to be compact) to a compact Lie group G, the traces are computed in some linear representation of G and q_4 is a constant depending on the representation; see Section 4.1; B is a compact three manifold such that its boundary is M. The right-hand-side of (10.0.1) is independent modulo integers from the continuation of the map $g: M \to G$ to the three manifold B, and therefore the classical field equations, derived by the variational principle, are well-defined. If the space-time metric is Minkowskian $x_0^2 - x_1^2$, defining the light-cone variables $x_\pm = x_0 \pm x_1$ we obtain the field equations

$$\left(\Lambda + \frac{3\pi q_4}{2}\right) \partial_-(g^{-1}\partial_+ g) + \left(\Lambda - \frac{3\pi q_4}{2}\right) \partial_+(g^{-1}\partial_- g) = 0. \tag{10.0.2}$$

In particular, when $\Lambda = \pm\frac{3\pi q_4}{2}$ the equation can be easely solved. For example, at $\Lambda = \frac{3\pi q_4}{2}$ we have $\partial_-(g^{-1}\partial_+ g) = 0$ and the general solution is of the form

$$g(x_+, x_-) = A(x_+)B(x_-),$$

where A and B are arbitrary smooth functions of the variable indicated, with values in G.

In the Hamiltonian approach the phase space of the σ model is the space of smooth maps from the one-dimensional physical space (which is assumed to be compactified as the circle S^1) to the group G. The first step is to determine the symplectic form in the phase space. The natural choice is the 2-form on LG which we studied in Section 4.2. In geometric quantization one constructs a complex line bundle over the phase space such that the curvature of the bundle is equal to the symplectic form. In this case the bundle has a holomorphic structure which determines a natural choice of *polarization*: The Schrödinger wave functions will be the holomorphic sections of the line bundle. This is exactly the same situation we met in Section 9.5 in another context.

One of the interesting features of the model is that in certain cases, despite the apparent nonlinearity, it actually describes fermions. For example, when the gauge group G is the orthogonal group $O(2N)$ we can compare the current algebra (8.0.4) with the current algebra of the σ model; but the latter is the Lie algebra of $\widehat{LG}$ and so the current algebras of both models are isomorphic. The same is true for the energy momentum tensors: The commutation relations of (half of the components of) the energy momentum tensor of free fermions are given by (7.0.3) and the energy momentum T_{++} of the σ model is obtained by the Sugawara construction from the currents $J(x_+)$ [the components T_{--} are obtained similarly from the currents $J(x_-)$] [Witten, 1984].

The explicit construction of the highest weight states (σ model vacua) in Section 10.2 is based on Felder, Gawedzki, and Kupiainen [1988]; the construction of the representations of $\widehat{LG}$ in the spirit of Borel-Weil theory is from Pressley and Segal [1986]. For further reading on σ models see, e.g., the lectures by Forger [1988], Rajeev [1988], and Witten [1988] and references therein.

10.1. The two-dimensional σ model

Consider the system described in the previous chapter: The phase space is the space LG of smooth loops on a compact simple group G. The (degenerate) "symplectic" form is given by

$$\omega(X,Y) = k\frac{\theta^2}{4\pi}\int_{S^1}(X, dY) \tag{10.1.1}$$

where k is an integer. As a classical Hamiltonian we take

$$H(g) = \frac{\theta^2}{8\pi}\int_{S^1}(dgg^{-1}, dgg^{-1})d\phi.$$

As shown in Section 9.5 the Hamiltonian $H = G_0$ generates the rotations $\phi \mapsto \phi + \phi_0$ in the loop space LG. For quantizing the classical system we fix a complex line bundle E^k with a connection over LG such that the first Chern class of the bundle corresponds to (10.1.1) (i.e., the curvature of the bundle is equal to ω.) From the discussion in Section 4.2 follows that the sections of the bundle E^k can be realized as functions $\psi : DG \to \mathbf{C}$ such that (in the notation of Section 4.2)

$$\psi(fg) = \psi(f) \cdot e^{2\pi i k[\gamma(f,g)+C(g)]} \tag{10.1.2}$$

where $g \in \mathcal{G}$ and $f \in DG$. The central extension $\widehat{LG}$ acts in the space $\Gamma(E^k)$ of sections from the left through

$$[T(f_0, \lambda)\psi](f) = \lambda^{-k} e^{-2\pi i k \gamma(f_0^{-1}, f)} \psi(f_0^{-1} f) \tag{10.1.3}$$

and from the right through

$$[T_r(f_0, \lambda)\psi](f) = \lambda^{-k} e^{-2\pi i k \gamma(f, f_0)} \psi(f f_0). \tag{10.1.4}$$

The connection in E^k is defined by the covariant derivatives

$$(\nabla_X \psi)(f) = \frac{d}{dt}\psi(e^{-t\tilde{X}} f)|_{t=0} + i\alpha_f(\tilde{X})\psi(f) \tag{10.1.5}$$

where $\tilde{X} : D \to \mathrm{g}$ is an extension of the tangent loop $X : S^1 \to \mathrm{g}$ and

$$\alpha_f(Z) = -k\frac{\theta^2}{8\pi} \int (df f^{-1}, dZ). \tag{10.1.6}$$

Note the difference between (4.2.9) and (10.1.6) which is due to the fact that here we are defining the connection in terms of left action corresponding to right invariant vector fields on LG, whereas in Section 4.2 the Lie algebra element Z corresponds to left invariant vector field. The section $\nabla_X \psi$ does not depend on the choice of $\tilde{X}$. If Y is another extension then

$$\psi(e^{-tY} f) = \psi(e^{-t\tilde{X}} f) \cdot e^{2\pi i k(\gamma(\exp(-t\tilde{X})f, g) + C(g))},$$

where $g = f^{-1}\exp(t\tilde{X})\exp(-tY)f$. Taking the derivative with respect to t at $t = 0$ of both sides we get

$$\begin{aligned}\frac{d}{dt}\psi(e^{-tY} f)|_{t=0} = {} & \frac{d}{dt}\psi(e^{-t\tilde{X}} f)|_{t=0} \\ & + ik\frac{\theta^2}{8\pi}\psi(f) \int (df f^{-1}, Y - \tilde{X}).\end{aligned}$$

The second term on the right is precisely the difference $[\alpha_f(Y) - \alpha_f(\tilde{X})] \times \psi(f)$ and thus $\nabla_X \psi$ is well-defined.

The commutators give

$$[\nabla_X, \nabla_Y] - \nabla_{[X,Y]} = i\omega(X, Y) \tag{10.1.7}$$

so that ω is indeed the curvature of E^k.

We shall proceed as in the string theoretic discussion in Section 9.2 and choose a polarization on LG. A natural splitting is again given by the splitting of the complexified tangent space to positive and negative Fourier components. Strictly speaking, this defines a polarization only in the space ΩG of based loops because of the zero modes on LG. In any case, we are free to impose the condition

$$\nabla^r_X \psi = 0, \quad \text{for } X = \sum_{n>0} X_n e^{in\phi} \tag{10.1.8}$$

for wave functions $\psi \in \Gamma(E^k)$. Here ∇^r_X means the covariant differentiation to the direction of the left invariant vector field on LG corresponding to the loop $X : S^1 \to \mathbf{g}$, i.e., ∇^r_X is computed using the right action

$$\frac{d}{dt} \psi(f e^{tX})|_{t=0}.$$

By fixing a Cartan subalgebra $\mathbf{h} \subset \mathbf{g}$ and a system Φ^+ of positive roots for the Lie algebra $\mathbf{g}$ we can impose the additional condition

$$\nabla^r_{x_\alpha} \psi = 0, \quad \alpha \in \Phi^+. \tag{10.1.9}$$

Since the left and right actions commute, the operators $T(g, \lambda)$ commute with the ∇^r_X's and therefore the subspace $\Gamma_{an}(E^k) \subset \Gamma(E^k)$ of sections satisfying (10.1.8) and (10.1.9) is invariant under the left group action.

10.2. The σ model vacua in two dimensions

We ask now what are the σ model "vacua", that means the highest weight vectors for the left action, in $\Gamma_{an}(E^k)$. Let us first consider the case $G = SU(2)$ for the sake of simplicity.

We shall make use of the *Birkhoff decomposition*

$$f = f_- f_0 f_+ \tag{10.2.1}$$

of loops $f : S^1 \to SL(2, \mathbf{C})$ into loops $f_+(f_-)$, which can be analytically continued to holomorphic (antiholomorphic) functions $D \to SL(2, \mathbf{C})$, and to a homomorphism $f_0 : S^1 \to SL(2,\mathbf{C})$. The decomposition is unique up to a conjugation of f_0 by an element $g \in SL(2,\mathbf{C})$; see Pressley and Segal [1986] for detailed discussion of this and other useful factorizations of LG. We shall use the same symbol $f_\pm$ for a loop and its (anti)holomorphic continuation into the disc D. For technical reasons (related to the Birkhoff decomposition) it is more convenient to consider the line bundle over the bigger space LG_c than the compact form LG $[G_c = SL(2,\mathbf{C})]$. Sections in $\Gamma(E^k)$ are defined in the same way as in the compact case: A section ψ is a function $\psi : DG_c \to \mathbf{C}$ which satisfies the cocycle condition (10.1.2) where now both f and g take values in G_c.

A section over LG_c gives of course by restriction a section over LG. But a section over LG can also be uniquely extended to a section over LG_c by the following observation. According to Pressley and Segal [1986], Theorem 8.1.1, any loop in G_c can be written uniquely as

$$f = f_u \cdot f_+,$$

where $f_u : S^1 \to G$ with $f_u(1) = 1$ and f_+ extends to a holomorphic map $f_+ : D \to G_c$. Using the Cartan decomposition of a complex nonsingular matrix to a unitary matrix and an upper triangular matrix with positive entries on the diagonal we can write f uniquely as $f_u \cdot f_+$ such that $f_+(0)$ is an upper triangular matrix with positive entries on the diagonal and $f_u \in LG$. Thus one gets a well-defined map $LG_c \to LG$ by $f \mapsto f_u$ and we can pull back sections in the line bundle E^k over LG to sections of the corresponding bundle over LG_c.

Suppose $\psi \in \Gamma_{an}$ is a vacuum vector; this means that it is annihilated by all generators T^a_n, $n < 0$ and $a = 1,2,3$, corresponding to the left action of $\widehat{LG}$, and by T^-_0. We use the basis

$$T^- = \begin{pmatrix} 0 & 0 \\ 1 & 0 \end{pmatrix} \quad T^+ = \begin{pmatrix} 0 & 1 \\ 0 & 0 \end{pmatrix} \quad T^3 = \begin{pmatrix} 1 & 0 \\ 0 & -1 \end{pmatrix}$$

for the complexified Lie algebra of $SU(2)$. The vacuum should be an eigenvector of T^3 corresponding to an eigenvalue $-j$. The condition (10.1.8) means that ψ is invariant under the right action of elements $(f_{++}, 1)$, where f_{++} is in the subgroup N_+ of holomorphic loops $1 + \sum_{n>0} A_n z^n$ with $A_0 = 1$; we define similarly N_-. The vacuum is invariant under the left transformations by $(f_{--}, 1)$ with $f_{--} \in N_-$.

Invariance under the right action by elements f_{++} and left action by f_{--} tells us that

$$\begin{aligned} \psi(f) &= \psi(f_{--} g_1 f_0 g_2 f_{++}) \\ &= e^{2\pi i k \gamma(f_{--}^{-1}, g_1 f_0 g_2 f_{++})} e^{2\pi i k \gamma(g_1 f_0 g_2, f_{++})} \psi(g_1 f_0 g_2), \end{aligned} \tag{10.2.2}$$

where g_1, g_2 are constant loops. It can be shown that the set Σ_0 of loops $f_- f_+$ form a dense subset in LG_c and therefore any section is uniquely defined by its restriction to the subspace of the loops with $f_0 = 1$. Thus, by (10.2.2), the vacuum vector ψ is uniquely defined by its restriction to the space of constant loops.

Again using the commutativity of left and right actions we can diagonalize the generator T^3 simultaneously for the left and right actions. Let j^r be the eigenvalue for the right action. The condition (10.1.9) together with the eigenvalue equation for the right T^3 means that ψ, when restricted to the constant loops, is a *holomorphic section* in the complex line bundle over $SL(2,\mathbf{C})/B^- = SU(2)/S^1 = S^2$ (B^- is the subgroup of lower triangular matrices) indexed by the Chern class j^r. Holomorphic sections exist only when j^r is a non-negative integer (see Chapter 3) and there is a unique ray of lowest weight vectors with the lowest weight $-j = -j^r$. Thus the problem reduces to deciding which values of j^r actually occur for a given k.

Let Σ_n denote the set of loops $f_- e_n f_+$, where e_n is the loop

$$e_n(\phi) = \begin{pmatrix} e^{in\phi} & 0 \\ 0 & e^{-in\phi} \end{pmatrix}.$$

By the Birkhoff factorization theorem LG_c is a disjoint union of the sets Σ_n, $n \in \mathbf{Z}$. There is another factorization which we shall use: The group LG_c is a union of the open sets U_n consisting of loops $e_n f_- f_+$ [Pressley and Segal, 1986, Theorem 8.7.2]. Let $n < 0$. From the discussion in Pressley and Segal [1986], Section 8.7, it follows that any element in U_n can be written as $f = f_- f(c) e_n f_+$ where

$$f(c) = \begin{pmatrix} 1 & 0 \\ cz^n & 1 \end{pmatrix},$$

and c is a complex parameter.

By a simple computation $f(c)e_n = g_- g_+$ on the circle $|z| = 1$, where

$$g_+ = \begin{pmatrix} 1 & z^n/c \\ 0 & 1 \end{pmatrix}, \qquad g_- = \begin{pmatrix} \bar{z}^n & -1/c \\ c & 0 \end{pmatrix}.$$

The maps $g_\pm$ extend to (anti)holomorphic functions $D \to SL(2,\mathbf{C})$ for all values of the parameter $c \neq 0$. We can write $f_+ = a_1 f_{++}$ and $f_- = f_{--} a_2$ where a_1 and a_2 are constants in G_c. Then

$$\psi(f_- f(c) e'_n f_+) = \psi(a_2 g_- g_+ a_1) e^{2\pi i k(\gamma(f,g) + C(g))}, \tag{10.2.3}$$

where we have fixed an extension e'_n of the loop e_n, $f = f_- f(c) e'_n f_+$ and $g = f^{-1} a_2 g_+ g_- a_2$. Writing

$$g_- = \begin{pmatrix} 1/c & 0 \\ 0 & c \end{pmatrix} \begin{pmatrix} c\bar{z}^n & -1 \\ 1 & 0 \end{pmatrix}$$

we observe that $\psi(a_2 g_- g_+ a_1)$ can be written as $h_1(c) \cdot c^{-j}$, where h is a regular function of c with $h_1(0) = \psi(a_2 a_0 a_1)$, a_0 being the constant loop $\begin{pmatrix} 0 & -1 \\ 1 & 0 \end{pmatrix}$.

Taking the derivative of $h_2 = \exp\{2\pi i k[\gamma(f,g) + C(g)]\}$ with respect to c one obtains

$$h'_2 = \left[\frac{k}{c} + q(c)\right] h_2$$

where $q(c)$ is a regular function of c at the point $c = 0$. Thus the solution of the differential equation can be written as

$$h_2(c) = c^k \cdot e^{s(c)}, \tag{10.2.4}$$

where $s(c)$ is a regular function of c at $c = 0$. Summarizing the discussion above we get

$$\psi(f_- f(c) e'_n f_+) = c^{k-j} \cdot h_1(c) e^{s(c)}, \tag{10.2.5}$$

and thus $\psi(f)$ is a regular analytic function of the parameter c at $c = 0$ if and only if $k - j$ is a non-negative integer. For a fixed non-negative integer k the spectrum of highest weight representations of $\widehat{LG}$ is parametrized by $j = 0, 1, \ldots, k$ and each representation (k, j) occurs with multiplicity one.

10.3. The σ model in four dimensions

The quantization of the nonlinear σ model in higher dimensions than two differs in an essential way from the two-dimensional case. In two dimensions the configuration space of the quantum system was the loop group LG; this is almost a symplectic manifold. In fact, the group ΩG of based loops is a symplectic manifold and we strongly exploited this fact in determining the commutation relations of the quantum observables. However, in four dimensions the corresponding object is the mapping space $S^3 G$ (in the Schrödinger picture) which does not have any natural symplectic structure.

In the two-dimensional case the symplectic structure was given by the commutator anomaly in the Lie algebra $L\mathfrak{g}$. It would be natural to try to use the commutator anomaly also in four dimensions to define a symplectic form on S^3G. The formula (4.3.16) indeed defines a closed 2-form on S^3G but this is highly nondegenerate. For example, at the point $g = 1$ it vanishes identically! In order to proceed we shall take S^3G as the configuration space of a quantum system but we shall apply the method of geometric quantization in the cotangent bundle (phase space) $M = T^*(S^3G)$. A cotangent bundle has a natural symplectic form $\omega_0 = d\theta$, where θ is the 1-form defined by

$$\theta(g,\xi;u,p) = \xi(u),$$

where $g \in S^3G, \xi \in T_g^*(S^3G), u \in T_g(S^3G)$ and the component of the tangent vector (u,p) along the fiber of M at g is identified as an element p of the fiber T_g^* since the latter is a linear space. Identifying right invariant vector fields as the elements of the Lie algebra $S^3\mathbf{g}$ in the usual way we obtain by an exterior differentation from θ

$$\omega_0(g,\xi;(u,p),(u',p')) = -\xi([u,u']) + p(u') - p'(u). \tag{10.3.1}$$

Instead of ω_0 we shall consider the form

$$\omega = \omega_0 + \omega_1$$

where ω_1 is the curvature form on S^3G,

$$\omega_1(g;u,u') = 2\pi i q_6 \int_{S^3} \mathrm{tr}(dgg^{-1})^2(udu' - u'du), \tag{10.3.2}$$

where q_6 is a normalization coefficient (see Section 4.3) which in the case $G = SU(N)$ (in the defining representation) is equal to $-i/48\pi^3$. It will turn out that ω is the correct symplectic form for describing the σ model chiral fermions.

Exercise 10.3.3. Show that the form ω is nondegenerate.

We shall identify $S^3\mathbf{g}$ with its dual using the inner product

$$(u,v) = \int_{S^3} \mathrm{tr}\, uv\, d(vol\, S^3).$$

For any $u \in S^3\mathbf{g}$ define the real valued function f_u on M:

$$f_u(g,\xi) = \xi(u). \tag{10.3.4}$$

The Hamiltonian vector field on M corresponding to the function f_u is

$$X_u(g,\xi) = \delta_u - 2\pi i q_6\, (0, du(dgg^{-1})^2 + (dgg^{-1})^2 du) \tag{10.3.5}$$

where the first term is the vector field on M induced by the natural left action of the infinitesimal generators of S^3G and the second component w of the second term is the cotangent vector which sends the tangent vector v to $\int \mathrm{tr}\, wv\, d(vol\, S^3)$. We can check (10.3.5) by the following computation:

$$\begin{aligned}
&\omega(g,\xi; X_u, (v,\xi')) \\
&\quad = \int \mathrm{tr}\, wv\, d(vol\, S^3) + \xi'(u) + 2\pi i q_6 \int \mathrm{tr}(dgg^{-1})^2(vdu - udv) \\
&\quad = \xi'(u) = \mathcal{L}_{(v,\xi')} f_u
\end{aligned}$$

Thus we obtain the Poisson bracket relations

$$(10.3.6)\quad \{f_u, f_v\} = X_u \cdot f_v = f_{[u,v]} - 2\pi i q_6 \int \mathrm{tr}(dgg^{-1})^2(udv - vdu).$$

We have here the same commutation relations which we already discussed in Chapter 4, and which we showed to be capable of describing both fermions and bosons.

In the Lagrangian approach the σ model is described by the Wess-Zumino-Witten action

$$(10.3.7)\qquad L = \Lambda \int \mathrm{tr}\, \partial_\mu g g^{-1} \partial^\mu g g^{-1} + \pi q_6 \int_{M_5} \mathrm{tr}(dgg^{-1})^5$$

where m_0 is a real constant and M_5 is a five-dimensional manifold which has the four-dimensional space-time M_4 as its boundary. The field equations obtained by the variational principle are

$$(10.3.8)\quad \partial_\mu(\partial^\mu g g^{-1}) + \frac{\pi q_6}{\Lambda} \epsilon^{\alpha\beta\gamma\delta} \partial_\alpha g g^{-1} \partial_\beta g g^{-1} \partial_\gamma g g^{-1} \partial_\delta g g^{-1} = 0.$$

We would like now to construct the Hamiltonian function H on the cotangent bundle which reproduces these field equations. As a first step we try the ansatz $H_0 = \frac{\Lambda}{2}\mathrm{tr}(\partial_k g g^{-1}\, \partial_k g g^{-1}) + \frac{\Lambda}{2}\mathrm{tr}\, \xi^2$; this is the Hamiltonian one would obtain by a Legendre transformation from the quadratic piece of the Lagrangian, replacing in the kinetic energy term $\partial_0 g g^{-1} \partial^0 g g^{-1}$ the momentum $\partial_0 g g^{-1}$ by ξ. We can add the free parameter Λ to the symplectic form, $\omega^{(\Lambda)} = \Lambda\omega_0 + \omega_1$. The Hamilton equations of motion on $(M, \omega^{(\Lambda)})$ are

$$\begin{aligned}
\dot{\xi} &= \partial_k(\partial_k g g^{-1}) + \frac{\pi q_6}{\Lambda}[(dgg^{-1})^2 d\xi + d\xi(dgg^{-1})^2] \\
\dot{u} &= \xi,
\end{aligned}$$

where $\dot{u} = \partial_0 g g^{-1}$. Inserting ξ from the second equation into the first we get

$$\partial_0(\partial_0 g g^{-1}) = \partial_k(\partial_k g g^{-1}) + \frac{\pi q_6}{\Lambda}[d(\partial_0 g g^{-1})(dgg^{-1})^2 + (dgg^{-1})^2 d(\partial_0 g g^{-1})].$$

This not quite what we want because the second term on the right is not Lorentz invariant. However, by keeping the Hamiltonian but modifying the symplectic structure we can recover the equations (10.3.8). Define

$$\omega_2 = -\pi q_6 \int \mathrm{tr}\ \{(dgg^{-1})^3[u,v] \\ +(dgg^{-1})dv(dgg^{-1})u - (dgg^{-1})du(dgg^{-1})v\}.$$

The difference $\omega_1 - \omega_2$ is the exterior derivative of the 1-form

$$u \mapsto \pi q_6 \int \mathrm{tr}(dgg^{-1})^3 u.$$

Using the cohomologically equivalent form $\Lambda\omega_0 + \omega_2$ instead of $\Lambda\omega_0 + \omega_1$ the Hamilton equations of motion give (10.3.8).

CHAPTER 11 THE KP HIERARCHY

11.0. Introduction

As an application of the theory of infinite-dimensional Grassmannians and the representation theory of $\mathbf{gl}_1$ we shall study in this chapter certain nonlinear "exactly solvable" systems of differential equations. Exactly solvable means here that the nonlinear system can be transformed to an (infinite-dimensional) linear problem. A prototype of the equations is the Korteweg-de Vries equation

$$\frac{\partial u}{\partial t} = \frac{3}{2} u \frac{\partial u}{\partial x} + \frac{1}{4} \frac{\partial^3 u}{\partial x^3}.$$

It turns out that it is more natural to consider an infinite system of equations like that above, for obtaining explicit solutions. The set of equations is called the KdV hierarchy and it can be derived from another set of equations, the KP (Kadomtsev-Petviashvili) hierarchy. The Grassmannian approach can be more directly applied to the KP hierarchy and therefore we shall mainly consider the KP case.

The basic idea we want to explain is the following. The nonlinear KP system when written in the *Hirota bilinear form* can be interpreted as the condition that a Grassmannian plane W lies on the orbit of the group $\widehat{GL}_1$ through the vacuum in a Fock space. This condition can be expressed as a system of nonlinear relations for the Plücker coordinates of the plane W; these relations give the equations of the KP hierarchy. The linear action of $\widehat{GL}_1$ in the Fock space can be used for constructing new (solitonic) solutions of the KP hierarchy from a given solution. Using the boson-fermion correspondence of Chapter 8 the group action can be written more directly in terms of the (bosonic) variables of the solutions. One can generate rather explicit forms of the solutions using vertex operators.

In Chapter 12 we shall explain how the solutions of the KP hierarchy can be used for parametrization of certain moduli spaces which are relevant in applications to two-dimensional conformal field theory.

This chapter is based mainly on the articles by Kac and Peterson [1986], and Segal and Wilson [1985]; the author has profited also from the articles by Date, Kashiwara, Jimbo, and Miwa [1983] as well as Sato and Sato [1982]. The idea of using the Plücker relations on Grassmannians for describing solutions of the soliton equations is due to M. Sato but the approach has long roots in the inverse scattering method: see, e.g., Gardner, Green, Kruskal, and Miura [1967]; Lax [1968]; Novikov [1974]; Zakharov and Faddeev [1971]; Zakharov and Shabat [1974]; Gelfand and Dikii [1976], and references in these papers.

11.1. The Plücker relations and the Hirota bilinear equation

Let $\mathcal{F}$ be the fermionic Fock space consisting of linear combinations of the holomorphic sections ψ_S, $S \in \mathcal{S}$, of the dual determinant bundle DET_1^* over Gr_1 (we use the notation of Chapter 6). We recall that these sections form an orthonormal basis in the Fock space and the vacuum vector corresponding to $S = \{1, 2, 3, \ldots\}$ is denoted by ψ_0. In general, in the notation of Section 6.3,

$$\psi_S = f_{i_1} \wedge \ldots f_{i_p} \wedge f_{j_1} \wedge f_{j_2} \wedge \cdots = a_{i_1}^* \ldots a_{i_p}^* a_{j_1} \ldots a_{j_q} \psi_0 \tag{11.1.1}$$

where $i_1 > i_2 > \cdots > i_p > 0$ are the positive integers in the set $\mathbf{Z} \setminus S$ and $0 \geq j_1 > \cdots > j_q$ are the nonpositive integers in S and the a_n's and a_n^*'s are the fermionic annihilation and creation operators with the only nonzero anticommutators

$$a_n^* a_m + a_m a_n^* = \delta_{n,m}.$$

Let X be the orbit $\widehat{GL_1} \cdot \psi_0$, where the action of $\widehat{GL_1}$ on sections of DET_1^* is as usual

$$[T(g, q, \lambda)\psi](w) = \lambda^{-1}\psi(g^{-1}wq).$$

In fact, as we saw in Section 6.5, the orbit X can be identified as the space $DET_1^\times$ of nonzero vectors in DET_1 by the Plücker embedding $u = (z, \lambda) \mapsto \xi(u)$, $\xi(u)(w) = \overline{\lambda}\det(z_-^* w_- + z_+^* w_+)$.

If ψ is any section of DET_1^* the Plücker coordinate corresponding to $S \in \mathcal{S}$ is

$$\psi^S = \psi(w_S),$$

where $w_S \in St_1$ consists of the vectors $e_{i_1}, e_{i_2}, \ldots$, where $i_1 < i_2 < \ldots$ is the set S of integers. The Fock space inner product is

$$< \psi, \psi' >= \sum_{S \in \mathcal{S}} \psi^S \overline{\psi'^S}. \tag{11.1.2}$$

We introduce operators B and B^* acting in $\mathcal{F} \otimes \mathcal{F}$,

$$B = \sum_{n \in \mathbf{Z}} a_n \otimes a_n^*, \quad B^* = \sum_{n \in \mathbf{Z}} a_n^* \otimes a_n.$$

Under the diagonal action $T(g,q,\lambda)\psi \otimes T(g,q,\lambda)\psi'$ the components a_n^* transform like the components of a vector and a_n transforms like a co-vector and therefore both B and B^* commute with T. (Remember that the Lie algebra of $\widehat{GL}_1$ is represented by the operators $E_{ij} = a_i a_j^*$, when $i \neq j$ or $i = j \leq 0$, $E_{ii} = a_i a_i^* - 1$ when $i > 0$.) Since clearly $B\psi_0 = B^*\psi_0 = 0$ we have

$$(11.1.3) \qquad B(\tau \otimes \tau) = \sum_{n \in \mathbf{Z}} a_n \tau \otimes a_n^* \tau = 0$$

for any $\tau \in X$. Writing τ in terms of the Plücker coordinates,

$$\tau = \sum_{S \in \mathcal{S}} \tau^S \cdot \psi_S,$$

we can rewrite (11.1.3) as

$$(11.1.4) \qquad \sum_{n \in S \setminus S'} \tau^{S \setminus \{n\}} \cdot \tau^{S' \cup \{n\}} = 0,$$

for all $S, S' \in \mathcal{S}$ such that $S \setminus S'$ is nonempty; note that the sum is in fact finite since the intersection of S with the complement of S' is always a finite set. The equations (11.1.4) are known as *Plücker relations.* In the finite-dimensional case these are the classical Plücker relations. Namely, consider a p-dimensional plane W in an N-dimensional vector space H. Let $e_1, \ldots, e_N$ be a basis of H and $w_1, \ldots, w_p$ a basis of W. We can write

$$(11.1.5) \qquad w_1 \wedge w_2 \wedge \cdots \wedge w_p = \sum_{1 \leq i_1 < \cdots < i_p \leq N} \xi_{(i)} e_{i_1} \wedge \cdots \wedge e_{i_p},$$

where the $\xi_{(i)}$'s are complex numbers which satisfy the Plücker relations

$$(11.1.6) \qquad \sum_{\nu=1}^{p+1} (-1)^\nu \xi_{i_1 \ldots i_{p-1} j_\nu} \cdot \xi_{j_1 \ldots \hat{j}_\nu \ldots j_{p+1}},$$

where the caret means that the corresponding index has been deleted. It is a classical result that if a collection $\{\xi_{(i)}\}$ of complex numbers satisfies (11.1.6) then it corresponds to some plane W via (11.1.5).

If the basis vectors w_i are transformed by a matrix A, $w_i' = \sum w_j A_{ji}$, then from the basic properties of the exterior product it follows that the Plücker coordinates $\xi_{(i)}$ will be multiplied by the determinant of the matrix A, in the same way as in the infinite-dimensional case in

Section 6.5. Note that the Plücker coordinates appear in two different ways. First, they give projective coordinates for the Grassmannian planes. Secondly, they can be used to parametrize holomorphic sections of the dual determinant bundle. However, these are two different aspects of the same thing: The Plücker embedding maps projectively the Grassmannian planes to sections of the dual determinant bundle such that the Plücker coordinates of the plane correspond (projectively) to the Plücker coordinates of the section.

We can now show that $\tau \in \mathcal{F}$ belongs to X if the relations (11.1.4) [or (11.1.3)] are satisfied. Let $S \in \mathcal{S}$. We define the *Young diagram* Y_S to be the box diagram consisting of i_1 boxes in the first row, i_2 boxes in the second row ... i_n boxes in the nth row, $-j_1$ boxes in the $(n+1)$th row ..... $-j_m$ boxes in the $(n+m)$th row, where $i_1 > i_2 > \ldots i_n$ are the positive integers in the complement of S, $j_1 > \ldots j_m$ are the negative integers in S, and the boxes labelled by the i's are placed on the right-hand-side of a vertical axis and the boxes labelled by the j's on the left-hand-side. Let D be a rectangle of height s and breadth t (the vertical axis in the middle). Let $\tau \in \mathcal{F}$ satisfy (11.1.4) and let $\tau(D)$ be the vector such that $\tau(D)^S = 0$ if Y_S is not contained in D and $\tau(D)^S = \tau^S$ otherwise. Then $\tau(D)$ has only a finite number of nonzero components, its Plücker coordinates satisfy (11.1.4) and using the classical result there is an operator $g = \begin{pmatrix} 1 & b \\ 0 & 1 \end{pmatrix} \in GL_1$ such that b is of finite-rank and $\tau(D) = T(g)\psi_0$. Letting s, t increase we get a sequence of $\tau(D)$'s converging to τ. Since the orbit X is closed we conclude that $\tau \in X$.

Next we shall use the boson-fermion correspondence of Section 8.2 for writing the Plücker relations (11.1.3) in terms of the bosonic coordinates $x_1, x_2, \ldots$. Note first that (11.1.3) is equivalent with

$$(11.1.7) \qquad z^0 - \text{term of} \qquad X(z)\psi_0 \otimes X^*(z)\psi_0 = 0.$$

Let us consider only the connected component X^0 of the orbit X corresponding to the action of the connected component of identity of GL_1 on the vacuum ψ_0 of index 0; any $\tau \in X^0$ is an element of $\mathcal{F}_0$. Further, in order to avoid discussing convergence questions, we shall consider only those points which can be reached by the action of $GL(\infty)$ (off diagonal blocks finite-rank operators) from ψ_0. Applying to (11.1.7) the map σ

from $\mathcal{F}_0$ to the bosonic Fock space $\mathcal{B}_0$ we obtain from Theorem 8.2.4

$$\text{z-term of } \exp\left[\sum_{j\geq 1} z^j(x'_j - x''_j)\right] \times$$

$$(11.1.8) \qquad \exp\left[\sum_{j\geq 1} \frac{z^{-j}}{j}\left(\frac{\partial}{\partial x'_j} - \frac{\partial}{\partial x''_j}\right)\right] \tau(x')\tau(x'') = 0,$$

where the function $\tau(x)$ is the image of τ under σ.

By the change of variables

$$x = \frac{1}{2}(x' + x''),\ y = \frac{1}{2}(x' - x'')$$

we can rewrite (11.1.8) as

$$z - \text{term of } \exp\left(\sum_{j\geq 1} 2z^j y_j\right)$$

$$(11.1.9) \qquad \exp\left(-\sum_{j\geq 1} \frac{z^{-j}}{j}\frac{\partial}{\partial y_j}\right)\tau(x+y)\tau(x-y) = 0.$$

Using the elementary Schur polynomials S_k of Section 8.2 we can reformulate this as

$$(11.1.10) \qquad \sum_{j\geq 0} S_j(2y)\tilde{S}_{j+1}\left(-\frac{\partial}{\partial y}\right)\tau(x+y)\tau(x-y) = 0,$$

where $\tilde{S}(x_1, x_2, \dots) = S(x_1, \frac{1}{2}x_2, \frac{1}{3}x_3, \dots)$. By a simple application of the Taylor expansion we see that the formula above is equivalent to

$$\sum_{j\geq 0} S_j(2y)\tilde{S}_{j+1}\left(-\frac{\partial}{\partial u}\right)$$

$$(11.1.11) \qquad \exp\left(\sum_{k=1}^{\infty} y_k \frac{\partial}{\partial u_k}\right)\tau(x+u)\tau(x-u)|_{u=0} = 0.$$

Using the notation

$$P(D)f\cdot f = P\left(\frac{\partial}{\partial u_1}, \frac{\partial}{\partial u_2}, \dots\right)f(x+u)f(x-u)|_{u=0},$$

where P is any polynomial, we can rewrite (11.1.11) as a system of *Hirota bilinear equations,*

$$P_{j_1 \dots j_k}(D)\tau \cdot \tau = 0. \tag{11.1.12}$$

The polynomials $P_{j_1 \dots j_k}$ are obtained by expanding (11.1.11) as a power series in the coordinates y_i and taking the coefficient of the monomial $y_1^{j_1} y_2^{j_2} \dots y_k^{j_k}$. For example,

$$P_j(x) = 2\tilde{S}_{j+1}(-x) - x_1 x_j,$$

with $P_1 = -2x_2$, $3P_2 = x_1^3$, $12P_3 = (x_1^4 - 4x_1x_3 + 3x_2^2) - 6(x_1^2 x_2 + x_4)$. The odd order terms of P in the Hirota bilinear form give a vanishing contribution and therefore the first nontrivial equation is

$$(D_1^4 - 4D_1D_3 + 3D_2^2)\tau \cdot \tau = 0. \tag{11.1.13}$$

Setting $x = x_1$, $y = x_2$, $t = x_3$, $u(x,y,t) = 2D_1^2 \log \tau(x,y,t,x_4,\dots)$ ($x_4, x_5, \dots$ are considered as parameters) one gets the classical Kadomtsev-Petviashvili equation

$$3u_{yy} = (4u_t - 6uu_x - u_{xxx})_x. \tag{11.1.14}$$

For this reason the totality of the equations (11.1.12) is called the *KP hierarchy.*

11.2. Soliton solutions of the KP hierarchy

We recall first the definition of the vertex operator from Section 8.2,

$$\Gamma(u,v) = \exp\left[\sum_{j>0}(u^j - v^j)x_j\right] \cdot \exp\left(-\sum_{j>0} \frac{u^{-j} - v^{-j}}{j} \frac{\partial}{\partial x_j}\right).$$

We shall consider the action of products of vertex operators on functions $f(x)$. For that purpose we first note that by Taylor's formula

$$\begin{aligned}
&\exp\left(-\sum \frac{u'^{-j} - v'^{-j}}{j} \frac{\partial}{\partial x_j}\right) \exp\left[\sum (u^j - v^j)x_j\right] \\
&= \exp\left[\sum (u^j - v^j)\left(x_j - \frac{u'^{-j} - v'^{-j}}{j}\right)\right] \\
&= \exp\left[\sum (u^j - v^j)x_j\right] \exp\left\{\sum \left[-\left(\frac{u}{u'}\right)^j \right.\right. \\
&\qquad \left.\left. + \left(\frac{u}{v'}\right)^j + \left(\frac{v}{u'}\right)^j - \left(\frac{v}{v'}\right)^j\right]/j\right\} \\
&= \exp\left[\sum (u^j - v^j)x_j\right] \cdot \\
&\quad \cdot (1 - u/u')(1 - u/v')^{-1}(1 - v/u')^{-1}(1 - v/v')
\end{aligned} \tag{11.2.1}$$

where we have used the expansion $\log(1-z) = -\sum_{j>0} \frac{z^j}{j}$ for $|z| < 1$; the above formula is valid in the range $|u|, |v| < min\{|u'|, |v'|\}$. Using (11.2.1) and once more Taylor's formula we get

$$\Gamma(u',v')\Gamma(u,v)f(x) =$$

$$\frac{(u'-u)(v'-v)}{(v'-u)(u'-v)} \times \exp\left[\sum_{j>0}(u^j - v^j + u'^j - v'^j)/j\right] \cdot f(y) \tag{11.2.2}$$

where $y_j = x_j - (u^j - v^j + u'^{-j} - v'^{-j})/j$. We want to prove the following theorem:

THEOREM 11.2.3. *Let τ be a solution of the KP hierarchy. Then $(1 + a\Gamma(u,v))\tau$ is also a solution for any $a \in \mathbf{C}$.*

First we prove the following lemma:

LEMMA 11.2.4. *Let L_{high} be the irreducible representation of $\mathbf{gl}(\infty)$ with the highest possible weight $2\omega_0$ contained in the tensor product $\mathcal{F}_0 \otimes \mathcal{F}_0$, or equivalently in $\mathcal{B}_0 \otimes \mathcal{B}_0$ (see Section 8.2). Then τ is a solution of the KP hierarchy if and only if $\tau \otimes \tau \in L_{high}$.*

PROOF: Because of the unitarity relations $E_{ij}^* = E_{ji}$ the representation of $\mathbf{gl}(\infty)$ in the tensor product space decomposes to a direct sum $L_{high} \oplus L_{high}^{\perp}$ of invariant subspaces. The highest weight vector in L_{high} is $\psi_0 \otimes \psi_0$ corresponding to the weight $2\omega_0$. It follows that the orbit $X = \widehat{GL}(\infty) \cdot (\psi_0 \otimes \psi_0)$ is contained in L_{high}. Since the orbit X consists of those points $\tau \otimes \tau$ for which τ satisfies the equations of the KP hierarchy we conclude that $\tau \otimes \tau \in L_{high}$ whenever τ is a solution of the KP hierarchy.

Assume then that $\tau \otimes \tau \in L_{high}$. From the irreducibility of L_{high} it follows that

$$\tau \otimes \tau = \sum a_i T(g_i)(\psi_0 \otimes \psi_0). \tag{11.2.5}$$

Since T commutes with the operator S we reduce from (11.2.5) that $S(\tau \otimes \tau) = 0$ which implies that $\tau \otimes \tau \in X$ or in other words τ satisfies the KP hierarchy.

PROOF OF 11.2.3: By the assumption and the lemma the vector $\tau \otimes \tau$ is in L_{high}. Because L_{high} is invariant under the action of $\mathbf{gl}(\infty)$ we conclude from Theorem 8.2.5 that also

$$\Gamma(u,v)\tau \otimes \tau + \tau \otimes \Gamma(u,v)\tau \in L_{high}. \tag{11.2.6}$$

Applying $\Gamma(u',v')$ (with $|u|,|v| < min(|u'|,|v'|)$) to (11.2.6) and then letting $u' \to u$, $v' \to v$ we see that also

$$\Gamma(u,v)\tau \otimes \Gamma(u,v)\tau \in L_{high}. \tag{11.2.7}$$

combining (11.2.6) and (11.2.7) we obtain

$$(1 + a\Gamma(u,v))\tau \otimes (1 + a\Gamma(u,v))\tau \in L_{high}$$

and the assertion in (11.2.3) follows now from (11.2.4).

Note that by induction on N we obtain from (11.2.2)

$$\Gamma(u_N, v_N)\Gamma(u_{N-1}, v_{N-1}) \dots \Gamma(u_1, v_1) \cdot 1$$

$$= \exp\left[\sum_{k>0}\sum_{j=1}^{N}(u_j^k - v_j^k)x_k\right] \cdot \prod_{1\le i<j\le N} \frac{(u_j - u_i)(v_j - v_i)}{(u_j - v_i)(v_j - u_i)}. \tag{11.2.8}$$

Since $\tau(x) = 1$ is a solution we obtain from (11.2.3) and (11.2.8) that

$$\tau(x) = \sum_{r=0}^{N} a_1 a_2 \dots a_r \exp(\sum_{k\ge 1}\sum_{\nu=1}^{r}(u_{j_\nu}^k - v_{j_\nu}^k)x_k)\times$$

$$\times \prod_{1\le\nu<\mu\le r} \frac{(u_{j_\nu} - u_{j_\mu})(v_{j_\mu} - v_{j_\nu})}{(u_{j_\nu} - v_{j_\mu})(v_{j_\nu} - v_{j_\mu})}$$

is a solution of the KP hierarchy for all $a_1, \dots, a_N \in \mathbf{C}$; this is called the *N-soliton solution* of the KP hierarchy. In the case $N = 1$ we obtain through the transformation $U(x,y,t) = 2D_1^2 \log\tau$ the classical 1-soliton solution

$$U(x,y,t) = \tfrac{1}{2}(u-v)^3\{cosh\tfrac{1}{2}[(u-v)x + (u^2 - v^2)y + (u^3 - v^3)t + \alpha]\}^{-2},$$

where α is a constant. It describes the motion of a solitary wave in the xy plane moving in the direction of the vector $(u - v, u^2 - v^2)$ (when u, v are real), t being the time.

11.3. The Lax formulation of the KP hierarchy

Let $\mathcal{D}$ be the space of formal differential operators

$$P = \sum_{j=-\infty}^{m} a_j(x) D^j,$$

where $D = \frac{d}{dx}$ and the coefficients a_j are formal Laurent series in the variable x; m is an arbitrary integer. The product of two formal differential operators is defined by the generalized Leibnitz rule

$$\left[\sum_j a_j(x)D^j\right]\left[\sum_k b_k(x)D^k\right] = \sum_n \left[\sum_{j;l\geq 0} \binom{j}{l} a_j(x)\frac{d^l b_{n+l-j}}{dx^l}\right] D^n.$$

Note that for fixed n the sum over j and l is actually finite since $a_j = 0$ and $b_k = 0$ for $j, k >> 0$. We denote by $\mathcal{D}^+$ the space of ordinary differential operators (with $a_j = 0$ for $j < 0$). The order ord$P$ is the maximal index j for which $a_j \neq 0$. We denote by $\mathcal{D}^-$ the space of formal differential operators with negative order; it is clear that

$$\mathcal{D} = \mathcal{D}^+ \oplus \mathcal{D}^-.$$

Both $\mathcal{D}^\pm$ are subalgebras of $\mathcal{D}$ with respect to the product defined above. A Lie algebra structure in $\mathcal{D}$ is introduced naturally as $[P,Q] = PQ - QP$. For $P \in \mathcal{D}$ we denote its projections in $\mathcal{D}^\pm$ by $P_\pm$. The subspace $\mathbf{C}\frac{d}{dx} \oplus \mathcal{D}^-$ is denoted by $\mathcal{K}$.

Consider a family $L = L(t) \in \mathcal{K}$ depending on an infinite set of parameters $t = (t_1, t_2, \dots)$ and set $B_n = (L^n)_+$. We consider a Lax system

$$\frac{\partial L}{\partial t_n} = [B_n, L], \quad n = 1, 2, \dots. \tag{11.3.1}$$

We want to show that the system (11.3.1) is equivalent to the system of equations of the KP hierarchy.

PROPOSITION 11.3.2. *For a fixed n the system* (11.3.1) *is equivalent to the infinite system of equations*

$$\frac{\partial a_i}{\partial t_n} = f_i(a), \quad i = -1, -2, \dots$$

where the f_i's are certain polynomials of the coefficients $a_j(x)$ and their derivatives; if we define the weight of $D^j a_i$ to be $j - i + 1$ then f_i is a homogeneous polynomial of weight $n - i + 1$ (the weight of a product being by definition the sum of the weights of the factors).

PROOF: Write $L^n = B_n + (L^n)_-$; since L^n commutes with L we have

$$[B_n, L] = -[(L_n)_-, L]$$

which shows that the right-hand side of (11.3.1) can be written as $\sum_{i<0} f_i(u)D^i$. Comparing the coefficients of each D^i on both sides of (11.3.1) we get the infinite system of equations in (11.3.2). From the product rule follows that the coefficient of D^j in L^n is a homogeneous polynomial in the a_i's of weight $n-j$ and therefore the coefficient of D^i in the commutator $[(L^n)_-, L]$ has weight $(n-j)+(-i+j+1) = n-i+1$.

In order to connect (11.3.2) with the KP system for the τ functions we introduce as an intermediate step the *Baker function* ϕ_W which is a function of variables $(t_1, t_2, \dots)$ depending also on a parameter W.

Let H be again the space of complex valued square integrable functions on the circle S^1 with the splitting $H = H_+ \oplus H_-$ to non-negative and negative Fourier components and let Gr_1 be the corresponding Hilbert-Schmidt Grassmannian. Denote by Γ_+ the group of holomorphic maps from the unit disc to $\mathbf{C}^\times$. Let Gr' be the subset of Gr_1 consisting of planes $W \subset H$ such that $W + H_- = H$ and denote

$$\Gamma_+^W = \{g \in \Gamma_+ \mid g^{-1}W \in Gr'\}.$$

PROPOSITION 11.3.3. *For each $W \in Gr_1$ there is a unique function $\phi_W(g,z)$ of the variable $g \in \Gamma_+^W$, $z \in S^1$ of the form*

$$\phi_W(g,z) = g(z)\left[1 + \sum_{i=1}^{\infty} b_i(g)z^{-i}\right]$$

such that $z \mapsto \phi_W(g,z)$ is an element of W for each $g \in \Gamma_+^W$.

PROOF: Let $\pi_g : g^{-1}W \to H_+$ be the orthogonal projection. Since for $g \in \Gamma_+^W$ we have $g^{-1}W \cap H_- = 0$ the projection is bijective. The vector $\sum b_i(g)z^{-i}$ is simply the inverse image of the function $1 \in H_+$ under the projection π_g.

One can show that the coefficients b_i are holomorphic functions of g and they extend to meromorphic functions on Γ_+; see Segal and Wilson [1985]; Section 5. We shall write a general element $g \in \Gamma_+$ as

$$g = \exp(xz + t_2 z^2 + t_3 z^3 + \dots)$$

so that we can write $\phi_W = \phi_W(x,t,z)$ as function of the variables t_i (with $t_1 = x$.)

PROPOSITION 11.3.4. *For each $n = 2, 3, \dots$ there is a unique ordinary differential operator*

$$P_n = D^n + p_{n,2}D^{n-2} + \cdots + p_{n,n-1}D + p_{n,n}$$

such that

$$\frac{\partial \phi_W}{\partial t_n} = P_n \phi_W,$$

where the coefficients p_{nj} are certain polynomials of the coefficients a_i and their derivatives.

PROOF: Denoting by $O(z^m)$ terms such that $O(z^m)/z^m \to$ *const.* as $|z| \to \infty$ we get from Proposition 11.3.3

$$\frac{\partial \phi_W}{\partial t_n} = g(z)[z^n + a_1 z^{n-1} + O(z^{n-2})]$$

and also

$$D^m \phi_W = g(z)[z^m + O(z^{m-1})].$$

Comparing the coefficients of the powers z^m with $m \geq 0$ of the expressions in brackets in the above two equations we conclude that there is a unique ordinary differential operator, with coefficients as stated in the proposition, such that

$$\frac{\partial \phi_W}{\partial t_n} - P_n \phi_W = g(z) \cdot O(z^{-1}). \tag{11.3.5}$$

The right-hand-side of (11.3.5) is in W (as a function of z) for all (x, t) and so the factor behaving like $O(z^{-1})$ must be in $g^{-1}W$; but $O(z^{-1})$ belongs to H_- and because of $H_- \cap g^{-1}W = 0$ for all $g \in \Gamma_+^W$ we conclude that the right-hand-side of (11.3.5) is equal to zero.

PROPOSITION 11.3.6. *For any formal differential operator $P = \sum p_i D^i \in \mathcal{K}$ there exists an operator $K \in \mathcal{D}$ of the form*

$$K = 1 + \sum_{j=-\infty}^{-1} a_j(x) D^j$$

such that $K^{-1}PK = D$. K is uniquely defined modulo right multiplication by operators of the form $1 + \sum_{j=-\infty}^{-1} c_j D^j$, where the c_j's are constants.

PROOF: First, the operator $K \in 1 + \mathcal{D}_-$ has an inverse: if we use the ansatz $K^{-1} = 1 + \sum_{j<0} b_j D^j$, multiply K with the ansatz from the left and put the product equal to 1 we get the infinite system of equations

$$a_j + b_j + \sum_{i+k \geq j} b_i \binom{i}{i+k-j} D^{i+k-j} a_k = 0. \tag{11.3.7}$$

The third term involves only the coefficients b_i with $i < j$ and therefore the system (11.3.7) has a unique solution $b = (b_1, b_2, \dots)$.

The solution K is found in the same way by comparing the coefficients of D^m on both sides of the equation $PK = KD$. The coefficient on the right is a_{m-1} whereas on the left it is

$$p_m + \sum_{m \le i+j} p_i \binom{i}{i+j-m} D^{i+j-m} a_j.$$

The sum is of the form $a_{m-1} + Da_m +$ terms involving a_j's with $j > m$. This set of equations for $m < 1$ can be used to solve successively the a_j's. The uniqueness follows from the fact that only the constant coefficient operators commute with D.

Denote by K the formal differential operator defined by the coefficients of the Baker function ϕ_W,

$$K = 1 + \sum_{i<0} a_i(x,t) D^i.$$

It is easy to check that the equation $\frac{\partial \phi_W}{\partial t_n} = P_n \phi_W$ is equivalent to

$$\frac{\partial K}{\partial t_n} = P_n K - KD^n. \tag{11.3.8}$$

Since the derivatives of K belong to $\mathcal{D}_-$ it follows that $P_n = (KD^nK^{-1})_+$ and therefore the operator $Q = KDK^{-1}$ is of the form

$$Q = D + \sum_{i<0} q_i(x,t) D^i.$$

Note also that $(Q^n)_+ = (KDK^{-1})^n_+ = (KD^nK^{-1})_+ = P_n$.

PROPOSITION 11.3.9. *The coefficients q_i of Q are meromorphic functions of all the variables (x,t) and Q satisfies the differential equation*

$$\frac{\partial Q}{\partial t_n} = [(Q^n)_+, Q].$$

PROOF: From the definition of Q we get

$$\frac{\partial Q}{\partial t_n} = \frac{\partial K}{\partial t_n} DK^{-1} - KDK^{-1} \frac{\partial K}{\partial t_n} K^{-1}. \tag{11.3.10}$$

From (11.3.8) we have

$$\frac{\partial K}{\partial t_n}K^{-1} = P_n - KD^nK^{-1} = P_n - Q^n = (Q^n)_+ - Q^n.$$

Inserting this into (11.3.10) the right-hand-side reduces to $[(Q^n)_+, Q]$.

We recognize in (11.3.9) again the Lax system (11.3.1) with $L = Q$.

Exercise 11.3.11. Show that the system of partial differential equations in (11.3.9) is equivalent to the *Zakharov-Shabat* system

$$\frac{\partial P_n}{\partial t_m} - \frac{\partial P_m}{\partial t_n} + [P_n, P_m] = 0,$$

where $P_n = (Q^n)_+$. (This "zero curvature condition" is of course the integrability condition for the linear system $\frac{\partial \phi}{\partial t_n} = P_n\phi$.)

In order to connect (11.3.9) with the τ function formulation of the KP system we need still one lemma. For $W \in Gr'$ we define the associated τ function on Γ_+ as $\tau_W(g) = \det(1 + a^{-1}bw_-)$, where we have chosen the admissible basis $w = (w_+, w_-)$ of W such that $w_+ = 1$ in the fixed basis of H_+ defined by the Fourier modes and we have written as usual

$$g^{-1} = \begin{pmatrix} a & b \\ 0 & d \end{pmatrix}.$$

Note that the value of τ_W at g is simply the value of the function on Γ_+ at the point g obtained via the local trivialization $W \mapsto w = (1, w_-)$ from the section $\psi(w) = T(g) \cdot \psi_0$, defined by g action on the vacuum section $\psi_0(w) = \det w_+$; this means that τ_W represents a point on the orbit $X \subset DET_1^*$.

LEMMA 11.3.12. *For each* $0 \neq \alpha \in \mathbf{C}$ *let* $\xi_\alpha(z) = 1 - z/\alpha$ *and for* $W \in Gr'$ *let* f_0 *be the unique element of* H_- *such that* $1 + f_0 \in W$. *Then we have*

$$\tau_W(\xi_\alpha) = 1 + f_0(\alpha), \text{ for } |\alpha| > 1.$$

PROOF: Clearly $\xi_\alpha \in \Gamma_+$ when $|\alpha| > 1$ and the inverse is given by the power series

$$\xi_\alpha^{-1} = 1 + \sum_{i>0} (z/\alpha)^i.$$

We can write ξ_α^{-1} in the block form

$$\xi_\alpha^{-1} = \begin{pmatrix} a & b \\ 0 & d \end{pmatrix}.$$

Since

$$b(z^{-p}) = \sum_{n\geq p} \alpha^{-n} z^{n-p} = \alpha^{-p} \cdot \xi_\alpha^{-1}(z)$$

for $p > 0$, we have $a^{-1}bz^{-p} = \alpha_{-p}$ or in other words $a^{-1}b$ sends an element $f \in H_-$ to the constant function $f(\alpha)$ in H_+. The map $a^{-1}bw_-$ [where $w = (1, w_-)$ represents the plane W] must then also have rank one and therefore

$$\begin{aligned}\tau_W(\xi_\alpha) &= \det(1 + a^{-1}bw_-) \\ &= 1 + \operatorname{tr}(a^{-1}bw_-) = 1 + f_0(\alpha),\end{aligned}$$

where we have used the fact that $1 + f_0$ is the first vector w_1 of the basis w (which in turn follows from $1 + f_0 \in W$).

PROPOSITION 11.3.13. *Let* $|\alpha| > 1$, $W \in Gr'$ *and* $g \in \Gamma_+^W$. *Then the modified Baker function* $\tilde{\phi}_W = g^{-1} \cdot \phi_W$ *can be expressed as*

$$\tilde{\phi}_W(g, \alpha) = \tau_W(g \cdot \xi_\alpha)/\tau_W(g).$$

PROOF: From the definition of τ_W follows that

$$\tau_W(g\xi_\alpha)/\tau_W(g) = \tau_{g^{-1}W}(\xi_\alpha).$$

Applying the lemma to $g^{-1}W$ we see that $\tau_{g^{-1}W}(\xi_\alpha)$ is equal to $1+f_0(\alpha)$, where $f_0 \in H_-$ is the unique function such that $1 + f_0 \in g^{-1}W$. On the other hand, from (11.3.3) we know that $\tilde{\phi}_W(g, \cdot)$ is the unique function of the form $1 + \sum_{i<0} a_i(x,t)z^i$ which belongs to $g^{-1}W$.

Using the power series expansion for $\log(1 - z/\alpha)$ (for $|\alpha| > 1$) we can write $g \cdot \xi_\alpha = \exp[(x - 1/\alpha) + (t_2 - 1/2\alpha^2) + (t_3 - 1/3\alpha^3) + \ldots]$ and thus

$$\tilde{\phi}_W(x, t, \alpha) = \frac{\tau_W(x - 1/\alpha, t_2 - 1/2\alpha^2, t_3 - 1/3\alpha^3, \ldots)}{\tau_W(x, , t_2, t_3, \ldots)}.$$

Recalling the definition of the vertex operator $\Gamma(u)$ from 8.2 (with $x_1 = x$ and $x_n = t_n$ for $n > 1$) and using $\phi_W(x,t,z) = \exp(xz + t_2z^2 + \ldots)\tilde{\phi}_W(x,t,z)$ we can write compactly

$$\phi_W(x, t, \alpha) = \frac{\Gamma(\alpha)\tau_W(x,t)}{\tau_W(x,t)}. \tag{11.3.14}$$

Let us summarize the discussion above. Starting from a solution τ of the KP hierarchy corresponding to a point $\tau \in \Gamma_+\psi_0 \subset X$ we can construct the function τ_W through the canonical local trivialization in DET_1 over Gr'. The function τ_W in turn determines the modified Baker function $\tilde{\phi}_W$ through (11.3.13) which can be used to define the formal differential operator K. The associated operator $Q = KDK^{-1}$ satisfies the system of differential equations in (11.3.9). Conversely, we may start from a solution Q of (11.3.9) and construct K using (11.3.6). Then we construct the (modified) Baker function from the coefficients $a_i(x,t)$ of K. Next we would like to construct a plane $W \in Gr'$ such that our Baker function is *the* Baker function ϕ_W associated to the plane W. However, we can directly derive the τ function from ϕ_W by the formula

$$\frac{\partial}{\partial t_n}\log\tau_W(x,t) = \frac{1}{2\pi i}\oint \frac{1}{z^n}\left(z^2\frac{\partial}{\partial z}+\sum_{j>0} z^{j+1-n}\frac{\partial}{\partial t_n}\right)\log\tilde{\phi}_W(x,t,1/z)dz,$$

where the integral is performed around a small circle at the origin. By a direct computation one can show that if τ_W is a solution of the above equation (which is unique modulo a multiplicative constant), then ϕ_W is indeed given by the formula (11.3.14).

The Plücker coordinates $W_S = \det e_S^* w$ (cf. Section 6.5) of the plane W corresponding to a given τ function can be obtained as follows. Let $g^{-1} \in \Gamma_+$ be given in the block triangular form as in (11.3.12). Then $\tau_W(g) = \det(1 + a^{-1}bw_-)$, where w is the basis of W such that $w_+ = 1$. Using (6.5.2) we have

$$\tau_W(g) = \sum_{S\in\mathcal{S}} W_S \det(1 + a^{-1}b)_S, \tag{11.3.15}$$

where $(1 + a^{-1}b)_S$ denotes the matrix obtained from the $\mathbf{N} \times \mathbf{Z}$ matrix $(1, a^{-1}b)$ by selecting the columns labelled by the indices $i \in S$.

Exercise 11.3.16. Using the results of Section 8.3 show that the coefficient of W_S in (11.3.15) for $S = \{i_1, i_2, \dots\}$ is equal to the Schur polynomial $S(\lambda)(x, t_2, t_2, \dots)$ where λ is given by $\lambda_k = k - i_k$. (Remember that $i_k = k$ for $k >> 0$.)

Since the Schur polynomials form an orthonormal basis in the bosonic Fock space this exercise shows that the Plücker coordinates of the plane W are indeed uniquely determined by the τ function. Note in particular that τ_W is a polynomial if and only if only finitely many of the Plücker coordinates are nonzero; this latter condition is equivalent to $W \in Gr_0$, where Gr_0 is the Grassmannian corresponding to the orbit $GL_0 \cdot H_+$ of the group GL_0 consisting of matrices with finite rank off-diagonal blocks.

11.4. The KdV equation and the reduced KP hierarchy

Let ℓ be an integer > 1 and $\mathbf{g}_\ell$ the Lie algebra

$$\mathbf{g}_\ell = \{u \in \mathbf{gl}(\infty) \mid [u, h_j] = 0 \,\forall j \equiv 0 \mathrm{mod} \ell\},$$

where $h_j = \sum E_{n,n+j}$. In the previous sections we saw how to associate a solution (the function τ_W) to each plane W in the Γ_+ orbit of a highest weight vector in a Fock space. The *ℓth reduced KP hierarchy* KP_ℓ is obtained by restricting the symmetry group Γ_+ to the subgroup corresponding to the Lie algebra $\mathbf{g}_\ell$. In the bosonic realization of the Fock space h_j is represented by jx_j for $j < 0$ and by $\frac{\partial}{\partial x_j}$ for $j > 0$. It follows that the functions $\tau(x)$ corresponding to points in the reduced orbit do not depend on $x_{n\ell}$, $n \in \mathbf{Z}$. From the relation between the Baker function and the τ function (Proposition 11.3.13) we see that the Baker function ϕ for the KP_ℓ hierarchy satisfies

$$\frac{\partial \phi(x,t,z)}{\partial t_{n\ell}} = z^{n\ell}\phi(x,t,z)$$

for $n = 1, 2, \ldots$. Thus the KP_ℓ hierarchy can be viewed as the integrability condition for the following linear system (cf. Exercise 11.3.11):

$$(11.4.1) \qquad \begin{aligned} \frac{\partial \phi}{\partial t_j} &= z^j \phi, \ j = \ell, 2\ell, 3\ell \ldots \\ \frac{\partial \phi}{\partial t_j} &= (Q^j)_+ \phi, \ j \not\equiv 0 \mathrm{mod} \ell \end{aligned}$$

The KP_2 hierarchy is called the KdV hierarchy since it contains the Korteweg-de Vries equation

$$\frac{\partial u}{\partial t} = \frac{3}{2} u \frac{\partial u}{\partial x} + \frac{1}{4} \frac{\partial^3 u}{\partial x^3}.$$

This equation is obtained from the KP equation (11.1.14) when u does not depend on the variable $y = t_2$. One can show that in the Hirota bilinear form the KP_ℓ system is

$$\sum_{j \geq 0} S_j(2y) \tilde{S}_{j+\ell+1}\left(-\frac{\partial}{\partial u}\right)\left(\exp \sum_{n>0} y_n \frac{\partial}{\partial u_n}\right) \tau(x+u)\tau(x-u)|_{u=0} = 0.$$

Let us study the action of the Lie algebra $\mathbf{g}_\ell$ in the Fock space; we shall see that this gives a construction of the affine Lie algebra A_ℓ^1. Let $u = \sum_{i,j} a_{ij} E_{ij} \in \mathbf{g}_\ell$. Then

$$a_{i+\ell, j+\ell} = a_{i,j}$$

for all $i, j \in \mathbf{Z}$ and therefore there is a linear one-to-one relation between elements of $\mathbf{g}_\ell$ and Laurent polynomials

$$A(z) = \sum_{n \in \mathbf{Z}} A_n z^n$$

with coefficients in the algebra of $\ell \times \ell$ matrices, given by

$$A_n = (a_{i,j+n\ell})_{0 \le i,j < \ell}.$$

It is a straightforward computation to show that the commutation relations

$$[E_{ij}, E_{mn}] = \delta_{jm} E_{in} - \delta_{in} E_{mj} + c(E_{ij}, E_{mn}),$$

where the cocycle c is given in Exercise 6.3.6, imply the following commutation relations for the Laurent polynomials:

$$[A, B](z) = [A(z), B(z)] + \operatorname{Res} \operatorname{tr} \frac{dA(z)}{dz} B(z). \tag{11.4.2}$$

Restricting to the subalgebra of traceless $\ell \times \ell$ matrices A_n we recognize in (11.4.2) the commutation relations of the affine Lie algebra A_ℓ^1 in Section 2.2.

11.5. Vertex operators and kinks

In Chapter 8 we discussed the boson-fermion relation in terms of vertex operators. First we constructed a linear isomorphism $\sigma_m : \mathcal{F}_m \to \mathcal{B}_m$ and we saw that this map transports the linear operators $h_n = \sum_{k \in \mathbf{Z}} E_{k,k+n}$ in $\mathcal{F}_m$ to operators in $\mathcal{B}_m$ which can be represented by the bosonic creation and annihilation operators $x_n, \frac{\partial}{\partial x_n}$. In reverse, the fermionic creation and annihilation operators are the moments of the vertex operators $\Gamma(z)$ [multiplied by a power of z]. From the results of Section 11.3 we have an alternative formula for the vertex operator: Comparing Proposition 11.3.3 with the formula (11.3.14) we observe that the action of the vertex operator on bosonic state $\tau(x)$ can be written as

$$[\Gamma(z)\tau](x) = g(x)\tau(g \cdot \xi_z), \tag{11.5.1}$$

where $|z| > 1$ and $g(x) = \exp(x_1 x + x_2 z^2 + \dots) \in \Gamma_+$. Thus the action of the fermion field on $\tau_m \in \mathcal{B}_m$ (see Theorem 8.2.4) can be written as

$$\begin{aligned} \psi(z)\tau &= z^{m+1} \tau_{m+1}(g \cdot \xi_z) \\ \psi^*(z) &= z^{-m-1} \tau_{m-1}(g \cdot \xi_x^{-1}), \end{aligned} \tag{11.5.2}$$

where the lower index $m \pm 1$ reminds us that the resulting tau function should be considered as an element of $\mathcal{B}_{m\pm 1}$ instead of $\mathcal{B}_m$.

In this section we want to explain how the formula (11.5.2) for the fermion operators is related to solitonic structures on the circle. Keeping the notation of Section 11.3 we have a unitary action of the loop group $L\mathbf{C}^\times$ on H given by pointwise multiplication of vectors in $L_2(S^1, \mathbf{C})$ by the loops $g: S^1 \to \mathbf{C}^\times$. The group $L\mathbf{C}^\times$ splits naturally to disconnected components $L\mathbf{C}^\times_n$ indexed by the winding number $n \in \mathbf{Z}$ around the origin. The projective action of $L\mathbf{C}^\times_0$ in $\mathcal{F}$ is given by the embedding $LC_0 \subset GL_1^0$ and the action of the total group is defined naturally by setting

$$T(z^n) f_{i_1} \wedge f_{1_2} \wedge \cdots = f_{i_1+n} \wedge f_{i_2+n} \wedge \ldots \tag{11.5.3}$$

and observing that any $g \in L\mathbf{C}^\times_n$ can be written uniquely as $g(z) = g_0(z) z^n$, where $g_0 \in L\mathbf{C}^\times_0$. The choice (11.5.3) is natural because $T(z^n)$ maps the basis vector $e_i = z^i$ onto e_{i+n}; of course there are other alternatives to make $L\mathbf{C}^\times$ act projectively on $\mathcal{F}$. Note that $T(z^n)\mathcal{F}_m = \mathcal{F}_{m+n}$ and in particular the vacuum ψ_m is mapped onto ψ_{m+n}. In contrast to the case of loops in simple compact Lie groups the central extension of $L\mathbf{C}^\times$ corresponding to the embedding to GL_1 is topologically trivial [this is because of $\pi_2(L\mathbf{C}^\times) \simeq \pi_3(\mathbf{C}^\times) \equiv 0$]. This means that we can choose a global section $\chi: L\mathbf{C}^\times \to \widehat{L}\mathbf{C}^\times$ and the extension can be described by an everywhere well-defined two-cocycle:

$$(g_1, \lambda_1)(g_2, \lambda_2) = (g_1 g_2, \lambda_1 \lambda_2 e^{c(f_1, f_2)}),$$

where we have written $g_k = \exp(i f_k)$ for loops $g_k : S^1 \to \mathbf{C}^\times$; the logarithm is uniquely defined up to a term $k \cdot 2\pi$. Using different choices of the trivialization χ there is a big freedom in the definition of c. Denote by $n(f)$ the winding number of the loop $\exp(if)$ and by $\hat{f}$ the mean value

$$\hat{f} = \frac{1}{2\pi} \int_0^{2\pi} f(\phi) d\phi,$$

where we have parametrized the loops using the angle $0 \leq \phi \leq 2\pi$. We shall use the following choice for the cocycle:

$$\begin{aligned} c(f_1, f_2) = {} & \frac{1}{4\pi} \int_0^{2\pi} f_1(\phi) f_2'(\phi) d\phi \\ & + \frac{1}{2}[\hat{f}_1 n(f_2) + \hat{f}_2 n(f_1)] - \frac{1}{2} n(f_1) f_2(0). \end{aligned} \tag{11.5.4}$$

The choice (11.5.4) is compatible with the action (11.5.3) of the discrete subgroup $\{z^n \mid n \in \mathbf{N}\}$ since c vanishes on this subgroup. From the cocycle formula we observe immediately:

PROPOSITION 11.5.5. *If f_1 and f_2 have disjoint supports then*

$$T(e^{if_1})T(e^{if_2}) = (-1)^{n_1 n_2} T(e^{if_2})T(e^{if_1}),$$

where n_1, n_2 are the winding numbers of the loops f_1 and f_2.

This result is of course analogous to the Proposition 4.4.12 which we proved for solitons in three space dimensions.

Consider the following two-parameter family of loops,

$$q_{u,v}(z) = \left(1 - \frac{\overline{u}}{z}\right)^{-1} (\overline{u}/z) \left(1 - \frac{z}{v}\right), \quad |v| > 1 > |u|.$$

These loops have all winding number $=+1$. Let ψ be any vector in the fermionic Fock space $\mathcal{F}_m$. The bosonized vector $\tau = \sigma_m(\psi) \in \mathcal{B}_m$ is then given by

$$\tau(x) = (\psi, T[\exp(x_1 z + x_2 z^2 + \dots)]\psi_m). \tag{11.5.6}$$

To check this formula we need only to show that when ψ on the right is replaced by $h_n\psi$ then $\tau \mapsto \frac{\partial \tau}{\partial x_n}$ for $n > 0$ and $\tau \mapsto n x_n \tau$ for $n < 0$; but this follows immediately from the definition of the h_n's and from the fact that the action of z^n in H is equal to the action of $\sum_k E_{k,k-n}$. We want to show that the action of the vertex operator

$$\Gamma_{u,v} \equiv \exp(\sum_{n>0} u^n x_n) \cdot \exp\left(-\sum_{n>0} v^{-n} \frac{\partial}{\partial x_n}\right)$$

can be written as

$$u^{m+1} z \Gamma_{u,v} \tau = (\psi, : T(q_{u,v}) : T(g(x))\psi_{m+1}). \tag{11.5.7}$$

Note that "z" has three different meanings: It is the loop parameter in $LC^\times$, it appears as the operator $T(z) : \mathcal{F}_m \to \mathcal{F}_{m+1}$ and as the bosonized operator $\sigma_{m+1} T(z) \sigma_m^{-1}$; this last operator is denoted for short as z (in accordance with the notation of Section 8.2). The normal ordering of the loop operator is defined as follows. First we decompose

$$\begin{aligned} q_{u,v}(z) &= \exp\left[\sum_{n>0} \frac{1}{n} (\overline{u}/z)^n\right] \cdot (\overline{u}/z) \cdot \exp\left[-\sum_{n>0} \frac{1}{n} (z/v)^n\right] \\ &\equiv q^-(z) u z q^+(z). \end{aligned}$$

The normal ordered operator is then

$$: T(q_{u,v}) := T(q^+)T(\overline{u}/z)T(q^-).$$

Using (11.5.4) we obtain

$$T(q^-)T(g(x)) = \exp\left(\sum_{n>0} x_n \overline{u}^n\right) \cdot T(g(x))T(q^-).$$

The vacuum ψ_m is invariant under $T(q^-)$. The operator $T(g(x))$ commutes with $T(\overline{u}/z)$ [simple computation from (11.5.4)],

$$T(\overline{u}/z) = \overline{u}T(\overline{u})T(1/z)$$

and thus the right-hand side of (11.5.7) is equal to

$$u^{m+1}\exp(\sum_{n>0} x_n u^n) \cdot (\psi, T(q^+)T(g(x))\psi_m),$$

where we have used $T(u)\psi_m = u^m\psi_m$. Finally,

$$T(q^+)T(g(x)) = T(q^+g(x)) = T\left(g\left(x_1 - \frac{1}{v}, x_2 - \frac{1}{2v^2}, \ldots\right)\right)$$

which proves (11.5.7).

Note that the vertex operator $\Gamma(u)$ is obtained from $\Gamma_{u,v}$ by setting $u = v$. However, one should keep in mind that in the case $u = v$ the vertex operator should be understood only as a formal power series in the indeterminate u. The coefficients of this power series are well-defined operators in the Fock space; comparing with (11.5.2) we see that these coefficients are the fermionic annihilation operators. If we had started from the complex conjugate of the loop $q_{u,v}$ we would have obtained the operator $\Gamma(u)^*$ which has as the coefficients the fermion creation operators.

For a geometric understanding of the construction of the fermion operators from the loops $q_{u,v}$ we look what happens when $\overline{u}$ and v converge towards a point $e^{i\alpha}$ on the unit circle. Writing $q^+ = \exp(if^+)$ and $q^- = \exp(if^-)$ we first observe that f^+ tends (in L^2 sense) to the function

$$f^+_\alpha = i\sum_{n>0} \frac{1}{n} e^{in(\phi-\alpha)}$$

and f^- converges towards

$$f_\alpha^- = -i \sum_{n>0} \frac{1}{n} e^{in(\alpha-\phi)}.$$

We conclude that the logarithm if of $q_{u,v}$ converges (in the L^2 sense) towards the function if_α,

$$f_\alpha(\phi) = -i\phi + \alpha - \pi + i \sum_{n\neq 0} \frac{1}{n} e^{in(\phi-\alpha)},$$

which is the Fourier expansion of the step function

$$s_\alpha(\phi) = \begin{cases} 0 & \text{when } 0 \leq \phi < \alpha \\ -2\pi & \text{when } \alpha \leq \phi \leq 2\pi. \end{cases}$$

Thus the fermion field at the point $e^{i\alpha}$ on the circle can be thought of as a infinitely narrow soliton placed at $e^{i\alpha}$, corresponding to a sudden "blip" of the angle $\log q_\alpha$ by the amount 2π.

CHAPTER 12 THE FOCK BUNDLE OF A DIRAC OPERATOR AND INFINITE GRASSMANNIANS

12.0. Introduction

In the earlier chapters we have studied representations of current algebras in fermionic Fock spaces. A (fermionic) Fock space is determined by a single Dirac operator D. To set up a Fock space we need a splitting of a complex Hilbert space H to the subspaces $H_\pm$ corresponding to positive and negative frequencies of D. However, in an interacting quantum field theory one really should consider a bundle of Fock spaces parametrized by different Dirac operators. For example, in Yang-Mills theory any smooth vector potential defines a Dirac operator and one must consider the whole bunch of these operators and associated Fock spaces if one wants to describe the interaction of the vector potential with Dirac spinor fields.

There are actually two different levels of complexity one can study. First, one can treat the vector potential as an external classical, nonquantized, field. Second, one tries to associate operators to the components of the vector potential and their canonically conjugate fields, acting in the space of sections of the *Fock bundle* and satisfying the canonical commutation relations. The sections of the Fock bundle are the Schrödinger wave functions in the Hamiltonian picture.

In Section 12.1 we shall study the $1+1$ dimensional case of massless fermions coupled to vector potentials. We construct the appropriate Fock bundle, the vacuum section and the quantized Dirac Hamiltonian for the external field problem. However, when trying to quantize the vector potentials we shall meet difficulties. The Hilbert space of the coupled system can be thought of as a tensor product of a pair of fermionic Fock spaces by a "fermionization" of the Yang-Mills system (this works only in space-time dimension 2); but the vacuum has infinite norm with respect to the inner product in the tensor product space.

In Section 12.2 we shall study the case of a Dirac operator parametrized by metrics on a Riemann surface. We shall explain (without proofs) the relation between solutions of the KP hierarchy and of certain moduli spaces related to metrics.

In Section 12.3 we shall consider the $3+1$ dimensional situation, for vector potentials coupled to Dirac spinors. This case is much more complicated already at the level of the external field problem. There are two alternative strategies to attack the problem. First, in order to have a well-defined action for the charge densities associated to the non-Abelian gauge group, we can define the Fock bundle such that the fibers are isomorphic to the space of sections of DET_2^*; this is the "generalized Fock space" we have studied in Chapter 6. The base is the Grassmannian Gr_2. The space of vector potentials A in the three-dimensional space is mapped into Gr_2 through the sign of the Dirac operator D_A. Gauge transformations act both in the fibers and the base. The total action does not have a commutator anomaly in the (twisted) Fock bundle. There is a well-defined vacuum vector. However, the action of the gauge group in the fibers is not unitarizable, as we saw in Section 6.5.

In the second alternative one trades the action of the charge densities in the fibers for the positive definite inner product using a standard construction of ordinary Fock spaces. Even though the current algebra is not represented here, there is an action of a smaller algebra $\widehat{\mathbf{gl}}_{2/3}$ in the space of holomorphic sections of the standard Fock bundle. This is the topic of the last subsections of Section 12.3.

In Section 12.4 we shall explain some ideas about a construction of a "universal Yang-Mills theory", due to Connes [1988] and Rajeev [1988b]. The point is that as the group GL_p contains all the gauge groups $S^d SU(N)$, $N = 1, 2, \ldots,$ (for $2p = d + 1$) there should be a way to formulate a gauge theory in terms of some infinite-dimensional manifolds related to the "universal gauge group" GL_2. In fact, this can be done, and what is interesting here, the method works precisely when the dimension $d + 1$ of the space time is four.

12.1. A two-dimensional example: Fermions coupled to a non-Abelian electric field

The Fock bundle

To warm up we start by considering an apparently simple example of massless fermions coupled to an external non-Abelian gauge field in 1+1 space-time dimensions. In local coordinates the Dirac equation takes the form

$$\gamma^\mu(\nabla_\mu + A_\mu)\psi + im\psi = 0 \tag{12.1.1}$$

where ∇ denotes the covariant space-time derivative, m is the mass of the particle, A_μ is a vector potential (one-form) taking values in the Lie algebra $\mathbf{g}$ of a compact Lie group G acting on the components of the field ψ, and $\gamma^{mu} = h^{\mu,i}\gamma_i$, $\mu = 0, 1$, is a pair of gamma matrices depending on the (Lorentzian) space-time metric through the components $h^{\mu,i}$ of a local orthonormal frame. We may take

$$\gamma_0 = \begin{pmatrix} 0 & 1 \\ 1 & 0 \end{pmatrix} \gamma_1 = \begin{pmatrix} 0 & 1 \\ -1 & 0 \end{pmatrix}.$$

When the mass $m = 0$ the two components of the Dirac spinor field decouple and we can solve the Dirac equation independently for the left and right components. We shall therefore concentrate on the one-component system

$$(\partial_x - \partial_t + A_x - A_t)\psi = 0.$$

We shall assume that the space-time is $S^1 \times \mathbf{R}$ (time is the real line) with Minkowskian metric and we shall move to the Hamiltonian formulation. We shall adopt the temporal gauge, i.e., we put $A_t \equiv 0$. The Hamiltonian is then

$$E = -i(\partial_x + A_x).$$

If the space were a real line we could solve the linear equation $\frac{d}{dx}\phi = A_x\phi$ and write $A_x = \phi'\phi^{-1}$. However, on the circle S^1 we have to take into account the holonomy and therefore we can only write

$$A_x = (\partial_x\phi)\phi^{-1} + b \tag{12.1.2}$$

where $b \in \mathbf{g}$ is a constant. The eigenvectors of E can be written as $\phi\psi$ where ψ is an eigenvector of $\partial_x + b$. The eigenvalues are $n + b(k)$, $n \in \mathbf{Z}$ and $1 \leq k \leq N$, where $b(1), \ldots, b(N)$ are the eigenvalues of the symmetric operator ib. In particular, the multiplicity of each eigenvalue is at most N.

For quantization of the fermions coupled to the vector potential A we want to construct for each A in the space $\mathcal{A}$ of smooth vector potentials the corresponding Fock space $\mathcal{F}_A$. Let H be the one-particle Hilbert space, that is, the space of square integrable one-component spinor fields ψ on the circle S^1. The Fock space carries an irreducible representation of the canonical anticommutation relations (CAR)

$$[a^*(u), a(v)]_+ = (u, v), \; [a(u), a(v)]_+ = 0 = [a^*(u), a^*(v)]_+ \tag{12.1.3}$$

where $u, v \in H$. To fix the representation we have to choose the vacuum vector. This means that we choose splitting $H = H_+ \oplus H_-$ and define the vacuum as the vector ψ_0 such that

$$a^*(u)\psi_0 = 0 \text{ for } u \in H_-, \; a(v)\psi_0 = 0 \text{ for } v \in H_+. \tag{12.1.4}$$

Concretely, the representation space can be identified as the quotient of the associative algebra generated by the creation and annihilation operators, modulo the left ideal generated by the operators $a^*(u)$ and $a(v)$ in (12.1.4).

For each $A \in \mathcal{A}$ we have the splitting $H = H_+(A) \oplus H_-(A)$ where $H_+(A)$ is the subspace spanned by the eigenvectors of E with eigenvalues $\lambda > 0$ and $H_-(A)$ is the orthogonal complement. This decomposition defines the Fock space $\mathcal{F}_A$. However, the mapping $A \mapsto \mathcal{F}_A$ is not continuous. If we have a continuous path $s \mapsto A(s) \in \mathcal{A}$ such that one of the eigenvalues of the energy operator $E(s)$ crosses the zero point at some point s_0 then there is a sudden jump in the splitting of H to positive and negative energy subspaces at the point $s = s_0$ which prevents us from considering the set of Fock spaces $\mathcal{F}_A$ as a smooth vector bundle over $\mathcal{A}$; but still we have a well-defined vector bundle $\mathcal{F}$ over the space of all subspaces H_+ of H. We must only keep in mind that $\mathcal{A}$ is not a smooth submanifold in the space of infinite-dimensional planes in H because of the discontinuity of $A \mapsto H_+(A)$.

If one wants to consider the Fock bundle really as a bundle over $\mathcal{A}$ there are at least two alternatives. The first one is via a construction of a projective Fock bundle and then using the fact that $\mathcal{A}$ is contractible and so the bundle of projective Fock spaces must be a projectivization of a trivial vector bundle over $\mathcal{A}$. We shall return to this in Section 12.3; we shall describe here the second alternative. However, both of these have drawbacks: (1) one must make some arbitrary choices of trivialization; (2) the Fock bundles over $\mathcal{A}$ do not admit the use of Fock space techniques for the quantization of the Yang-Mills field.

Consider the space $\hat{\mathcal{A}} = \{(A, \lambda) \mid A \in \mathcal{A},\ \lambda \text{ an eigenvalue of } E(A)\}$. The set $\hat{\mathcal{A}}$ is a bundle over $\mathcal{A}$ with fiber isomorphic with $\mathbf{Z}$ since the set of eigenvalues of the Hamiltonian $E(A)$ is discrete. For each $(A, \lambda) \in \hat{\mathcal{A}}$ we can construct the fermionic Fock space as follows. Let $H_+(A, \lambda) \subset H$ be the space spanned by the eigenvectors of $E(A)$ belonging to eigenvalues bigger than λ and let $H_-(A, \lambda)$ be its orthogonal complement. Choosing an orthonormal basis $\{e_n(A, \lambda)\}_{n \in \mathbf{N}_+}$ of $H_+(A, \lambda)$ we can define the Fock space $\mathcal{F}(A, \lambda)$ as described in Chapter 6. The choice of the basis defines the vacuum $\psi(A, \lambda)$ by

$$\psi(A, \lambda) = f_{-1} \wedge f_{-2} \wedge \ldots$$

where $\{f_n\}$ is the dual basis in H_+^*, $f_{-n}(e_m) = \delta_{n,m}$. The bundle $\hat{\mathcal{A}}$ is trivial since $\mathcal{A}$ is a vector space and thus for each $A \in \mathcal{A}$ one can choose $\lambda(A)$ in the spectrum of $E(A)$ in a smooth manner. We can also make a choice of the basis vectors $e_n(A, \lambda(A))$ as smooth functions of the vector potential and we have then a smooth bundle of Fock spaces $\mathcal{F}_A$ over $\mathcal{A}$.

The construction above is problematic (besides the unesthetical choice of a point on the spectrum and the basis) because there is no way to think of the sections of the Fock bundle as elements of the bosonic Fock space of the Yang-Mills field. The vacuum section is a nonholomorphic function of the background field A; on the other hand, elements of the bosonic Fock space should be holomorphic functions of A (they should be contained in the completion of the space of polynomial functions in the variable A).

Since the plane $H_+(A)$ is obtained from $H_+(0)$ by a gauge transformation ϕ followed by a shift operator due to the shifting of the momenta by the eigenvalues $b(k)$, we have $H_+(A) = g \cdot H_+(0)$ for $g \in GL_1$. Thus the set of planes under consideration is included in the Grassmannian Gr_1; this is an important characteristics of the 1+1 dimensional case. We shall denote in the following by $\mathcal{F}$ the bundle of Fock spaces over Gr_1.

The different Fock representations of CAR parametrized by points of Gr_1 are actually equivalent. Let $W' = g^{-1} \cdot W \in Gr_1$ for some $g \in GL_1$. Choose an element $\hat{g}$ in the central extension $\widehat{GL}_1$ of GL_1 which projects down to g. Let ψ_W be the vacuum vector of the Fock space $\mathcal{F}_W$. We define an equivalent representation of CAR in the same vector space $\mathcal{F}_W$ by

$$a'(u) = T(\hat{g})a(u)T(\hat{g}^{-1}), \;\; a^{*\prime}(u) = T(\hat{g})a^*(u)T(\hat{g}^{-1})$$

where T is the faithful representation of $\widehat{GL}_1$ in the Fock space $\mathcal{F}_W$ discussed in Chapter 6 (instead of $H = H_+ \oplus H_-$ we use the splitting $H = W \oplus W^\perp$ here). Now $a'(u) = a(gu)$ and $a^{*\prime}(u) = a^*(gu)$ and therefore the vector ψ_W is annihilated by $a^{*\prime}(u)$ for $u \in g^{-1} \cdot W$ and by $a'(v)$ for $v \in g^{-1} \cdot W^\perp$. Thus the primed operators form an irreducible Fock space representation of the CAR corresponding to the splitting $H = W' \oplus W'^\perp$. The equivalence of the two representations is *not natural*; it depends on the choice of $\hat{g}$ such that $W' = g^{-1} \cdot W$. For a fixed $W \in Gr_1$ it is not possible to make this choice as a continuous function of W'. In fact what we would need is only a choice of $\hat{g}$ modulo a right multiplication by the center of $\widehat{GL}_1$ because the center commutes with the CAR algebra; however, even this is not possible by the nontriviality of the principal bundle $GL_1 \to GL_1/\{\text{upper triangular operators}\} = Gr_1$. For this reason the Fock bundle $\mathcal{F}$ over Gr_1 is nontrivial.

Remark. The above point needs an explanation: Any Hilbert bundle over a manifold is trivial if the transition functions are allowed to take values in the general linear group of the fiber; this is simply because $GL(H)$ is contractible when H is infinite-dimensional. However, when considering the Fock bundle it is natural to consider it as a vector bundle such that the transition functions take values in an appropriate GL^p, or in other words, as an associated bundle to a certain principal GL^p bundle. The principal bundle is nontrivial.

The discussion above suggests that instead of the nontrivial Fock bundle $\mathcal{F}$ we could consider the trivial bundle $\mathcal{B} = Gr_1 \times \mathcal{F}_{H_+}$ where the vacuum ψ_W at the base point W would be defined as the line $T(\hat{g})\psi_{H_+}$ with $W = g \cdot H_+$. In this case the vacua at different points $W \in Gr_1$ define a nontrivial complex line bundle Vac over Gr_1. This line bundle is actually the dual determinant line bundle DET_1^*. Namely, by construction the group $\widehat{GL}_1$ acts transitively in $Vac^\times$ (the zero section deleted). The stability group N at the point ψ_{H_+} consists of the elements $(g, q, 1)$ with

$$g = \begin{pmatrix} a & 0 \\ c & d \end{pmatrix}, \quad q = a$$

and therefore $Vac^\times = \widehat{GL}_1/N$ as a manifold; the central elements $(1,1,\lambda)$ of $\widehat{GL}_1$ act as multiplication by λ^{-1} in the Fock space (=the space of holomorphic sections of the *dual* determinant bundle) and thus we may identify Vac as DET_1^*. The relation between $\mathcal{B}$ and $\mathcal{F}$ is this: Using the Fock space inner product we can associate to each $\psi \in Vac_W$ a linear form ψ^* on Vac_W, i.e., an element ψ^* of $DET_1^{**} = DET_1$, $u \mapsto (\psi, u) = \psi^*(u)$. The tensor product bundle $\mathcal{B} \otimes DET_1$ has a global section of the form

$$\xi(W) = \psi \otimes \psi^*, \ \psi \in Vac_W \text{ with } ||\psi|| = 1.$$

The value of ξ at $W \in Gr_1$ does not depend on the choice of ψ since multiplication of ψ by $e^{i\lambda}$ induces a multiplication of ψ^* by $e^{-i\lambda}$. The CAR algebra acts in the fibers of the twisted bundle $\mathcal{B} \otimes DET_1$ through the natural action on the first factor and therefore the vector $\xi(W) \neq 0$ is annihilated by the operators $a(u), a^*(v)$ for $u \in W$ and $v \in W^\perp$. It follows that $\mathcal{B} \otimes DET_1$ is in fact equivalent with the vector bundle $\mathcal{F}$. We shall take as a basic axiom the requirement that there should be a well-defined vacuum vector for each value W of the "external field" and therefore $\mathcal{F}$ is the "correct" Fock bundle.

Quantization of the vector potential

Up to this point we have only discussed the quantization of the fermions, treating the vector potential A as a classical background. Next we want to discuss briefly the quantization of A in the Grassmannian formulation (the vector potential being thought of as a point in Gr_1). Usually when quantizing the gauge field in the Schrödinger picture one starts from the canonical equal time commutation relations

$$[E_i^a, E_j^b] = 0 = [A_i^a, A_j^b]$$

$$(12.1.5) \qquad [E_i^a(x), A_j^b(y)] = \delta_{ij}\delta_{ab}\delta(x-y)$$

where the upper indices are the Lie algebra indices for **g** (in the adjoint representation) and the lower indices are the coordinate indices in space (in the present one-dimensional setting $i, j = 1$); the fields are not well-defined operators pointwise in space but one should understand them as operator valued distributions.

When considering the coupled Yang-Mills-Dirac system the canonical relations (12.1.5) are no longer consistent with the commutation relations of the quantized Dirac field in the Fock bundle. As we saw in the beginning there is no natural way to define the Fock bundle over $\mathcal{A}$ (because of the zero modes of the Dirac operator) and therefore we shall take the space of sections $\Gamma(\mathcal{F})$ as the space of Schrödinger wave functions for the coupled system. In the standard text book treatment of the (pure) Yang-Mills field the vector potential becomes just a multiplication operator after quantization,

$$\hat{A}_i^a(f)\Psi = (A_i^a, f) \cdot \Psi,$$

where $\Psi = \Psi(A)$ is a complex valued function in $\mathcal{A}$, $\hat{A}_i^a(f)$ is the smoothed operator by the smooth test function f and (A_i^a, f) denotes the L_2 inner product of a pair of real valued functions in the physical space. The electric field components are then represented by partial differential operators in the infinite-dimensional space $\mathcal{A}$, symbolically

$$(12.1.6) \qquad \hat{E}_i^a(f) = \int dx\, f(x) \frac{\partial}{\partial A_i^a(x)},$$

where the right-hand side means the derivative in the space $\mathcal{A}$ in the direction of the vector potential $A_j^b(x) = \delta_{ij}\delta_{ab} f(x)$.

In the Grassmannian formulation the vector potentials can still be represented by multiplication operators. If ϕ is any smooth function of the "vector potential" $W \in Gr_1$ then we associate to it the operator

$$(\hat{\phi}\Psi)(W) = \phi(W)\psi(W), \;\; \Psi \in \Gamma(\mathcal{F}).$$

Because of the twisting of the bundle $\mathcal{F}$ we must replace derivatives by covariant derivatives when acting on the sections of the bundle. In order to make sense of this we have to fix a connection in $\mathcal{F}$. Since $\mathcal{F}$ can be identified as $\mathcal{B} \times DET_1$ we can define a connection in $\mathcal{F}$ by taking the flat connection in the trivial bundle $\mathcal{B}$ and the natural connection in DET_1; the curvature of the latter is

$$\omega_W(X, Y) = \mathrm{tr} FXY, \tag{12.1.7}$$

where $F : H \to H$ is the operator which is $+1$ on W and -1 on $W^{\perp}$ and X, Y are tangent vectors at W (that is, they are Hermitian operators in $\epsilon + L_2$ which anticommute with F, ϵ being defined as in Section 6.1). Because of the nonvanishing of the curvature the electric field operators ∇_X do not commute. For example, if X denotes the vector field on Gr_1 induced by the action of a Lie algebra element $X \in \mathbf{gl}_1$ then

$$[\nabla_X, \nabla_Y] = \frac{1}{8} \mathrm{tr} F[F, X][F, Y]. \tag{12.1.8}$$

The Hamiltonian of the Yang-Mills field

We have discussed the kinematics of the Dirac-Yang-Mills system. In order to define the quantum dynamics we have to fix a Hamiltonian. The complete Hamiltonian consists of the Yang-Mills (YM) Hamiltonian, the free Dirac Hamiltonian, and an interaction Hamiltonian. We shall first discuss the YM Hamiltonian.

The classical YM Hamiltonian in $n+1$ space-time dimensions is

$$\int \left(E_i^a E_i^a + \frac{1}{2} B_{ij}^a B_{ij}^a \right) d^n x$$

where $B_{ij}^a = \partial_i A_j^a - \partial_j A_i^a + [A_i, A_j]^a$ are the components of the non-Abelian magnetic field. In 1+1 dimensions we have only the electric field components. In the Grassmannian formulation the electric field components are quantized as the directional covariant derivatives in the bundle $\mathcal{F}$. Thus it is natural to quantize the YM Hamiltonian as the *covariant Laplace operator*. The definition of the Laplace operator Δ in an infinite-dimensional space requires some care. Since the curvature of $\mathcal{F} = \mathcal{B} \otimes DET_1$ is entirely due to the second factor it is sufficient to define the action of Δ on sections of DET_1.

Formally at least, we expect the Laplace operator to be something like the second-order Casimir invariant of the symmetry group U_1 of the Grassmannian. For example, on the unit sphere $S^2 = SU(2)/U(1) = U(2)/[U(1) \times U(1)]$ (which is a Grassmannian manifold of complex dimension one) the Laplace operator can be written as $J_1^2 + J_2^2 + J_3^2$, where $\{J_1, J_2, J_3\}$ is an orthonormal basis of the Lie algebra of $SU(2)$. Let $\{e_{ij}\}_{i,j\in\mathbf{Z}}$ be the Weyl basis of the Lie algebra of $\mathbf{gl}_1$. The second-order Casimir of $\mathbf{gl}_1$ can be written as $e_{ij}e_{ji}$ (sum over repeated indices). To make sense of this infinite sum, when acting in $\Gamma(DET_1)$, we shall use a normal ordering prescription: We write $\mathbf{gl}_1 = \mathbf{g}_+ + \mathbf{h} + \mathbf{g}_-$, where the Cartan subalgebra $\mathbf{h}$ consists of all diagonal matrices, $\mathbf{g}_+$ consists of upper triangular and $\mathbf{g}_-$ of lower triangular matrices in $\mathbf{gl}_1$. We define

$$(12.1.9) \qquad : xy := \begin{cases} yx, \text{ when } x \in \mathbf{g}_-, \, y \in \mathbf{g}_+ \\ xy \text{ otherwise when } x, y \in \mathbf{g}_\pm, \mathbf{h} \end{cases}$$

We define the normal ordered Casimir as

$$(12.1.10) \qquad \Delta =: e_{ij}e_{ji} :$$

The domain of definition of Δ includes at least the dual of finite energy vectors in a Fock space by identifying the space of *antiholomorphic* sections of DET_1 as the dual of the standard Fock space $\mathcal{F}_{H_+}$. Remember that the generator e_{ij} is represented in the Fock space as

$$(12.1.11) \qquad E_{ij} = \begin{cases} a_i^* a_i - 1 \text{ when } i = j \leq 0 \\ a_i^* a_j \text{ otherwise} \end{cases}$$

where $a_i = a(e_i)$, $a_i^* = a^*(e_i)$. The basis vector $a_{i_1}^* \ldots a_{i_n}^* a_{j_1} \ldots a_{j_m} \psi_0$ is an eigenvector of Δ corresponding to the eigenvalue $(i_1 + \ldots i_n) - (j_1 + \ldots j_m)$ as can be seen by a substitution from (12.1.11) to (12.1.10), taking account of the defining relations $a_i\psi_0 = 0 = a_{-i}^*\psi_0$ for $i > 0$ and $a_0^*\psi_0 = 0$ of the Fock vacuum. Thus in the subspace of antiholomorphic sections of DET_1 the Laplace operator reduces to the first-order operator

$$\mathcal{H} = \sum_i iE_{ii}.$$

But $\mathcal{H}$ is the quantized free Dirac operator in one space dimension! The energy of the YM field is the Hamiltonian of the free Dirac field. This is again a manifestation of the intriguing relations between bosonic and fermionic theories in 1+1 dimensions. Note however that we have selected a small subspace in $\Gamma(\mathcal{F})$ as the domain of definition of the YM Hamiltonian Δ. [This has the consequence that the multiplication operators $\phi(W)$ are defined only for constant functions ϕ; this is not a serious drawback because we can treat the electric field, not the vector potential, as a fundamental field and there is no need to quantize A separately.] The subspace consists of vectors of the fermionic Fock space with coefficients in the space $\Gamma_{antihol}(DET_1)$. Since we identified the antiholomorphic sections of DET_1 as vectors in the fermionic dual Fock space, it would be natural to think of the Hilbert space of the coupled quantized Dirac-Yang-Mills system as a tensor product $\mathcal{F}_{H_+} \otimes \mathcal{F}^*_{H_+}$. However, the inner product in $\Gamma_{antihol}(\mathcal{F})$ cannot be chosen as the inner product of a tensor product of two Hilbert spaces. The reason is that the vacuum section would not have a finite norm. The vacuum in $\mathcal{B} \otimes DET_1$ is the section

$$\psi_0(f,w) = \det w^{(f)} \tag{12.1.12}$$

where $w^{(f)}$ denotes the matrix representing the orthogonal projection of an admissible basis w to another admissible basis f. For any fixed f the function $w \mapsto \psi_0(f,w)$ is a holomorphic section of DET_1^*, that is, an element of the Fock space $\mathcal{F}_{H_+}$. As a function of $f \in St_1$, ψ_0 can be viewed as an antiholomorphic section of DET_1 with coefficients in the Fock space $\mathcal{F}_{H_+}$. For a fixed f, ψ_0 is a vector annihilated by the operators $a^*(v)$, $a(u)$ with u in the plane W spanned by the basis f and v in its complement and therefore it is the vacuum at the point W in the base space Gr_1.

Restricting to unitary basis and using the notation of Section 6.5 we have

$$\psi_0(f,w) = \sum_S \overline{\psi_S(f)}\psi_S(w). \tag{12.1.13}$$

Since the sections ψ_S form an orthonormal basis in the Fock space the sum above is a sum over an infinite set of orthogonal vectors with norm of each vector equal to one if we use the tensor product norm in $\Gamma_{antihol}(\mathcal{F})$. There seems to be no natural inner product which would make ψ_0 a vector of finite norm. We have traded the existence of an inner product in the "physical space" to the existence of the vacuum section. If we had taken the trivial Fock bundle as our play ground we could have quantized the system in a Hilbert space using the standard rules of QFT [Paranjape, 1988], but then we would have lost the vacuum vector!

The complete Hamiltonian

Next we ask what is the operator $\hat{E}$ in the Fock bundle representing the Dirac Hamiltonian $E(A) = -i(\partial_x + A_x)$. As we have seen, formally the space of Schrödinger wave functions for the Dirac-Yang-Mills system is $\mathcal{F}_{H_+} \otimes \mathcal{F}^*_{H_+}$. The YM Hamiltonian acts in the second component as the free Dirac operator. There is another free Dirac operator which acts on the first factor; we shall argue that this is $\hat{E}$.

In one space dimension the classical Hamiltonian $E(A)$ has the property that its spectrum is obtained from the spectrum of the free Dirac operator $E(0)$ by a translation by the vector $(b(1),\ldots,b(N))$ [i.e., the eigenvalues corresponding to the index k of the internal symmetry are shifted by $b(k)$] and the set of eigenvectors is obtained by a gauge transformation $\phi(x)$. However, because of the normal ordering prescription in the definition of the Fock space $\mathcal{F}_A$ only the differences of the eigenvalues matter. The energy of the vacuum is always normalized to zero and the quantized Hamiltonian in the fiber $\mathcal{F}_A$ is

$$\begin{aligned}\hat{E}(A) &= \sum_{n\in\mathbf{Z};1\leq i\leq N} (\lambda_{ni}-\lambda_{0i})E_{ni,ni}\\ &= \sum_{n,i} nE_{ni,ni}\end{aligned} \tag{12.1.14}$$

where n enumerates the Fourier modes and i is the internal symmetry index. The elements $e_{ni,mj}$ of the Lie algebra $\mathbf{gl}_1$ have been defined with respect to a basis of eigenvectors of $E(A)$: The eigenvector belonging to the eigenvalue $\lambda_{nj}=n+b(j)$ is $\phi\cdot e^{inx}v_j$, where ϕ is a gauge transformation such that $A=(\partial_x\phi)\phi^{-1}+diag(b(1),\ldots,b(N))$ and $\{v_1,\ldots,v_N\}$ is the standard basis in $\mathbf{C}^N$.

The formula (12.1.14) shows that the fermionic Hamiltonian, as a distribution of operators $\hat{E}(A)$ in the fibers of the Fock bundle, can indeed be interpreted as the free quantized Dirac operator acting on the first factor in $\mathcal{F}_{H_+}\otimes\mathcal{F}^*_{H_+}$.

We want to stress that the results depend essentially on the fact that we have chosen the mass equal to zero. The spectrum of the massive Dirac operator consists of the eigenvalues $\pm\sqrt{[n+b(k)]^2+m^2}$, $n\in\mathbf{Z}$ and $1\leq k\leq N$ and therefore the spectrum of $E(A)$ is not in general obtained by a shift from the set of eigenvalues of $E(0)$.

12.2. Dirac operator on a Riemann surface

Let R be a Riemann surface without a boundary, equipped with a spin structure. Let L be the complex line bundle of one-component spinors of degree j on R (of chirality $+$, for example); in local coordinates a spinor of degree j transforms like $(dz)^j$. Fix a point $p\in R$ and a local holomorphic coordinate z on a disc D_0 surrounding p. Let D_∞ be the complement of D_0, slightly extended such that $D_0\cap D_\infty$ is an open collar containing a circle S^1. We can choose the coordinate z such that

$z = 0$ at p and $|z| = 1$ on the boundary S^1 of D_0. In addition, fix a local trivialization ξ of L (i.e., a local holomorphic section without zeros) on D_0. A section of L on D_0 can then be given by a complex valued function ϕ. Let W be the completion (with respect to L_2 inner product) of the space of smooth complex valued functions f on the circle S^1 such that $f\xi$ extends to a holomorphic section over D_∞. We shall again use the splitting of the Hilbert space H of square integrable complex valued functions on S^1 to the subspaces $H_\pm$ consisting of positive and nonpositive Fourier modes, respectively. The following theorem is proven in Segal and Wilson [1985]:

THEOREM 12.2.1. *The mapping* $(R, p, z, L, \xi) \mapsto W$ *maps the given data into the Grassmannian* Gr_1 *with respect to the splitting* $H = H_+ \oplus H_-$ *and the Fredholm index of the Grassmannian plane* W *is equal to* $(2j-1)(g-1)$, *where* g *is the genus of* R *and* $j \in \mathbf{Z}/2$ *is the spin.*

The map which assigns a point on the Grassmannian to the local data is called the Krichever map because it was first used by Krichever for constructing solutions of the KP equations by algebro-geometric methods [Krichever, 1977].

The Krichever map is not bijective but one can show that it is one-to-one between equivalence classes of local data (equivalence with respect to biholomorphic maps between surfaces) and a certain subset of the Grassmannian.

In the previous chapter we saw how solutions of the KP hierarchy could be obtained from points $W \in Gr_1$ as τ functions. Here we have a correspondence between points on the Grassmannian and Riemann surfaces. The triangle can be completed. This is the solution to the classical *Schottky problem* : How to characterize the points in the Siegel upper half space which are period matrices associated to Riemann surfaces. To explain the solution we need a few basic definitions.

For given positive integer g the *Siegel upper half space* X_g consists of symmetric complex $g \times g$ matrices Ω such that the imaginary part of Ω is positive definite. To each $\Omega \in X_g$ we associate the complex function

$$\vartheta(\mathbf{z}, \Omega) = \sum_{\mathbf{n} \in \mathbf{Z}^g} \exp(\pi \mathbf{n}^t \cdot \Omega \mathbf{n} + 2\pi i \mathbf{n} \cdot \mathbf{z}) \tag{12.2.2}$$

of the complex vector variable $\mathbf{z} \in \mathbf{C}^g$; $\mathbf{x} \cdot \mathbf{y} = x_1 y_1 + \dots x_g y_g$. If Ω is the *period matrix* of a Riemann surface R then $\vartheta(\mathbf{z}, \Omega)$ is the (generalized) *theta function* associated to R. The period matrix is defined as follows. For a given Riemann surface R of genus g there are $2g$ loops $A_1, \dots, A_g$ and $B_1, \dots, B_g$ on R such that each starts from the same base point p

and the number of intersection points of the loops (other than the base point) is given by

$$I(A_i, A_j) = 0 = I(B_i, B_j)$$
$$I(A_i, B_j) = \delta_{ij}.$$

Furthermore, the loops can be chosen such that after removing the loops R becomes a contractible set (open cell). There is a basis of holomorphic one-forms ω_i on R such that

$$\int_{A_j} \omega_i = \delta_{ij};$$

see, e.g., Mumford [1983]. The period matrix is then (with respect to the chosen *homology basis* $\{A_i, B_i\}$)

$$\Omega_{ij} = \int_{B_i} \omega_j.$$

One can show that Ω is symmetric and $\operatorname{Im}\Omega$ is positive definite. The Schottky problem is this: What characterizes the period matrices among all points in X_g? The following solution has been given in Shiota [1986] (see also Mulase [1984] for a related formulation):

THEOREM 12.2.3. *The following two conditions are equivalent:*

(1) *There exists a $g \times \mathbf{N}_+$ matrix $A = (A_{ij})$ of rank g and a quadratic form $Q(t) = \sum_{i,j \in \mathbf{N}_+} Q_{ij} t_i t_j$ ($Q_{ij} = Q_{ji} \in \mathbf{C}$) such that for any $\zeta \in \mathbf{C}^g$ the function*

$$\tau(t) = \exp(Q(t)) \cdot \vartheta(At + \zeta, \Omega)$$

is a solution of the KP hierarchy.

(2) *Ω is the period matrix for a Riemann surface of genus g.*

We have now two different ways to associate a τ function to a Riemann surface. First, we have the construction by the Theorem 12.2.3. Second, we can associate a Grassmanian plane W to the local data (R, p, z, ξ) as explained above. To each W there corresponds a function τ_W as described in Section 11.3. One might ask whether these apparently different constructions are related. The answer is yes and we shall describe the relation without giving the proofs; for the proofs we refer to Segal and Wilson [1985], Shiota [1986], and Mulase [1984].

The matrix A is determined by the expansion

$$\omega_i = (A_{i1} + A_{i2}z + A_{i3}z^2 + \dots)dz, \qquad i = 1, 2, \dots, g.$$

For each $n \in \mathbf{N}_+$ there is a unique 1-form η_n such that it has a pole of order $n+1$ at the point $p \in R$ and no other singularities, and normalized such that

$$\int^q \eta_n = z^{-n} + O(z) \text{ in a neighborhood of } p.$$

We can then expand

$$\int^q \eta_n = z^{-n} - 2\sum_{j=1}^{\infty} Q_{nj}\frac{z^j}{j}$$

in a neighborhood of $q = p$. The coefficients define the matrix Q in Theorem 12.2.3. Assuming now that $j = 1/2$, the τ function of the theorem is the *tau* function constructed in Chapter 11 from the plane W. We have to assume that the Fredholm index of W is zero in order to apply the construction of Chapter 11 and therefore we have assumed that the spin is 1/2.

12.3. Dirac operator in 3+1 space-time dimensions

Construction of the Fock bundle

Let M be a compact three dimensional spin manifold with a fixed Riemann metric and an associated spin bundle S. Let $E(0)$ be the Dirac operator in S. Next let V be a trivial vector bundle over M with a unitary action of a compact finite-dimensional gauge group G in the fibers V_x. Let $\mathbf{g}$ be the Lie algebra of G and $\mathcal{A}$ the vector space of all smooth $\mathbf{g}$-valued 1-forms A (vector potentials) on M. To each $a \in \mathcal{A}$ we associate a Dirac Hamiltonian $E(A)$ operating on sections of $S \times V$ in the standard way. The assumption about the triviality of V could be relaxed in the following discussion but we shall keep it for the sake of a technical simplification.

Let H be the space of square-integrable sections of the bundle $S \times E$. As in Section 12.1, for each $A \in \mathcal{A}$ we can write

$$H = H_+(A) \oplus H_-(A)$$

where $H_+(A)$ is the sum of eigenspaces $H(A, \lambda)$ of the Dirac Hamiltonian $E(A)$ with $\lambda > 0$ and $H_-(A)$ is the sum of the eigenspaces with $\lambda \leq 0$. We denote the subspaces $H_\pm(A)$ for $A = 0$ by $H_\pm$ and the L_4 Grassmannian defined by the splitting $H = H_+ \oplus H_-$ is Gr_2. Using the techniques of Section 6.1 one can show that the off-diagonal blocks of the operator $E(0) - E(A)$ with respect to the decomposition $H = H_+ \oplus H_-$ belong to the Schatten ideal L_4.

The mapping $\mathcal{A} \to Gr_2$, $A \mapsto H_+(A)$ is not continuous; it has discontinuities at the points in $\mathcal{A}$ where an eigenvalue of $E(A)$ crosses the point zero. This is a potential cause of trouble in the construction of the bundle of Fock spaces $\mathcal{F}_A$, $A \in \mathcal{A}$. The discontinuity can be overcome by the following trick. First define $H_\pm(A, \lambda)$ for each real number λ as the direct sum of energy eigenspaces $H(A, \mu)$ with $\mu > \lambda$ (in the case of plus sign) or with $\mu \leq \lambda$ (in the case of minus sign). Let $\mathcal{F}_{A,\lambda}$ be the Fock space defined by the splitting

$$H = H_+(A, \lambda) \oplus H_-(A, \lambda).$$

Denote by $\psi_{A,\lambda}$ the Fock vacuum of $\mathcal{F}_{A,\lambda}$. It has the property that it is annihilated by the creation operators belonging to energy levels $\mu \leq \lambda$ and by annihilation operators for energy levels $\mu > \lambda$.

PROPOSITION 12.3.1. *The Fock spaces $\mathcal{F}_{A,\lambda}$ and $\mathcal{F}_{A,\mu}$ are canonically isomorphic up to a phase factor.*

PROOF: Let $\lambda < \mu$. Choose a basis $\{e_n\}_{n \in \mathbf{Z}}$ of H such that e_n with $n \leq 0$ belongs to $H_-(A, \lambda)$, e_n with $n > N$ belongs to $H_+(A, \mu)$ and the rest form a basis of the finite-dimensional space $U = H_+(A, \lambda) \ominus H_+(A, \mu)$ and let a_i^*, a_i be the corresponding creation and annihilation operators. A general element in the Fock space is a linear combination of ordered monomials in creation and annihilation operators acting on the vacuum vector. The isomorphism $\mathcal{F}_{A,\lambda} \to \mathcal{F}_{A,\mu}$ is obtained by sending the vector

$$a_{i_1}^* \ldots a_{i_p}^* a_{j_1}^* \ldots a_{j_q}^* a_{k_1} \ldots a_{k_r} \psi_{A,\lambda},$$

where $i_1 > \ldots i_p > N \geq j_1 > \cdots > j_q > 0 \geq k_1 > \cdots > k_r$, to the vector

$$a_{i_1}^* \ldots a_{i_p}^* a_{l_1} \ldots a_{l_{N-q}} a_{k_1} \ldots a_{k_r} \psi_{A,\mu},$$

where $l_1 > \cdots > l_{N-q}$ is the complement of $\{j_1, \ldots, j_q\}$ in $\{1, 2, \ldots, N\}$. In a change of basis this isomorphism is multiplied by $\det T$ where T is the matrix representing a change of orthonormal basis in U.

For any $A_0 \in \mathcal{A}$ we can choose λ such that it is not an eigenvalue of $E(A_0)$; it is also not an eigenvalue of $E(A)$ when A is in some open neighborhood O of A_0. It follows that the projective space $P(\mathcal{F}_A)$ of complex lines in $\mathcal{F}_A$ is well-defined and varies continuously as a function of the parameter $A \in O$. Thus there is no problem in defining the bundle of projective Fock spaces over $\mathcal{A}$. Since $\mathcal{A}$ as a vector space is contractible one can show that there is a trivial vector bundle $\mathcal{F}$ over $\mathcal{A}$ and a projective isomorphism between $P(\mathcal{F})$ and the projectivization of $\mathcal{F}$.

In the following we want to avoid working with projective Fock spaces and therefore, as in Section 12.1, we consider the Fock spaces parametrized by the elements of Gr_2 and not by vector potentials; we would be tempted to take as a basic object the trivial Fock bundle over Gr_2, with each fiber isomorphic to the Fock space $\mathcal{F}_0$, say. Then the discontinuities would be buried in the discontinuity of the mapping $\mathcal{A} \rightarrow Gr_2$; the pull-back bundle over $\mathcal{A}$ should be considered as a bundle of projective Fock spaces as explained above. However, as in the case in Section 12.1 we shall see that this is wrong: The correct object will be a twisted Fock bundle over Gr_2. (Again, as in Section 12.1, the non-triviality of a Fock bundle must be understood with respect to the structure group GL^2.)

Let $F \in Gr_2$. A (generalized) Fock space $\mathcal{F}_F$ relative to the splitting $H = F \oplus F^\perp$ is defined by a choice of basis $\{f_1, f_2, \dots\}$ in F. We want the different Fock spaces to be in a certain sense comparable. For that reason we fix once and for all a basis $\{e_1, e_2, \dots\}$ in H_+ and we shall require that the basis $\{f_n\}$ be admissible relative to $\{e_n\}$, that is, the orthogonal projection $f^{(e)}$ of the f-basis onto the e-basis is in $1 + L_2$,

$$pr_{H_+}(f_n) = \sum_j f_{jn}^{(e)} e_j. \tag{12.3.2}$$

For each given $\{f_n\}$ we can define the Fock space $\mathcal{F}_F$ to consist of sections of the correspoding dual determinant bundle DET_2^*. A section is a function $\psi(w)$ in the space of all admissible basis St_2 satisfying

$$\psi(wt) = \psi(w) \cdot \omega_2(w^{(f)}, t), \ t \in GL^2, \tag{12.3.3}$$

where $w^{(f)}$ is the matrix relating the F-projection to the basis $\{f_n\}$, i.e., $w^{(f)}$ is defined analogously to (12.3.2),

$$pr_F(w_n) = \sum_j w_{jn}^{(f)} f_j. \tag{12.3.4}$$

Exercise 12.3.5. Show that a basis w of a plane $U \in Gr_2$ is admissible with respect to the basis e of H_+ if and only if it is admissible with respect to any $f \in St_2$.

The problem with the above construction of the Fock spaces $\mathcal{F}_F$ is that it depends of a choice of admissible basis f in each $F \in Gr_2$. What we have constructed is in fact a bundle over St_2 and not over Gr_2. If we could choose a basis f_F of F as a continuous function of F we would have a bundle over Gr_2; however, that is not possible since the GL^2 bundle St_2 over Gr_2 is nontrivial. Since the definition of a section ψ depends on f we shall write explicitly $\psi = \psi(w,f)$. Thinking of ψ also as a function of f we have introduced additional degrees of freedom to the wave function; these new degrees of freedom (at a fixed F) are parametrized by the fiber of the projection $\pi : St_2 \to Gr_2$ at $F \in Gr_2$. The requirement $\psi(w,f) = \psi(w,f')$ for $\pi(f') = \pi(f)$ would be in conflict with (12.3.3). What we can require is that

$$\psi(w,ft) = \psi(w,f) \cdot \omega'(w,f;t), \;\; t \in GL^2, \tag{12.3.6}$$

where ω' is some cocycle. What is the precise form of ω'?

Comparing

$$\psi(wt_1, ft_2) = \psi(wt_1,f)\omega'(wt_1,f;t_2) = \psi(w,f)\omega_2(w^{(f)},t_1)\omega'(wt_1,f;t_2)$$

with

$$\psi(wt_1,ft_2) = \psi(w,ft_2)\omega_2(w^{(ft_2)},t_1) = \psi(w,f)\omega_2(w^{(ft_2)},t_1)\omega'(w,f;t_2)$$

we observe that for consistency ω' must satisfy

$$\omega'(w,f;t_2)\omega_2(w^{(ft_2)},t_1) = \omega'(wt_1,f;t_2)\omega_2(w^{(f)},t_1) \tag{12.3.7}$$

for all $t_1, t_2 \in GL^2$ and $w, f \in St_2$. We claim that

$$\omega'(w,f;t) = \omega_2(w^{(f)},t^{-1}) \tag{12.3.8}$$

is a solution of (12.3.7). To prove this all we need is to recall the definition

$$\omega_2(w^{(f)},t) = \frac{\det_2 w^{(f)}t}{\det_2 w^{(f)}}$$

and to observe that $w^{(ft)} = t^{-1}w^{(f)}$ and $(wt)^{(f)} = w^{(f)}t$; (12.3.7) is now seen to hold by a substitution from (12.3.8).

Next we want to show that the functions $\psi : St_2 \times St_2 \to \mathbf{C}$ satisfying (12.3.3) and (12.3.6) can be thought of as sections of a vector bundle $\mathcal{F}'$ over Gr_2 which is a tensor product of the determinant bundle DET_2 and a trivial Fock bundle $\mathcal{B}$ (with fiber $\mathcal{F}_{H_+}$) over Gr_2; the situation here is parallel to what we had in Section 12.1 but we want to go through the details because there are some extra complications arising from the generalized determinants.

To prove the claim above we construct a function $\xi : St_2 \times St_2 \to \mathbf{C}^\times$ such that for any ψ satisfying (12.3.3) and (12.3.7) the product $\psi' = \xi\psi$ satisfies the following conditions

$$\begin{aligned} \psi'(wt, f) &= \psi(w, f) \cdot \omega_2(w_+, t) \\ \psi'(w, ft) &= \psi'(w, f) \cdot \omega_2(f_+, t)^{-1}, \end{aligned} \tag{12.3.9}$$

where we have written, in order to conform with the notation of Section 6.2, $w_+ = w^{(e)}$, $f_+ = f^{(e)}$. The first condition says that for a fixed f the function ψ is a section of DET_2^* and therefore can be identified as an element of the Fock space $\mathcal{F}_{H_+}$. The second condition is just the defining relation for a section of the bundle DET_2.

For finding ξ we make first the following observations. Let us for simplicity work with unitary bases only ($\{e_n\}, \{w_n\}, \{f_n\}$ orthonormal). Let $\{e_n\}_{n\in -\mathbf{N}}$ be a basis of H_-. As in Chapter 6, the matrix elements of w_- are determined by

$$pr_{H_-} w_n = \sum_{j<0} (w_-)_{jn} e_j$$

and similarly for f_-. Then

$$w^{(f)} = f_+^* w_+ + f_-^* w_- = f^* w$$

and the function $\psi_0(w, f) = \det_2 f^* w$ is easily seen to satisfy the conditions (12.3.3) and (12.3.6). Consider the following function, familiar from Section 6.5,

$$\phi(w, f) = \det_2(f^* w) \cdot \exp[\operatorname{tr}(f^* w - f_+^* - w_+ + 1 - \tfrac{1}{4} F_{21}^* W_{21})],$$

where

$$W = \begin{pmatrix} W_{11} & W_{12} \\ W_{21} & W_{22} \end{pmatrix}$$

is the Hermitian operator representing the plane $W = \operatorname{Span}(w_1, w_2, \dots)$ and F is the corresponding matrix for the plane spanned by the f_n's. ϕ satisfies the following conditions:

$$\phi(wt, f) = \phi(w, t)\omega_2(w_+, t), \;\; \phi(w, ft) = \phi(w, f)\overline{\omega_2(f_+, t)}. \tag{12.3.10}$$

The ratio $\phi(w,f)/\det_2(f^*w)$ is essentially the function we are looking for. Namely, the cocycle $\overline{\omega_2(f_+,t)}$ in (12.3.10) is cohomologous to the cocycle $\omega_2(f_+,t)^{-1}$ in (12.3.9).

Exercise 12.3.11. Show that there is a function $\beta : St_2 \to \mathbf{C}^\times$ such that

$$\omega_2(f_+,t)^{-1} = \frac{\beta(ft)}{\beta(f)}\overline{\omega_2(f_+,t)}\,\forall\, t \in GL^2.$$

A solution for ξ is equal to $\phi(w,f)\beta(f)/\det_2(f^*w)$, where β is a solution of the preceding exercise.

Group actions in the Fock bundle

The section $\psi_0 = \det_2(f^*w)$ is the vacuum vector in the space of sections $\Gamma(\mathcal{F}')$. At the point $(f_+,f_-) = (1,0)$ (which is the basis $\{e_n\}$ of H_+) we have $\psi_0 = \det_2 w_+$; this is the unique ray in $\Gamma(DET_2^*)$ which is left invariant by the subgroup $N \subset \widehat{GL}_2$ consisting of triples (n,a,λ) with

$$n = \begin{pmatrix} a & 0 \\ c & d \end{pmatrix} \in GL_2.$$

The matrix n is written with respect to the decomposition $H = H_+ \oplus H_-$. When f is at the general position the section $w \mapsto \psi(w,f)$ of $DET_2^*(F)$ defines the ray which is invariant under the conjugated group gNg^{-1}, where g is an element of GL_2 which maps the plane H_+ onto F. In other words, gNg^{-1} consists of triples (n,a,λ), where n is of the triangular form as above but now with respect to the decomposition $H = F \oplus F^\perp$.

We have in fact two different groups $\widehat{GL}_2$ acting in $\mathcal{F}'$. The first one, to be denoted by G_l, acts only on the variable w in $\psi(w,f)$,

$$[T_l(g,q,\lambda)\psi](w,f) = \alpha_l(g,q;w,f)\psi(g^{-1}wq) \tag{12.3.12}$$

and the second one, to be denoted by G_r acts on the second variable,

$$[T(g,q,\lambda)\psi](w,f) = \alpha_r(g,q;w,f)\psi(w,g^*fq). \tag{12.3.13}$$

Exercise 12.3.14. Construct the factors α_l and α_r such that the right-hand-sides in (12.3.12) and (12.3.13) satisfy the conditions (12.3.3) and (12.3.6). Hint: Use (12.3.10) and the transformation $\psi \mapsto \xi\psi$.

There is a group $G \simeq GL_2$ which acts "diagonally" in $\mathcal{F}'$. For any $g \in GL_2$ choose $q \in GL(H_+)$ such that $aq^{-1} - 1 \in L_2$ and put

$$[T(g)\psi](w,f) = \psi(g^{-1}wq, g^*fq). \tag{12.3.15}$$

By (12.3.3) and (12.3.6) the right-hand side does not depend on the choice of q and therefore we have a representation of G in $\mathcal{F}'$.

The physical interpretation of the groups G_l, G_r and G is the following. For a fixed value of the "background" field f [the vector potentials are represented by points $F = \pi(f)$ on the Grassmannian] the elements of G_l represent the effect of a gauge transformation in the fermionic Fock space $\Gamma(DET_2^*)$ at f. The action of G_r gives then naturally the action of gauge transformations on vector potentials. The diagonal group G is the symmetry group of the coupled Dirac-Yang-Mills system.

We can now check that our vacuum section $\psi_0 \in \mathcal{F}'$ is invariant under the group G. The matrix $w^{(f)}$, representing the orthogonal projection of a system of vectors $w_1, w_2, \ldots$ to a system $f_1, f_2, \ldots$, is related to the orthogonal projection $w'^{(f')}$ of the transformed system $w_n' = (g^{-1}w_j)q_{jn}$ to the system $f_n' = (g^* f_j)q_{jn}$ by $w'(f') = q^{-1}w^{(f)}q$, and thus $\psi_0(w,f) = \det_2 w^{(f)}$ is invariant.

The vacuum ψ_0 is the only vector which is invariant under G. First, because GL_2 acts transitively on Gr_2 the values of an invariant function ψ are completely determined by giving the function $w \mapsto \psi(w,e)$. On the other hand the basis e is left fixed by the transformation $e \mapsto g^*eq$ when

$$g = \begin{pmatrix} a & b \\ 0 & d \end{pmatrix}$$

and $q = (a^*)^{-1}$. Therefore we must have $\psi(g^{-1}wq, e) = \psi(w,e)$ for these transformations; but we can choose $w \in St_2$ such that the orbit of the triangular group through w is dense and therefore ψ is completely determined by its value at the point (w,e). However, if we restrict to the unitary subgroup $U_2 \subset G$ then the space of U_2 invariant vectors in $\mathcal{F}'$ is infinite-dimensional. For example, all vectors of the form $\psi = h(W,F)\psi_0(w,f)$ are invariant, where h is any U_2 invariant function of F and W. The space of invariant functions is infinite-dimensional. For example, all polynomials of the function $\det_4 WF^*$ are invariant, since $W \mapsto g^{-1}Wg$ and $F \mapsto g^*Fg*^{-1}$ for $g \in U_2$. The infinite determinant converges because $W = \epsilon + W'$ and $F = \epsilon + F'$ where $W', F' \in L_4$; now $WF^* = 1 + W'\epsilon + \epsilon F'^* + W'F'^* \in 1 + L_4$. The interest in the unitary subgroup $U_2 \subset GL_2$ lies in the fact that the gauge transformations are contained in U_2.

The space of holomorphic sections

A basis for the sections $\psi(w,f)$ of $\mathcal{F}'$ which are holomorphic in w and *antiholomorphic* in f is constructed as follows. Let $S : H \to H$ be any linear operator of which differs from 1 type a finite rank operator. The function

$$\psi_S(w,f) = \det_2(f^*Sw)\cdot \exp \operatorname{tr}(f^*Sw - f^*w) = \det(f^*Sw)\cdot \exp \operatorname{tr} f^*w$$

satisfies (12.3.3) and (12.3.6) and therefore $\psi_S \in \mathcal{F}'$. Taking linear combinations of the ψ_S's with complex coefficients one obtains a dense subspace of the space of bi(anti)holomorphic sections, with respect to the topology of uniform convergence on compact subsets. The vacuum section is obtained by setting $S = 1$.

Let $\mathcal{F}_{hol}$ be the subbundle of $\mathcal{F}'$ such that the fiber of $\mathcal{F}_{hol}$ at $F \in Gr_2$ consists of all holomorphic sections of the bundle $DET_2^*(F)$. A section of $\mathcal{F}_{hol}$ is a linear combination of functions

$$\psi(w,f) = \lambda(F)\cdot \psi_S(w,f) \tag{12.3.16}$$

where λ is an arbitrary complex valued smooth function on Gr_2. The group U_2 acts in $\mathcal{F}_{hol}$ as follows. Let $g \in U_2$, $F \in Gr_2$, and $f \in St_2$ be an admissible basis of F. Let $w \mapsto \psi_S(w,f)$ be a typical element in the fiber over F in $\mathcal{F}_{hol}$. By (12.3.15) this vector over F is mapped to the function

$$\psi'(w,h) = \psi_S(g^*wq, g^*hq) = \det(q^*h^*gSg^*wq)\cdot \exp(-\operatorname{tr} h^*w)$$

$$= \psi_{gSg^*}(h,w) \tag{12.3.17}$$

where $h = gfq^{-1}$ is an admissible basis of gFg^{-1}.

The range of S can be extended beyond 1+ finite rank operators. Clearly ψ_S is well-defined if and only if $K = f^*Sw - f^*w$ is a trace class operator. However, we can improve the convergence of the trace by subtracting from K the operator $S_{11} - 1$; this is due to the fact that w_+ and f_+ are of the form 1+ Hilbert-Schmidt operator. We have decomposed the operators into blocks with respect to the splitting $H = H_+ \oplus H_-$. Now

$$K = f_+^*(S_{11}-1)w_+ + f_+^*S_{12}w_- + f_-^*S_{21}w_+ + f_-^*(S_{22}-1)w_-.$$

For a given S the four different terms are linearly independent functions of w and f and therefore their traces should be finite separately. The first term is finite after subtraction of $S_{11} - 1$ if and only if $S_{11} - 1$ is a Hilbert-Schmidt operator whereas the middle terms are finite if and only if S_{12} and S_{21} are in $L_{4/3}$. The trace-class condition on the last term means that S_{22} has to be Hilbert-Schmidt. Thus the "renormalized" holomorphic section of $DET_2^*(F)$ takes the form

$$\det(f^* S w) \cdot \exp[-\mathrm{tr}(f^* w + S_{11} - 1)]$$

which really should be read as

$$\det{}_2(f^* S w) \cdot \exp[\mathrm{tr}(f^* S w - f^* w - S_{11} + 1)]. \tag{12.3.18}$$

In the following we shall denote by ψ_S the renormalized section in (12.3.18). Note that the multiplication of $\psi_S(\cdot, w)$ by the factor $\exp[\mathrm{tr}(1 - S_{11})]$ has an effect on the U_2 action in the fibers of $\mathcal{F}_{hol}$: in addition to the transformation of the parameter $S \mapsto gSg^*$ as the fiber $\mathcal{F}_{hol}(F)$ is mapped onto the fiber $\mathcal{F}_{hol}(gFg^*)$ there is a multiplication of the sections by the factor $\exp[(gSg^*)_{11} - S_{11}]$.

Exercise 12.3.19. Show directly that $(gSg^*)_{11} - S_{11}$ is a trace-class operator for all $g \in U_2$ and $S - 1 \in \begin{pmatrix} L_2 & L_{4/3} \\ L_{4/3} & L_2 \end{pmatrix}$.

Group actions in $\mathcal{F}_{hol}$

In the case of DET_1^* the holomorphic sections (i.e., vectors in the fermionic Fock space) in the $\widehat{U}_1$ orbit through the vacuum, modulo the action of the center, formed a Grassmannian of the type Gr_1. In the situation at hand each fiber of $\mathcal{F}_{hol}$ is isomorphic to the space of holomorphic sections of the bundle DET_2^*. Since DET_1^* is a dense subspace of DET_2^* the space $\Gamma_{hol}(DET_2^*)$ can be identified as a subspace of $\Gamma_{hol}(DET_1^*)$. As we have seen in Chapter 6 the group $\widehat{GL_2}$ does not act in $\Gamma_{hol}(DET_2^*)$. However, there is a smaller group whichs acts holomorphically; this group is (a central extension of) $GL_{2/3}$. Remember that the reason for the break-down of the holomorphicity is the regularization term $\mathrm{tr} q^{-1} b F_{21}$ in the formula

$$(g, q, 1) \cdot (w, \lambda) = (gwq^{-1}, \lambda e^{-\mathrm{tr}[(1-q^{-1}a)(w_+ - 1) + q^{-1} b(\frac{1}{2} F_{21} - w_-)]})$$

defining the action of $\widehat{GL_2}$ in DET_2. Assuming now that

$$g = \begin{pmatrix} a & b \\ c & d \end{pmatrix}$$

is in $GL_{2/3}$, i.e., $b, c \in L_{4/3}$, we can remove the regulator since it is finite ($F_{21} \in L_4$ and therefore bF_{21} is a trace-class operator). Thus we have a representation of the *central* extension $\widehat{GL}_{2/3}$ in the space $\Gamma_{hol}(DET_2^*)$ by the formula

$$(12.3.20) \qquad [T(g,q,\lambda)\psi](w) = \lambda(g^{-1}Fg) \cdot e^{-\mathrm{tr}[(1-q\alpha)(w_+-1)-q^{-1}bw_-]}$$

where $\begin{pmatrix} \alpha & \beta \\ \gamma & \delta \end{pmatrix} = g^{-1}$. Clearly $GL_{2/3}$ is the maximal GL_p group which acts in the space of holomorphic sections.

The stability subgroup in $\widehat{GL}_{2/3}$ which leaves the vacuum $\psi(w) = \det_2 w_+$ invariant consists of the group elements $(g,q,1)$ with $b=0$, $q = a$. It follows that the $\widehat{GL}_{2/3}$ orbit through the vacuum can be projectively parametrized by the Grassmannian $Gr_{2/3}$. Despite the fact that each fiber of $\mathcal{F}_{hol}$ is isomorphic as a vector space with the space $\Gamma_{hol}(DET_2^*)$, the group $\widehat{GL}_{2/3}$ does not act in $\mathcal{F}_{hol}$. The reason is that the bundle $\mathcal{F}_{hol}$ is a twisted bundle over Gr_2. For each $F \in Gr_2$ there is the group $\widehat{GL}_{2/3}(F)$ which is the central extension of the group $GL_{2/3}(F)$ consisting of matrices $h = gh'g^{-1}$ where $h' \in GL_{2/3}$ (the group $GL_{2/3}$ is defined with respect to the fixed splitting $H = H_+ \oplus H_-$) and $g \in GL_2$ is any element such that $F = g\epsilon g^{-1}$; in other words, $GL_{2/3}$ consists of bounded invertible operators in H such that $[h,F] \in L_{4/3}$. The group $GL_{2/3}(F)$ acts in the fiber $\mathcal{F}_{hol}(F)$ by the formula (12.3.20), but there is no natural way to identify $GL_{2/3}$ with $GL_{2/3}(F)$ since the isomorphism of the groups depends on the choice of g.

Let $GL_{2,2/3}$ be the subgroup of $GL_{2/3}$ consisting of elements $g = \begin{pmatrix} a & b \\ c & d \end{pmatrix}$ such that $a-1$ and $d-1$ are Hilbert-Schmidt operators. The central extension of this subgroup acts in $\mathcal{F}_{hol}$. Namely, it is easily seen that $GL_{2,2/3}$ is a *normal subgroup* of GL_2 and therefore $GL_{2,2/3} \subset GL_{2/3}(F)$ for *any* $F \in Gr_2$.

Exercise 12.3.21. Show that $GL_{2,2/3} = \bigcap_{F \in Gr_2} GL_{2/3}(F)$.

We have now a representation of $GL_2 \times \widehat{GL}_{2,2/3}$ in $\Gamma(\mathcal{F}_{hol})$. The interpretation of the various spaces and groups is the following. For a given sign operator $F \in Gr_2$, associated to some Dirac operator in three space dimensions, we have a "squeezed" Fock space $\mathcal{F}_{hol}(F)$; the Fock space is squeezed because there are less holomorphic sections in DET_2^* than in DET_1^*. The elements of the group GL_2 represent the combined gauge transformations acting both in the base Gr_2 (which can be thought of roughly as the space of vector potentials) and in the fibers. The elements of $GL_{2/3}$ correspond to "pure fermionic currents"; the parentheses refer to the fact that the (time) components of the smooth currents are really elements of the Lie algebra of the bigger group GL_2.

The standard Fock bundle $\mathcal{F}^{(2)}$

The "standard" twisted Fock bundle $\mathcal{F}^{(2)}$ over Gr_2 is constructed as follows. The fibers of $\mathcal{F}^{(2)}$ are isomorphic with the standard Fock space $\mathcal{H}$ which is the space of holomorphic sections of DET_1^*. We define the fiber $\mathcal{F}_F^{(2)}$ at $F \in Gr_2$ to be the Fock space created from the vacuum by the canonical anticommutation relations, the vacuum being a vector v_F which is annihilated precisely by the operators $a(u)$ for $u \in F$ and by the creation operators $a^*(u)$ for $u \in F^\perp$. The canonical anticommutation relations are

$$[a^*(u), a(u')]_+ = (u, u'), \ [a(u), a(u')]_+ = [a^*(u), a^*(u')]_+ = 0,$$

where (u, u') is the inner product of vectors $u, u' \in H$. It is known that the representations defined by the planes F and $g \cdot F$ are equivalent if and only if $g \in GL_1$; see Araki [1987] for a detailed discussion of representations of the CAR algebra. By construction, the bundle $\mathcal{F}^{(2)}$ has a nowhere vanishing section of the vacuum subbundle. However, this bundle is twisted in the same way as the bundles $\mathcal{F}'$ and $\mathcal{F}_{hol}$. In fact, this happens already in the case of Gr_1; in that case the various Fock spaces parametrized by elements of Gr_1 are equivalent as representation spaces of CAR but the equivalence is canonically determined only up to a phase (the phase being related to the phase arising from the central extension of GL_1). The all important difference in the case Gr_2 as compared to Gr_1 is the fact that the various Fock spaces are no longer even projectively equivalent.

Next we want to translate the above remarks about the construction of $\mathcal{F}^{(2)}$ to the language of determinant bundles in order to obtain a more geometric understanding of $\mathcal{F}^{(2)}$. Let $St_{p,q}$ denote the set of pairs (w, f) where $f \in St_q$ (relative to the fixed basis $e_1, e_2, \ldots$ of H_+) and w is an admissible basis relative to f in the L_p sense; this means that

$$pr_F \, w_i = \sum_j a_{ji} f_j$$

where F is the Grassmannian plane spanned by the vectors f_i and $a - 1$ is an operator in L_p. (As before, we denote $a = w^{(f)}$.) We define the bundle $\mathcal{F}^{(2)}$ indirectly by proclaiming that the space of sections of $\mathcal{F}^{(2)}$ consists of complex valued functions ψ on $St_{1,2}$ which are holomorphic in the variable w and such that

$$\begin{aligned} &\psi(wt, f) = \psi(w, f) \cdot \det t, \ t \in GL^1 \\ (12.3.22) \qquad &\psi(wq, fq) = \psi(w, f), \ q \in GL^2 \end{aligned}$$

An example of such a section is the function $\psi_0(w, f) = \det w^{(f)}$. The group GL_2 acts (without the extension) in the space $\Gamma(\mathcal{F}^{(2)})$ by

$$[T(g)\psi](w, f) = \psi(g^{-1}wq, g^* fq), \tag{12.3.23}$$

where $q \in GL(H_+)$ such that $a - q \in L_2$. The right-hand side of (12.3.23) does not depend on the choice of q by the second relation in (12.3.22).

For any given $F \in Gr_2$, after choosing an admissible basis f of F, the fiber of $\mathcal{F}_{hol}$ is isomorphic with the space of holomorphic sections of DET_1^*. However, the isomorphism $\Gamma_{hol}(DET_1^*) \simeq \mathcal{F}_F^{(2)}$ depends on the choice of f. For this reason we cannot trivialize the bundle $\mathcal{F}^{(2)}$. For a similar reason the group $\widehat{GL_1}$ does not act in the total space of the bundle; the group acts in a fiber $\mathcal{F}_F^{(2)}$ only after choosing an admissible basis f of F. The group $GL_1(F)$ consisting of operators with L_2 off-diagonal blocks with respect to the decomposition $H = F \oplus F^\perp$ is isomorphic with the group GL_1 (which is defined with respect to the standard splitting $H = H_+ \oplus H_-$) but the isomorphism $GL_1 \to GL_1(F)$, $h \mapsto ghg^{-1}$ depends on the choice of the element $g \in GL_2$ such that $F = g \cdot H_+$; the central extension of $GL_1(F)$ acts in the fiber $\mathcal{F}_F^{(2)}$ but there is no way to choose g as a continuous function of $F \in Gr_2$ and therefore $\widehat{GL_1}$ does not act in $\mathcal{F}^{(2)}$.

The CAR algebra in $\mathcal{F}^{(2)}$

Next we define a representation of the CAR in the space of sections $\Gamma(\mathcal{F}^{(2)})$. Let $F \in Gr_2$ and let $f \in St_2$ be an admissible basis of F. For any frame w' such that $(w', f) \in St_{1,2}$ we can construct a holomorphic section of $DET_1^*(F)$ by

$$\psi_{w'}(w) = \det w^{(w')}, \ (w, f) \in St_{1,2}. \tag{12.3.24}$$

The fiber $\mathcal{F}_F^{(2)}$ is then the completion of the space of linear combinations of the vectors $\psi_{w'}$ with respect to the inner product

$$(\psi_{w'}, \psi_{w''}) = \det(w'^* w'').$$

The construction of the fiber $\mathcal{F}_F^{(2)}$ does not depend essentially on the choice of f; if $\hat{f} = fq$ for some $q \in GL^2$ then the section $\psi_{w'}$ at f is mapped to the section $\hat{\psi}$,

$$\hat{\psi}(w) = \psi_{w'q}(w), \ (w, \hat{f}) \in St_{1,2}.$$

The action of the creation operator on a vector $\psi_{w'} \in \mathcal{F}_F^{(2)}$ is now defined by

$$a^*(u)\psi_{w'} = \psi_{u\cup w'} \tag{12.3.25}$$

where for any $u \in H$, $u \cup w'$ denotes the frame $\{u, w_1, w_2, \dots\}$. The annihilation operators are defined by

$$a(u)\psi_{w'} = \sum_{i=1,2,\dots} (-1)^{i-1}(u, w_i') \cdot \psi_{w'(i)} \tag{12.3.26}$$

where $w'(i)$ is the frame obtained from w' by deleting the vector w_i'.

Exercise 12.3.27. Show that the operators defined above indeed satisfy the canonical anticommutation relations.

In the case when $f = \{e_1, e_2, \dots\}$ we can recover some familiar formulas. First, setting $w' = f$ the formula (12.3.24) gives the standard form $\psi(w) = \det w_+$ of the vacuum vector ψ_0 in the fermionic Fock space. Denoting $a_i = a(e_i)$ and $a_i^* = a^*(e_i)$ then the Fock space vector

$$a_{i_1}^* \dots a_{i_n}^* a_{j_1} \dots a_{j_m} \psi_0$$

represents the state with holes in the negative energy sea at the levels $0 \geq j_1 > \dots > j_m$ and the positive energy states filled corresponding to the energy levels $i_1 > \dots > i_n$. This vector is represented by the determinant $\det w_S$, where $S = \{j_1, \dots, j_m\} \cup \{i > 0 \mid i \neq i_k; 1 \leq k \leq n\}$ and w_S is the matrix obtained from the $\mathbf{Z} \times \mathbf{N}$ matrix w (in the basis e) by selecting the rows labelled by S.

We know that the representations of CAR in the fibers $\mathcal{F}_F^{(2)}$ and $\mathcal{F}_{F'}^{(2)}$ are not equivalent when F and F' are not related by a GL_1 transformation; on the other hand, any element $g \in GL_2$ which maps F to F' induces a linear isomorphism of the fiber $\mathcal{F}_F^{(2)}$ onto $\mathcal{F}_{F'}^{(2)}$. There is no contradiction since the action of GL_2 in $\mathcal{F}^{(2)}$ is *not equivariant* with respect to the action of CAR; it is *not true* that $T(g)a(u)T(g^{-1}) = a(g \cdot u)$ for $u \in H$ and $g \in GL_2$.

As in the case of $\mathcal{F}_{hol}$ the subgroup $\widehat{GL}_{2/3} \subset \widehat{GL}_1$ acts in the total space of $\mathcal{F}^{(2)}$. Namely, let $(w, f) \in St_{1,2}$ and $h \in \widehat{GL}_{2,2/3}$. Then also $(hwq^{-1}, f) \in St_{1,2}$, when q is chosen such that $a - q \in L_1$; we have written $h \in \begin{pmatrix} a & b \\ c & d \end{pmatrix}$ in the splitting $H = H_+ \oplus H_-$. We shall prove this only in the case when f is orthonormal, the general case being technically slightly more complicated but essentially straightforward. We have to show that $f^* hwq^{-1} - 1$ is a trace-class operator; by assumption $f^* w - 1$ is in L_1 and therefore

$$f^* hwq^{-1} - 1 = f_+^*(aw_+ q^{-1} - w_+) + f_+^* bw_- q^{-1} + f_-^* cw_+ q^{-1} + f_-^*(dw_- q^{-1} - w_-) \bmod L_1.$$

Since f_- and w_- are in L_4 and $b, c \in L_{4/3}$ the two terms in the middle are trace-class operators. The first term is in L_1 since $aq^{-1} - 1 \in L_1$. By $d - 1 \in L_2$ the last term is of the form $f_-^* w_-(q^{-1} - 1)$ + a trace-class operator. But $f_-^* w_- \in L_2$ and $q^{-1} - 1$ is a Hilbert-Schmidt operator as a consequence of $a - 1 \in L_2$ and $a - q \in L_1$. We can now define the action of $\widehat{GL}_{2/3}$ in $\Gamma(\mathcal{F}^{(2)})$ by

(12.3.28) $$[T(h, q, \lambda)\psi](w, f) = \lambda^{-1}\psi(h^{-1}wq, f).$$

Concluding remarks

We emphasize the fact that the fermionic charge densities (= the infinitesimal gauge transformations acting on spinor fields) are not represented in the standard Fock bundle $\mathcal{F}^{(2)}$. These operators are represented in the space of sections of the bundle $\mathcal{F}'$. However, the particle creation and annihililation operators do not act in $\Gamma(\mathcal{F}')$, not even in the space $\Gamma(\mathcal{F}_{hol})$; from the point of view of the CAR algebra the topology of the fibers of $\mathcal{F}_{hol}$ is wrong: The CAR algebra acts in a Hilbert space but the fibers of $\mathcal{F}_{hol}$ are modelled by Schatten ideals L_4.

On the basis of our discussion it seems that we either have to abandon the particle interpretation of the states (sections of the generalized Fock bundle) or to abandon the idea that the charge densities should be represented as Hermitian operators.

We have not discussed at all the quantization of the Yang-Mills Hamiltonian in 3+1 dimensions. The techniques of Section 12.1 cannot be applied here. There seems to be no way to interpret the Hamiltonian as a quadratic Casimir operator. The action of the Casimir operator diverges even after normal ordering.

Another open problem which we have not touched is the question of generalizing the Sugawara construction of the Virasoro algebra to 3+1 dimensions. The Virasoro algebra is an extension of the Lie algebra of vector fields on the circle. In the case of the Hilbert-Schmidt group GL_1 we can define

$$L_n = -\sum_{k \in \mathbf{Z}} kE_{n+k,k}$$

as the generators of the Virasoro algebra acting in the fermionic Fock space. In fact, also in the case of GL_2 we can use the same formula to define a generalized Virasoro algebra. The commutation relations are now

$$[L_n, L_m] = (n-m)L_{n+m} + \eta(L_n, L_m) \tag{12.3.29}$$

where η is the cocycle defined in Section 6.3 and it is of the form $\operatorname{tr}(F-\epsilon)T_{nm}$, where T_{nm} is a matrix of finite rank. However, when living in S^3 instead of S^1, the algebra (12.3.29) is not really what we want. We want to construct a representation of the Lie algebra of vector fields on S^3, acting in the space $\Gamma(DET_2^*)$. In the one-dimensional case the success of the Virasoro construction can be traced back to the fact that $\mathit{Diff}\, S^1$ is a subgroup of GL_1. The group of diffeomorphisms of S^3 (or of any other higher dimensional manifold) is not a subgroup of any of the restricted groups GL_p. For this reason there is no natural way to represent the diffeomorphisms as linear operators in $\Gamma(DET_2^*)$.

12.4. A universal Yang-Mills theory

As we have seen, a smooth vector potential determines a point on a Grassmannian Gr_p. The plane $W \in Gr_p$ corresponding to a vector potential $A \in \mathcal{A}$ in a vector bundle over a d dimensional compact spin manifold M is the space of positive energy solutions of the corresponding Dirac operator. The Schatten index p is related to the dimension by $2p = d+1$. In this section we shall generalize this setting.

The off-diagonal blocks, with respect to the splitting $H = H_+ \oplus H_-$, of the sign operator F parametrizing a point in Gr_p are in L_{2p}. By $F^2 = 1$ the diagonal blocks of $F - \epsilon$ are in L_p, where ϵ is the sign operator associated to H_+. We shall formulate a "universal Yang-Mills Hamiltonian" as a functional in the cotangent bundle T^*M of an infinite-dimensional manifold M containing the Grassmannian Gr_p. The points of M are the operators $F-\epsilon$, but we omit the condition that F^2 should be equal to 1. Thus M is the linear space consisting of all linear Hermitian operators A in H such that the diagonal blocks of A are in L_p and the off-diagonal blocks are in L_{2p}. Points of M play the role of vector potentials.

The electric field E is the canonically conjugate variable to A, and thus E is interpreted as an element of the dual space M^*; since M is a linear space, the cotangent space at each point can be identified as M^*. The vectors $E \in M^*$ are linear operators in H such that the diagonal blocks are in $L_{p/p-1}$ and the off-diagonal blocks in $L_{2p/2p-1}$. The pairing between vectors $A \in M$ and the dual vectors is given by $(E, A) = \operatorname{tr} EA$.

The Hamiltonian of the Yang-Mills field is normally the integral of the trace of $\frac{1}{2}(E^2 + B^2)$. In the present setting $\int \mathrm{tr}$ will be replaced by the infinite-dimensional trace in H. Thus the electric part of the Hamiltonian is

$$\frac{1}{2}\mathrm{tr}E^2.$$

This converges for all $E \in T^*M$ if and only if $p \geq 2$.

A point F on the Grassmannian G_p can be written as $F = g\epsilon g^{-1}$ for some $g \in GL_p$. Thus

$$F - \epsilon = [g, \epsilon]g^{-1}.$$

Think of the elements of GL_p as generalized gauge transformations. If we interprete the commutator $[g, \epsilon]$ as dg, the equation above says that the vector potential $A = F - \epsilon$ is an exact gauge. With these definitions the points in $Gr_p \subset M$ are precisely all the exact gauges.

In Connes' noncommutative geometry the exterior derivative of a zero form (like g above) is the commutator with ϵ; the same is true for all forms of *even* degree. The exterior derivative of a form of odd degree is defined using the anticommutator. In particular, the covariant exterior derivative of a vector potential $A \in M$ is

$$dA + A^2 = \epsilon A + A\epsilon + A^2.$$

We can check that this makes sense by computing the curvature of an exact gauge $A = [g, \epsilon]g^{-1}$. Using $\epsilon^2 = 1$ we get

$$dA + A^2 = 0$$

as expected. In general, we shall interpret $B = dA + A^2$ as the magnetic field (in the Hamiltonian formulation of YM theory; A is thought of as the vector potential at time $t = 0$ in the temporal gauge, $A_0 = 0$).

The *gauge transformations* are defined by $A' = gAg^{-1} + dgg^{-1}$, as in an ordinary gauge theory. Note that if $A = dg_0g_0^{-1}$ then the gauge transformation corresponds to the left multiplication $g_0 \mapsto gg_0$. The magnetic field transforms homogeneously, $B \mapsto gBg^{-1}$. The gauge transformations in M induce the linear homogeneous transformation $E \mapsto gEg^{-1}$ in the dual M^*.

The complete Hamiltonian is

$$\mathcal{H} = \frac{1}{2}\mathrm{tr}(E^2 + B^2). \tag{12.4.1}$$

The diagonal blocks of the operator B are of the type L_p whereas the off-diagonal part is in $L_{2p/3}$. Thus the trace $\mathrm{tr}\, B^2$ converges if and only if $p \leq 2$. In order that the total Hamiltonian be convergent we must have $p = 2$. It follows that the universal YM theory exists only in space-time dimension $d = 3 + 1$. The Hamiltonian is gauge invariant by the symmetry of the trace.

Let ω be the natural symplectic structure in the cotangent bundle T^*M. The Hamiltonian equations of motion in (T^*M, ω) derived from the Hamiltonian $\mathcal{H}$ are then

$$\frac{\partial A}{\partial t} = E$$

$$\frac{\partial E}{\partial t} = -[\epsilon + A, B]. \qquad (12.4.2)$$

We shall not follow any further this road; see Rajeev [1988b] for some results on solutions of the model.

There is another related interesting approach to YM theory in terms of the noncommutative geometry, which is due to Connes [1988]. We shall explain it briefly. Any smooth function a on the $d+1$ dimensional space-time (which is supposed to be spin, with a given spin structure, and with a trivial vector bundle for the gauge group action) taking values in the Lie algebra $\mathbf{g}$ of the gauge group defines a bounded operator a in H through pointwise multiplication of the spinor fields by a. The operator $i[\epsilon, a]$ is an exact form of degree one, in terms of noncommutative geometry. The classical exact form associated to a is the ordinary exterior derivative da. Connes shows that one can uniquely extend this correspondence beyond exact forms: There is a linear map c from the space of 1-forms (in noncommutative geometry) to the space of ordinary $\mathbf{g}$ valued 1-forms such that for any Lie algebra valued function a we have $c(i[\epsilon, a]) = da$.

The map c is surjective but not injective. The Yang-Mills action $YM(A)$ of a vector potential A can be expressed using the trace in H. Let $\hat{A}$ be a 1-form in the noncommutative sense such that $c(\hat{A}) = A$. Let

$$\hat{B} = \epsilon\hat{A} + \hat{A}\epsilon + \hat{A}^2$$

be the corresponding curvature operator. One is tempted to write the (noncommutative) Yang-Mills action as a trace of the square of the curvature $\hat{B}$. However, there are some complications. First, the operator $\hat{B}^2$ is in general a trace class operator; it is "almost trace-class". There is a "renormalized trace", the *Dixmier trace,* which converges for $d + 1 = 4$. Connes shows that the operator $\hat{B}^2$ belongs to the class L_{1+} of operators T such that

$$\sup_N \frac{1}{\log N} \sum_1^N \lambda_n(T) < \infty,$$

where the $\lambda_n(T)$'s are the eigenvalues of T. The operators in L_{1+} have a "trace", to be denoted by $\text{Trace}_\omega T$, defined as an appropriate average of the sequence

$$\frac{1}{\log N} \sum_1^N \lambda_n(T).$$

The second complication is that $\hat{A}$ is not uniquely defined by A. In Connes [1988] is is shown that the correct way to construct the classical YM action is by the formula

$$YM(A) = 16\pi^2 \operatorname*{Inf}_{\hat{A}} \text{Trace}_\omega(\hat{B}^2),$$

where the infimum is taken over all 1-forms $\hat{A}$ such that $c(\hat{A}) = A$.

REFERENCES

Abraham, R. and J. Marsden [**1978**]: *Foundations of Mechanics.* Benjamin/Cummings, Reading, Mass., 1978.

Adler, S.L. [**1969**]: *Axial-Vector Vertex in Spinor Electrodynamics.* Phys. Rev. 177, p.2426, 1969.

Adler, S.L. and R. Dashen [**1968**]: *Current Algebras.* Benjamin, New York, 1968.

Ahlfors, L. [**1961**]: *Some Remarks on Teichmüller Spaces of Riemann Surfaces.* Ann. of Math. 74, p.171, 1961.

Alvarez, O. [**1985**]: *Topological Quantization and Cohomology.* Commun. Math. Phys. 100, p.279, 1985.

Alvarez, O., I. Singer, and B. Zumino [**1984**]: *Gravitational Anomalies and the Families Index Theorem.* Commun. Math. Phys. 96, p.409, 1984.

Alvarez-Gaume, L. and P. Ginsparg [**1984**]: *The Topological Meaning of Nonabelian Anomalies.* Nucl. Phys. B 243, p.449, 1984.

Araki, H. [**1987**]: *Bogoliubov Automorphisms and Fock Representations of Canonical Anticommutation Relations.* in: Contemporary Mathematics, American Mathematical Society vol. 62, 1987.

Atiyah, M.F. and I. Singer [**1971**]: *The Index of Elliptic Operators.* Ann. of Math. 87, p.484, 1968; 87, p.546, 1969; 93, p.119, 1971; 93, p.139, 1971.

————-[**1984**]: *Dirac Operators Coupled to Vector Potentials.* Proc. Natl. Acad. Sci. USA 81, p. 2597, 1984.

Bao, D. and V.P. Nair [**1985**]: *A Note on Covariant Anomaly and Equivariant Momentum Mapping.* Commun. Math. Phys. 101, p.437, 1985.

Bardeen, W.A. [**1969**]: *Anomalous Ward Identities in Spinor Field Theories.* Phys. Rev. 184, p.1848, 1969.

Bell, J. and R. Jackiw [**1969**]: *The PCAC Puzzle: $\pi^0 \to \gamma\gamma$ in the σ-Model.* Nuovo Cimento A60, p.47, 1969.

Benkart, G. [**1986**]: *Kac-Moody Bibliography and Some Related References.* Can. Math. Bull. 5, p.111, 1986.

Bernard, D. and J. Thierry-Mieg [**1987**]: *Level One Representations of the Simple Affine Kac-Moody Algebras in Their Homogeneous Gradation.* Commun. Math. Phys. 111, p.181, 1987.

Bers, L. [**1981**]: *Finite-dimensional Teichmüller Spaces and Generalizations.* Bull. Amer. Math. Soc. 5, p.131, 1981.

Blau, M. [1988]: *Group Cocycles, Line Bundles, and Anomalies.* Preprint, University of Vienna, UWThPh-1988-25.

Bonora, L. and P. Cotta-Ramusino [1983]: *Some Remarks on BRS Transformations, Anomalies and the Cohomology of the Lie Algebra of the Group of Gauge Transformations.* Commun. Math. Phys. 87, p.589, 1983.

Bonora, L., P. Cotta-Ramusino, M. Rinaldi, and J. Stasheff [1987]: *The Evaluation Map in Field Theory, Sigma Models and Strings -I.* Commun. Math. Phys. 112, p.237, 1987.

——— *The Evaluation...... -II.* Commun. Math. Phys. 114, p.381, 1988.

Bott, R. [1957]: *Homogeneous Vector Bundles.* Ann. of Math. 66, p.203, 1957.

Bott, R. and L.W. Tu [1982]: *Differential Forms in Algebraic Topology.* Springer-Verlag, New York, Heidelberg and Berlin, 1982

Bowick, M. and S. Rajeev [1987]: *String Theory as the Kähler Geometry of the Loop Space.* Phys. Rev. Lett. 58, p.535, 1987.

——— *Anomalies as Curvature in Complex Geometry* . Nucl. Phys. B 296, p.1007, 1987.

de Broglie, L. [1932]: *Sur le Champ Électromagnétique de l'Onde Lumineuse.* Comptes Rendus (Paris) 195, p.862, 1932.

Brown, R.W., C.C. Shi, and B.-L. Young [1969]: *Regularization and Ward Identity Anomalies.* Phys. Rev. 186, p.1491, 1969.

Carey, A. and K.C. Hannabuss [1987]: *Temperature States on Loop Groups, Theta Functions and the Luttinger Model.* J. Funct. Anal. 75, p.128, 1987.

Carey, A. and J. Palmer [1988]: *Gauge Anomalies on S^2 and Group Extensions.* Preprint, University of Adelaide; to be publ. in J. Math. Phys.

Carey, A. and S.N.M. Ruijsenaars [1987]: *On Fermion Gauge Groups, Current Algebras and Kac-Moody Algebras.* Acta Applicandae. Math. 10, p.1, 1987.

Chern, S.-S [1979]: *Complex Manifolds Without Potential Theory.* Springer-Verlag, New York, Heidelberg, Berlin, 1979. (second edition)

Chern, S.-S. and Simons [1974]: *Characteristic Forms and Geometric Invariants.* Ann. of Math. 99, p.48, 1974.

Coleman, S. [1975]: *Quantum sine-Gordon Equation as the Massive Thirring Model.* Phys. Rev. D 11, p.2088, 1975.

Connes, A. [1986]: *Non-Commutative Differential Geometry.* Publ. Math. IHES 62, p.81, 1986.

———**[1988]**: *The Action Functional in Non-Commutative Geometry.* Commun. Math. Phys. 117, p.673, 1988.

Cronström, C. and J. Mickelsson [1983]: *On Topological Boundary Characteristics in Nonabelian Gauge Theory.* J. Math. Phys. 24, p.2528, 1983.

Date, E., M. Kashiwara, M. Jimbo, and T. Miwa [1983]: *Transformation Groups for Soliton Equations.* in: Integrable Systems - Classical Theory and Quantum Theory. eds. M. Jimbo and T. Miwa. World Scientific Publ. Co., Singapore, 1983.

Faddeev, L. [1984]: *Operator Anomaly for the Gauss Law.* Phys. Lett. 145B, p.81, 1984.

Feingold, A.J. and I. Frenkel [1983]: *A Hyperbolic Kac-Moody Algebra and the Theory of Siegel Modular Forms of Genus 2.* Math. Ann. 263, p.87, 1983.

Felder, G., K. Gawedzki, and A. Kupiainen [1988]: *Spectra of Wess-Zumino-Witten Models with Arbitrary Simple Groups.* Commun. Math. Phys. 117, p.127, 1988.

Finkelstein, D. and J. Rubinstein [1968]: *Connection between Spin, Statistics, and Kinks.* J. Math. Phys. 9, p.1762, 1968.

Flato, M., J. Simon, H. Snellman, and D. Sternheimer [1972]: *Simple Facts about Analytic Vectors and Integrability.* Ann. scient. Ec. Norm. Sup. 4^e serie, t.5, p.423, 1972.

Floreanini, R. and R. Percacci [1988]: *Anomalous Gauss Law Algebras.* Preprint, S.I.S.S.A. 118 EP, Trieste, 1988.

Forger, M. [1988]: *Supersymmetric Sigma Models and Kähler Manifolds.* Preprint, Fakultät für Physik, University of Freiburg, 1988. To be publ. in Springer Lecture Notes in Physics, Proc. of the 7th Scheveningen Conference, 1987.

Frenkel, I. [1981]: *Two Constructions of Affine Lie Algebra Representations and Boson-Fermion Correspondence in Quantum Field Theory.* J. Funct. Anal. 44, p.259, 1981.

Frenkel, I., H. Garland, and G. Zuckerman [1986]: *Semi-Infinite Cohomology and String Theory.* Proc. Natl. Acad. Sci. USA 83, p.8442, 1986.

Frenkel, I. and V. Kac [1980]: *Basic Representations of Affine Lie Algebras and Dual Resonance Models.* Invent. Math. 62, p.23, 1980.

Friedan, D., Z. Qiu, and S. Shenkar [1984]: *Conformal Invariance, Unitarity, and Critical Exponents in Two Domensions.* Phys. Rev. Lett. 52, p.1575, 1984.

Fujikawa, K. [1979]: *Path Integral Measure for Gauge-Invariant Fermion Theories.* Phys. Rev. Lett. 42, p.1195, 1979.

Fujiwara, T., S. Hosono and S. Kitakado [1988]: *Chirally Gauged Wess-Zumino-Witten Model as a Constraint System.* Mod. Phys. Lett. A 3, no. 16, p.1585, 1988.

Fuks, D.B. [**1987**]: *Colomology of Infinite-dimensional Lie algebras.* Plenum Press, New York and London, 1987.

Gardner, C., J. Miura, M. Kruskal, and R. Miura [**1967**]: *Method for Solving the KdV Equation.* Phys. Rev. Lett. 19, p.1095, 1967.

Gel'fand, I.M. and L.A. Dikii [**1976**]: *Fractional Powers of Operators and Hamiltonian Systems.* Funct. Anal. Appl. 10, p.259, 1976.

Gilkey, P. [**1984**]: *Invariance Theory, the Heat Equation, and the Atiyah-Singer Index Theorem.* Publish or Perish, Inc., Wilmington, 1984.

Goddard, P., A. Kent, and D. Olive [**1986**]: *Unitary Representations of Virasoro and Super-Virasoro Agebras.* Commun. Math. Phys. 103, p.105, 1986.

Goddard, P., W. Nahm, A. Schwimmer, and D. Olive [**1986**]: *Vertex Operators for Non-Simply Laced Algebras.* Commun. Math. Phys. 107, p.179, 1986.

Goddard, P. and D. Olive [**1986**]: *Kac-Moody and Virasoro Algebras in Relation to Quantum Physics.* Internat. J. Mod. Phys. A 1, p.302, 1986.

Goodman, R. and N. Wallach [**1985**]: *Projective Unitary Positive-Energy Representations of Diff*(S^1). J. Funct. Anal. 63, p.299, 1985.

Green, M., J. Schwartz, and E. Witten [**1987**]: *Superstring Theory I-II.* Cambridge University Press, Cambridge, UK, 1987.

Greenberg, M. [**1966**]: *Lectures in Algebraic Topology.* W.A. Benjamin, New York, 1966.

Gross, D. and R. Jackiw [**1972**]: *Effect of Anomalies on Quasi-Renormalizable Theories.* Phys. Rev. D6, p. 477, 1972.

Halpern, M.B. [**1975**]: *Quantum "Solitons" which are* $SU(N)$ *Fermions.* Phys. Rev. D 11, p.1684, 1975.

Harada, K. and I. Tsutsui [**1987**]: *A Consistent Gauss Law in Anomalous Gauge Theories.* Prog. Theor. Phys. 78, no. 3, p.675, 1987.

Helgason, S. [**1978**]: *Differential Geometry, Lie Groups and Symmetric Spaces.* Academic Press, New York, 1978.

Hosono, S. and K. Seo [**1988**]: *Derivation of chiral anomalies and commutator anomalies in a fixed-time regularization method.* Phys. Rev. D38, p.1296, 1988.

Humphreys, J. [**1980**]: *Introduction to Lie algebras and Representation Theory.* Springer-Verlag, New York, Heidelberg, Berlin, 1980. (third printing)

Husemoller, D. [**1975**]: *Fiber Bundles.* Springer-Verlag, New York, Heidelberg, Berlin, 1975 (second edition).

Itzykson, C. and J.-B. Zuber [**1980**]: *Quantum Field Theory.* McGraw-Hill, New York, 1980.

Jackiw, R. [**1985**]: *Anomalies and Topology.* in: Symposium on Anomalies, Geometry, Topology. Eds. W.A. Bardeen and A.R. White. World Scientific, Singapore, 1985.

Jackiw, R. and K. Johnson [**1969**]: *Anomalies of the Axial Vector Current.* Phys. Rev. 182, p.1459, 1969.

Jackiw, R. and C. Rebbi [**1977**]: *Spinor Analysis of Yang-Mills Theory.* Phys. Rev. D 16, p.1052, 1977.

Jaffe, A., A. Lesniewski, and J. Weitsman [**1989**]: *Pfaffians on Hilbert Space.* Commun. Math. Phys. 1989.

———-[**1987**]: *Index of a Family of Dirac Operators on Loop Space.* Commun. Math. Phys. 112, p.75, 1987.

Jo, S.-G. [**1985**]: *Commutator of Gauge Generators in Non-Abelian Chiral Theory.* Nucl. Phys. B 256, p.616, 1985.

Kac, V. [**1985**]: *Infinite Dimensional Lie Algebras.* Cambridge University Press, Cambridge, UK, 1985. (second edition)

———[**1985b**]: *Constructing Groups Associated to Infinite-Dimens ional Lie Algebras.* in: Infinite-Dimensional Groups with Applications. Springer-Verlag, Berlin, Heidelberg, New York, 1985.

———[**1968**]: *Simple Irreducible Graded Lie Algebras of Finite Growth.* Math. USSR Izvestija 2, p.1271, 1968.

Kac, V., D. Kazhdan, J. Lepowsky, and R.L. Wilson [**1981**]: *Realization of Basic Representations of the Euclidean Lie Algeras.* Adv. in Math. 42, p.83, 1981.

Kac, V. and D. Peterson [**1986**]: *Lectures on Infinite Wedge Representations and MKP Hierarchy.* Seminaire de Math. Sup 102, p.141, Montreal University, 1986.

Kirillov, A.A. [**1987**]: *Kähler Structures on K-Orbits of the Group of Difeomorphisms of a Circle.* Funct. Anal. Appl. 21, p.122, 1987.

Kobayashi, S. and K. Nomizu [**1963; 1969**]: *Foudations of Differential geometry. I-II.* John Wiley & Sons, New York and London, 1963, 1969.

Kogut, J. and L. Susskind [**1975**]: *How Quark Confinement Solves the $\eta - 3\pi$ Problem.* Phys. Rev D 11, p.3594, 1975.

Kolokolov, I.V. and A.S. Yelkhovsky [**1987**]: *Schwinger Terms as a Source of Gauge Anomaly in Hamiltonian Approach.* Preprint, Inst. of Nuclear Physics, Novosibirsk, 1987.

Krichever, I.M. [**1977**]: *Integration of Non-Linear Equations By Methods of Algebraic Geometry.* Funct. Anal. Appl. 11, p.12, 1977.

Krichever, I.M. and S.P. Novikov [**1987**]: *Algebras of Virasoro type, Riemann Surfaces and Structures of the Theory of Solitons.* Funct.

Anal. Appl. 21, p.126, 1987.

Lax, P. [1968]: *Integrals of Non-Linear Equations of Evolution and Solitary Waves.* Commun. Pure Appl. Math. 21, p.467, 1968.

Macdonald, I.G. [1979]: *Symmetric Functions and Hall Polynomials.* Oxford University Press, Oxford, 1979.

Mandelstam, S. [1974]: *Soliton Operators for the Quantized sine-Gordon Model.* Phys. Rev. D 11, 3026, 1974.

Michor, P. [1980]: *Manifolds of Differentiable Mappings.* Shiva Math. Series 3, Orpington, Kent, 1980. Now available from: Holt-Saunders, Eastbourne, U.K.

Mickelsson, J. [1985a]: *Chiral Anomalies in Even and Odd Dimensions.* Commun. Math. Phys. 97, p.361, 1985. Also: Lett. Math. Phys. 7, p.45, 1983.

———[**1985b**]: *Two-Cocycle of a Kac-Moody group.* Phys. Rev. Lett. 55, p.2099, 1985.

———[**1987a**]: *Kac-Moody Groups, Topology of the Dirac Determinant Bundle and Fermionization.* Commun. Math. Phys. 110, p.173, 1987.

———[**1987b**]: *String Quantization on Group Manifolds and the Holomorphic Geometry of Diff* S^1/S^1. Commun.Math. Phys. 112, p.653, 1987.

———[**1988a**]: *Current Algebras, Gauge Group Extensions and Infinite-Dimensional Grassmannians.* Preprint, University of Freiburg, 1988.

———[**1988b**]: *Current Algebra Representation for the* 3+1 *Dimensional Dirac-Yang-Mills Theory.* Commun. Math. Phys. 117, p.261, 1988.

Mickelsson, J. and S. Rajeev [1988]: *Current Algebras and Determinant Bundles over Infinite-Dimensional Grassmannians.* Commun. Math. Phys. 116, p.365, 1988.

Milnor, J. and J. Stasheff [1974]: *Characteristic Classes.* Princeton University Press, Princeton NJ, 1974.

Moody, R.V. [1968]: *A New Class of Lie Algebras.* J. Algebra 10, p.211, 1968.

Mulase, M. [1984]: *Cohomological Structure in Soliton Equations and Jacobian Varieties.* J. Diff. Geom. 19, p.403, 1984.

Mumford, D. [1983]: *Tata Lectures on Theta I-II.* Birkhäuser, Boston, Basel, Stutgart, 1983 (I) and 1984 (II).

Murray, M.K. [1988]: *Another Construction of the Central Extension of the Loop Group.* Commun. Math. Phys. 116, p.73, 1988.

Nielsen, N.K., H. Römer and B. Schroer [1977]: *Classical Anomalies and the Local Version of the Atiyah-Singer Theorem.* Phys. Lett. 70B, p.445, 1977.

———[1978]: *Anomalous Currents in Curved Space.* Nucl. Phys. B 136, p.475, 1978

Nielsen, N.K. and B. Schroer [1978]: *Axial Anomaly and Atiyah-Singer Theorem.* Nucl. Phys. B 127, p.493, [1978].

Niemi, A.J. and G.W. Semenoff [1985]: *Quantum Holonomy and the Chiral Gauge Anomaly.* Phys. Rev. Lett. 55, p.927, 1985.

Novikov, S.P. [1974]: *A Periodic Problem for the Korteweg-de Vries Equations.* Funct. Anal. Appl. 8, p.54, 1974.

Omori, H. [1973]: *Groups of Diffeomorphisms and their Subgroups.* Trans. Amer. Math. Soc. 179, p.85, 1973.

Palais, R.S. [1965]: *On the Homotopy Type of Certain Groups of Operators.* Topology 3, p.271, 1965.

Paranjape, M.B. [1988]: *Quantization of the Anomalous, Chiral, Schwinger Model.* Nucl. Phys. B 307, p.649, 1988.

Pickrell, D. [1987]: *Measures on Infinite-Dimensional Grassmann Manifolds.* J. Funct. Anal. 70, 323, 1987.

——— **[1988]**: *On the Mickelsson-Faddev Extension and Unitary Representations.* Preprint, Dept. of Math., University of Arizona, Tucson, 1988.

Pilch, K. and Warner [1987]: *Holomorphic Structure of the Superstring Vacua.* Class. Quantum Grav. 4, p.1183, 1987.

Pressley, A. and G. Segal [1986]: *Loop Groups.* Clarendon Press, Oxford, 1986.

Quillen, D. [1988]: *Superconnection Character Forms and the Cayley Transform.* Topology 27, p.211, 1988.

Rajeev, S. [1984]: *Fermions from Bosons in 3+1 Dimensions through Anomalous Commutators.* Phys. Rev. D29, p.2944, 1984.

———**[1988]**: *Non-Abelian Bosonization without Wess-Zumino Terms I. New Current Algebra.* Preprint, Dept. of Physics, Univ. of Rochester, 1988.

———**[1988b]**: *An Exactly Integrable Algebraic Model for 3+1 Dimensional Yang-Mills Theory.* Preprint, Dept. of Physics, Univ. of Rochester, 1988.

Reyman, A. [1985]: *Differential Geometry, Groups and Mechanics. VII* (Russian) Akademie Nauk SSSR no. 146, 1985.

Reyman, A., M.A. Semenov-Tyan-Shanskii and L. Faddeev [1985]: *Quantum Anomalies and Cocycles of Gauge Groups.* Funct. Anal. Appl. 18, p.319, 1985.

Sato, M. and Y. Sato [1982]: *Soliton Equations as Dynamical Systems on Infinite Dimensional Grassmann Manifolds.* Lecture Notes in Num. Appl. Anal. 5, p.279, 1982.

Scherk, J. [1975]: *An Introduction to the Theory of Dual Models ans Strings.* Rev. Mod. Phys. 47, p.123, 1975.

Schwinger, J. [1959]: *Field Theory Commutators.* Phys. Rev. Lett. 3, p.296, 1959.

Segal, G. [1981]: *Unitary Representations of Some Infinite Dimensional Groups.* Commun. Math. Phys. 80, p.301, 1981.

Segal, G. and G. Wilson [1985]: *Loop Groups and Equations of KdV type.*Publ. Math. IHES N° 61, p.5, 1985.

Shiota, T. [1986]: *Characterization of Jacobian Varieties in Terms of Soliton Equations.* Invent. Math. 83, 333, 1986.

Simon, B. [1979]: *Trace Ideals and their Applications.* Cambridge University Press, Cambridge, UK, 1979.

Singer, I. [1985]: *Families of Dirac Operators with Applications to Physics.* Asterisque 323, 1985.

Skyrme, T.H.R. [1961]: *Particle States of a Quantized Meson Field.* Proc. Roy. Soc. A 262, p.237, 1961.

Sommerfield, C. [1968]: *Currents as Dynamical Variables.* Phys. Rev. 176, p.2019, 1968.

Spivak, M. [1979]: *A Comprehensive Introduction to Differential Geometry.* Publish or Perish, Inc., Berkeley, 1979 (second edition)

Stasheff, J. [1985]: *The de Rham Bar Construction as a Setting for the Zumino, Fadeev, etc. Descent Equations.* in: Symposium on Anomalies, Geometry, Topology. Eds. W.A. Bardeen and A.R. White, World Scientific, Singapore, 1985.

Sugawara, H. [1968]: *A Field Theory of Currents.* Phys. Rev. 170, p.1659, 1968.

Takasaki, K. [1988]: *Geometry of Universal Grassmann Manifold From Algebraic Point of View.* Preprint RIMS-623, Kyoto, 1988.

Treiman, S.B., R. Jackiw, B. Zumino, and E. Witten [1985]: *Current Algebras and Anomalies.* Princeton University Press, Princeton, NJ, 1985.

Vershik, A.M., I.M. Gelfand, and M.I. Graev [1980]: *Representations of the Group of Functions Taking Values in a Compact Lie Group.* Compositio Math. 42, p.217, 1980.

Warner, F.W. [1983]: *Foundations of Differential Manifolds and Lie Groups.* Springer-Verlag, New York, Berlin, Heidelberg, 1983. (second edition)

Warner, G. [1972]: *Harmonic Analysis on Semi-Simple Lie Groups I-II.* Springer-Verlag, New York, Heidelberg, Berlin, 1972.

Witten, E. [1983]: *Current Algebra, Baryons, and Quark Confinement.* Nucl. Phys. B223, p.433, 1983.

———-[**1984**]: *Non-Abelian Bosonization in Two Dimensions.* Commun. Math. Phys. 92, 455, 1984.

———-[**1988**]: *Topological Sigma Models.* Commun. Math. Phys. 118, p.411, 1988.

Yamagishi, H. [**1987**]: *A Space-Time Approach to Chiral Anomalies.* Preprint, Dept. of Physics, State University of New York at Stony Brook, 1987.

Zakharov, V.E. and L.D. Faddeev [**1971**]: *The Korteweg-de Vries Equation is a Fully Integrable Hamiltonian System.* Funct. Anal. Appl. 5, p.280, 1971.

Zakharov, V.E. and A.B. Shabat [**1974**]: *A Scheme for the Integration of the Non-Linear Equations of Mathematical Physics by the Method of the Inverse Scattering Problem.* Funct. Anal. Appl. 8, p.43, 1974.

Zoller, D. [**1986**]: *Cocycles and the Virasoro Algebra.* Preprint, The Enrico Fermi Institute, Chigaco, 1986.

Zumino, B. [**1984**]: *Chiral Anomalies and Differential Geometry.* in: Proc. of the Les Houches summer school, 1983. Eds. de Witt and R. Stora, North-Holland, Amsterdam, 1985.

INDEX